CENTRIFUGAL COMPRESSORS

A STRATEGY FOR AERODYNAMIC DESIGN AND ANALYSIS

Ronald H. Aungier

NEW YORK   ASME PRESS   2000

Library of Congress Cataloging-in-Publication Data

Aungier, Ronald H.
    Centrifugal compressors : a strategy for aerodynamic design and
    analysis / Ronald H. Aungier
        p.   cm.
    ISBN 0-7918-0093-8
    1. Compressors–Aerodynamics.   2. Compressors–Design and construction.
        I. Title.
    TJ267. 5. C5A96 2000
    621.5' 1—dc21   99-39156
                                                                              CIP

To
Anne, with love and gratitude.
Beth and Joan, from a proud father.
Dad and Mom, whose sacrifices made it possible.

TABLE OF CONTENTS

PREFACE

There are several excellent books that provide a general overview of centrifugal compressor aerodynamic technology. But, none of these books fully describe a working design and analysis system encompassing the interacting procedures, design guidelines and decision processes required. During the long and tedious process of developing my own design and analysis system, I have often wished such a book existed. The centrifugal compressor presents an extremely complex fluid dynamics problem. The conversion of basic technology into a working design and analysis system is an evolutionary process; its success depends on the continual validation and guidance that only results from the many triumphs and failures one encounters while developing these complex machines. In addition, the very best fluid dynamic technology available today is not sufficient, by itself. Success or failure of a compressor development program continues to be highly dependent on the judgments and basic design practice that comes only from past experience.

This book describes my own centrifugal compressor aerodynamic design and analysis system and the strategy I use while applying it. My intent was to provide a description sufficiently complete that both new and experienced compressor aerodynamicists can fully understand the methods I use. To this end, some care has been taken to present the basic thermodynamic and fluid dynamics principles, empirical models and key numerical methods that form the basis of these design and analysis methods. I have also done my best to describe the strategy or design practice I use; this was rather difficult, since it involves a process of reasoning rather than following an established set of principles.

This aerodynamic design and analysis system for centrifugal compressors is fairly comprehensive and it has produced significant performance improvements in recent years. It uses a very practical and efficient methodology, requiring minimal resources for its implementation. Indeed, a personal computer of rather modest capability is quite adequate to implement all of the procedures described in this book.

It should be obvious that my strategy for centrifugal compressor aerodynamic design and analysis is by no means the only one. The same basic functions accomplished by the methods described here have been addressed in alternate ways by other investigators. Indeed, I have developed and used many alternative methods over the past 28 years, only to discard them when a better approach was found. I have made no effort to draw comparisons or contrasts with alternate methods. My purpose was to describe methods that I found to be effective approaches to aerodynamic design and analysis—for whatever benefit the reader may derive from them.

I would like to express my sincere appreciation to The Elliott Company for permitting me to publish this book. I also want to express my gratitude to my long-time mentor and true friend, Mr. Frank J. Wiesner. Frank introduced me to centrifugal compressors, patiently guided me through the learning process and collaborated with me on this subject for many years. Many of the ideas in this book evolved directly from our mutual efforts to better understand centrifugal compressor aerodynamics. Finally, I wish to thank Dr. Naresh Amineni of The Elliott Company and Dr. Abraham Engeda of Michigan State University for their many helpful suggestions.

Chapter 1

INTRODUCTION

Centrifugal and axial-flow compressors are classified as dynamic compressors, or simply as turbomachines. In contrast to positive displacement compressors, the dynamic compressor achieves its pressure rise by a dynamic transfer of energy to a continuously flowing fluid stream. There is a substantial increase in radius across the rotating blade rows of the centrifugal compressor, which is its primary distinguishing feature from the axial-flow compressor. For this reason, the centrifugal compressor can achieve substantially higher stage pressure ratios than the axial-flow compressor. But the axial-flow compressor can achieve a much larger flow rate per unit frontal area. As illustrated in Fig. 1-1, these two types of compressors also have quite different performance characteristics. The centrifugal compressor approximates a constant head-variable flow machine, whereas the axial-flow compressor is closer to a constant flow-variable head machine. Neither description is strictly correct, but they serve to differentiate between the performance characteristics of these two types of dynamic compressors.

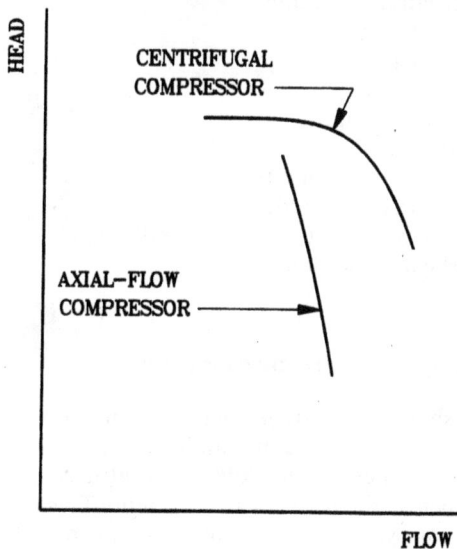

FIGURE 1-1. Comparison of Compressor Types

NOMENCLATURE

a = sound speed
C = absolute velocity
d_S = specific diameter
H = head
h = enthalpy
I = work input coefficient
M_U = rotational Mach number, U_2/a_{0t}
n_S = specific speed
$\dot{m}$ = mass flow
Q_0 = inlet volume flow ($\dot{m}/\rho_{0t}$)
Re = Reynolds number
r = radius
U = blade speed, ωr
η = efficiency
μ = head coefficient, gas viscosity
ρ = gas density
ϕ = stage flow coefficient, $\dot{m}/(\pi\rho_{0t}r_2^2 U_2)$
ω = rotation speed

Subscripts

B = a blade parameter
is = isentropic process
rev = thermodynamically reversible process
t = total thermodynamic condition
0 = impeller eye condition
1 = impeller blade inlet condition
2 = impeller tip condition
3 = vaned diffuser inlet condition
4 = diffuser exit condition
5 = crossover or volute inlet condition
6 = crossover or volute exit condition
7 = return channel vane or exit cone exit condition
8 = return channel exit condition

1.1 The Centrifugal Compressor Stage

Figures 1-2 and 1-3 show typical stage configurations for a single-stage centrifugal compressor or for the last stage in a multistage machine. The stage consists of a rotating impeller to energize the fluid and a diffuser to recover some of the fluid kinetic energy before the flow enters the volute. This diffusion process may be assisted by stationary vanes in the diffuser, as shown in the figures. This type of diffuser is called a vaned diffuser; alternatively, the diffuser may be a simple annular passage, which is known as a vaneless diffuser. Finally, a volute (or

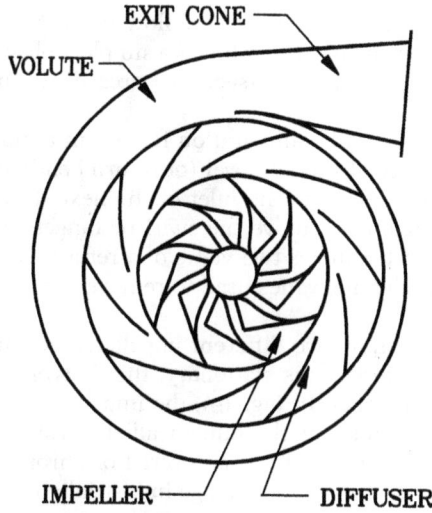

FIGURE 1-2. Front View of a Single Stage Compressor

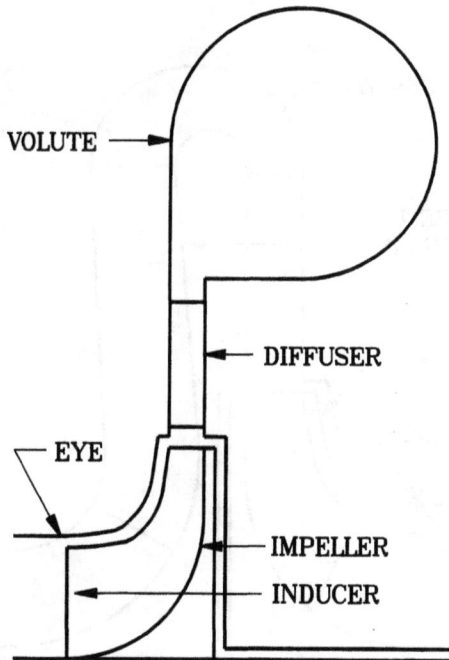

FIGURE 1-3. Side View of a Single Stage Compressor

scroll) is employed to smoothly collect the flow from the diffuser exit into the discharge pipe. An alternative to the volute is a simple "collector," which is similar to a volute but has a uniform cross-sectional area in its circumferential direction.

Figure 1-4 shows a typical configuration for use in a multistage compressor. Here the volute is replaced by a crossover (or return bend) and a return channel, which redirects the flow back to the inlet of the next stage. At the same time, vanes in the return channel remove the swirl or tangential velocity developed by the impeller. Figure 1-5 shows two views of a return channel to illustrate the typical type of return channel vanes used to remove the swirl component of the velocity.

Figures 1-3 and 1-4 show two different impeller styles. In Fig. 1-3, the outer or shroud wall of the impeller is stationary; this is typically referred to as an unshrouded or open impeller. In Fig. 1-4, the impeller blades have a shroud or cover attached to the blades at the outer wall that rotates with the impeller; this type of impeller is referred to as a covered or shrouded impeller. Covered impellers are commonly used in multistage industrial compressors where it can be difficult to hold acceptable tight clearances between the impeller blades and a stationary shroud for several stages. Usually an "eye seal" is included to reduce the flow leakage through the clearance gap between the cover and the casing wall. Also note that the impeller blade in Fig. 1-3 extends all the way into the axial portion of the passage, that is commonly called the impeller "eye." This portion of

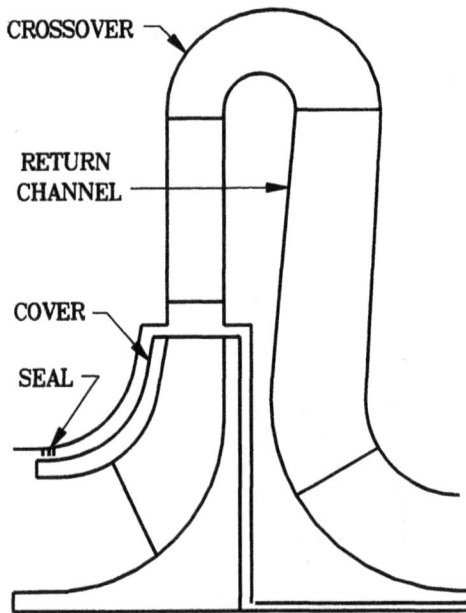

FIGURE 1-4. Multistage Compressor Stage

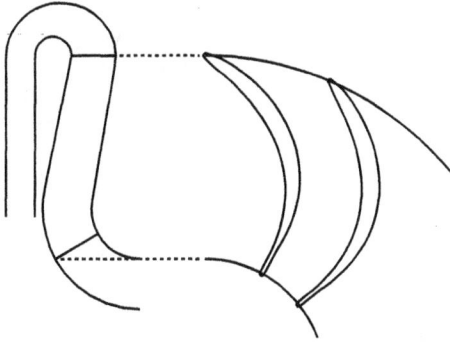

FIGURE 1-5. Return Channel Geometry

the impeller is referred to as the inducer, and this type of impeller is known as a full-inducer impeller. The impeller in Fig. 1-4 extends only partway into the axial portion of the passage and is called a semi-inducer impeller. A third type often used has impeller blades only in the radial portion of the passage, i.e., it has no inducer, and is sometimes called a radial impeller.

Figures 1-6 and 1-7 illustrate the station nomenclature used in this book. Typically, these station numbers are used as subscripts for various aerodynamic and geometrical parameters to designate values associated with these specific locations in the stage. For vaneless diffusers, station 3 (the vane leading edge) does not exist and stations 4 and 5 are identical. The locations for stations 0 through 5 for the volute and return channel stages are identical. Station 6 in a volute stage is the plane where all of the flow has been collected, "the full-collection plane," and station 7 is the exit cone discharge flange.

1.2 Dimensionless Parameters

From basic dimensional analysis, several groupings of dimensionless parameters that serve to characterize the stage can be developed. These parameters, by themselves, tell the aerodynamicist a great deal about the basic character of the stage, the performance levels it is likely to achieve and the type of design that will be most effective. There are many different dimensionless groupings used for this purpose. This book uses the stage flow coefficient, ϕ, the head coefficient, μ, the rotational Mach number, M_U and the Reynolds number Re.

$$\phi = \dot{m}/(\rho_{0t}\pi r_2^2 U_2) = Q_0/(\pi r_2^2 U_2) \qquad (1\text{-}1)$$

$$\mu = H_{rev}/U_2^2 \qquad (1\text{-}2)$$

$$M_U = U_2/a_{0t} \qquad (1\text{-}3)$$

where U_2 = impeller tip speed; r_2 = impeller tip radius; a_{0t} = sound speed based on inlet total thermodynamic conditions; ρ_{0t} = inlet total gas density; $\dot{m}$ = stage

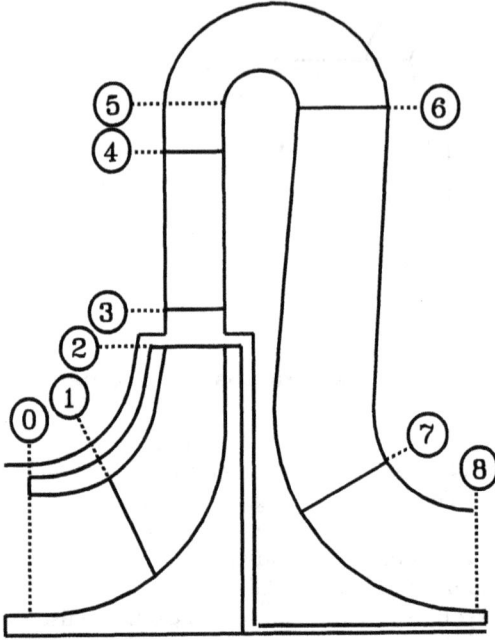

FIGURE 1-6. Computing Station Nomenclature

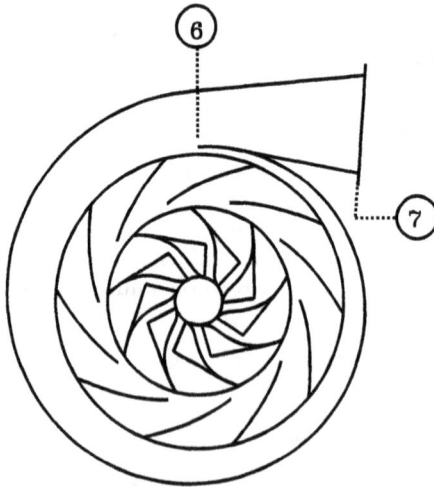

FIGURE 1-7. Computing Stations in Volutes

mass flow rate; $Q_0 = \dot{m}/\rho_{0t}$ is referred to as the inlet volume flow rate; H_{rev} = head (total enthalpy rise) required to produce the compressor's total pressure increase via an ideal or reversible thermodynamic process. The precise reversible process must be specified to fully define μ, but is not essential for the present discussion (see Chapter 2 for discussions on this and other thermodynamic parameters). A popular alternative to ϕ and μ is the specific speed, n_S and specific diameter, d_S (Balje, 1981), which are equally significant for characterizing centrifugal compressor stages, but lack the direct physical interpretation of ϕ and μ. They can be related to ϕ and μ by

$$n_S = 1.773\sqrt{\phi}/\mu_{is}^{0.75} \tag{1-4}$$

$$d_S = 1.128\mu_{is}^{0.25}/\sqrt{\phi} \tag{1-5}$$

where μ_{is} indicates that the reversible thermodynamic process chosen to define μ is an isentropic process. The stage flow coefficient characterizes the stage type and the efficiency level that can reasonably be expected. Figure 1-8, illustrates typical impeller styles used for various values of ϕ in multistage process compressors. Figure 1-9 illustrates efficiency levels that can reasonably be expected (a simplified presentation of results from Aungier, 1995). For low values of ϕ, passages are very narrow. A simple no-inducer impeller is commonly used, often with very simple two-dimensional blade shapes. Very high wall friction losses greatly limit the efficiency that can be expected. In contrast, at very high values of ϕ, the wide passages lead to large "curvature losses" to again limit the achievable efficiency levels. This can be alleviated by reducing curvature by using an impeller that discharges the flow somewhat less than radial, i.e., a "mixed-flow" impeller. Highly three-dimensional impeller blades are required to match the strong flow profile gradients. At intermediate values of ϕ, more conventional centrifugal impeller styles are used and rather good efficiency levels can be expected.

The stage head coefficient is another important characteristic parameter. Subsequent chapters will show that the impeller tip flow data (velocity and flow

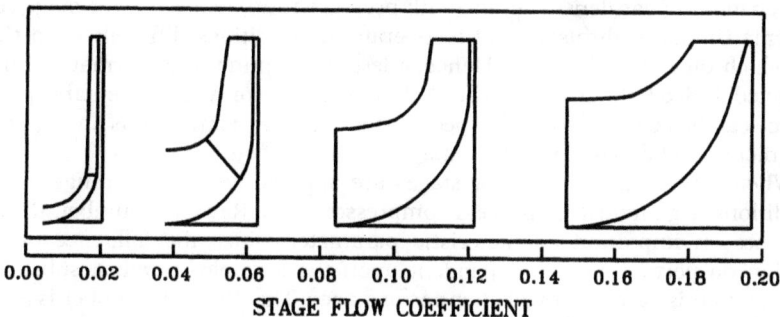

STAGE FLOW COEFFICIENT

FIGURE 1-8. Typical Impeller Styles

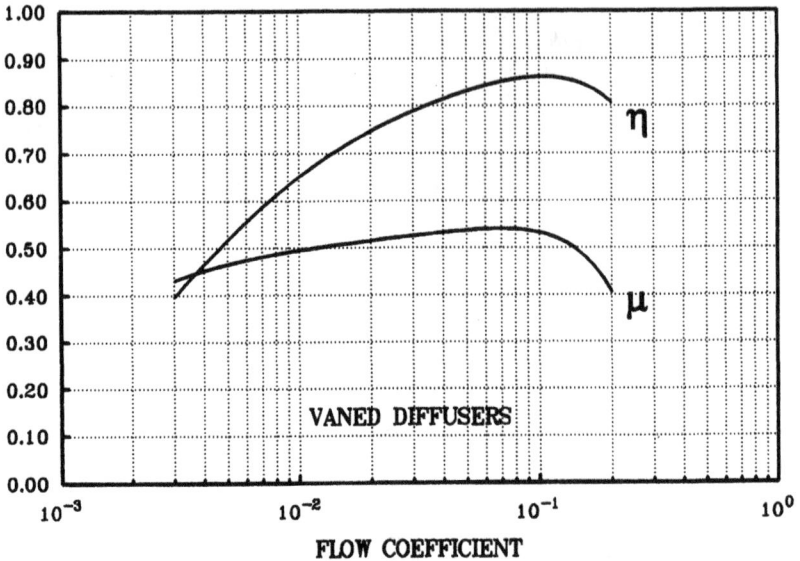

FIGURE 1-9. Typical Stage Performance Levels

angle) are direct consequences of the choice of μ. For this reason, μ greatly influences the efficiency levels that can be achieved by a centrifugal compressor. Figure 1-9 also shows typical values of μ that can be expected to yield the indicated stage efficiency levels (again, a simplified presentation of results from Aungier, 1995). These values simply represent "good" choices. Centrifugal compressors are designed for a wide range of values, depending on the specific design requirements. But, substantial deviations from the head coefficient levels shown can yield lower stage efficiencies than those shown in Fig. 1-9.

The rotational Mach number is a measure of the Mach number levels in the machine. The stage pressure ratio and temperature ratio are also directly related to this parameter. A centrifugal compressor stage will be designed for a specific value of M_U and will be optimized for that value. When operated at other values, changes in gas density levels (with pressure ratio) will cause the stage components (impeller, diffuser, etc.) to operate at conditions different from those for which they were designed. Hence, a less-than-optimum performance can be expected. Indeed, there will be a limited range of M_U for which acceptable performance can be achieved for each specific stage. Consequently, M_U is an important parameter to characterize centrifugal compressor stages.

When centrifugal compressor stages are applied over a wide range of inlet conditions (e.g., industrial process compressors), the Reynolds number, Re, also becomes an important characteristic parameter. Since the influence of wall friction on performance is dependent upon the Reynolds number, substantial changes in this parameter will alter performance. The Reynolds number is a measure of the ratio of inertia forces to viscous forces, Re = $\rho CL/\mu$. Here, μ is the gas viscosity, ρ is density, C is velocity and L is a characteristic length. It is now generally accepted that Re based on flow conditions and passage width at the diffuser entrance is the most representative value (Wiesner, 1979; Casey, 1985). But, the effect of the Reynolds number is different for each stage component and for each stage type. Thus, the Reynolds number effects are best dealt with using a performance analysis that handles wall friction losses in a comprehensive fashion.

1.3 Performance Characteristics

Figure 1-10 illustrates the basic performance characteristics of a centrifugal compressor stage. The impeller imparts energy to the fluid to raise the total enthalpy. Consistent with our head coefficient definition, we define the impeller work coefficient as

$$I = \Delta h_t / U_2^2 \qquad (1\text{-}6)$$

where Δh_t = actual increase in total enthalpy. The portion of the work input supplied by the impeller blades is designated as I_B and represents the useful work. The difference between the total work input coefficient, I, and the blade work

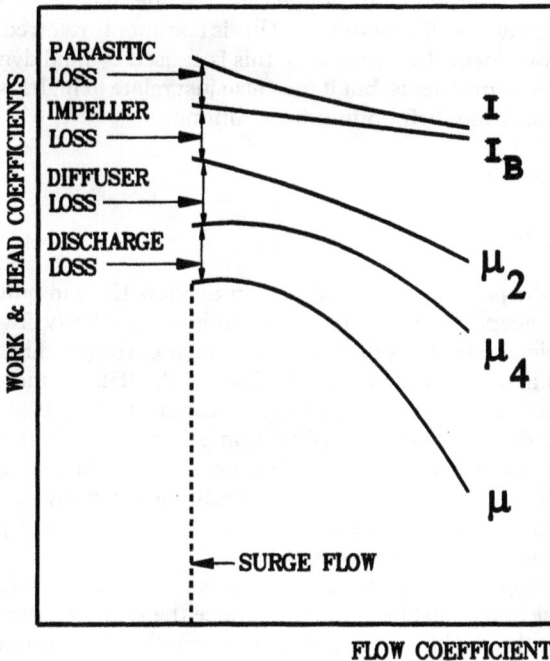

FIGURE 1-10. Stage Performance Characteristics

input coefficient is referred to as the parasitic loss. These are basically wasted energy, in the sense that they consume energy but do not contribute to the head developed in the compressor. Examples include the skin friction forces acting on the rotating disk and cover (disk friction loss), reverse flow at the impeller tip (recirculation loss) and cover seal leakage flows for covered impellers (leakage loss). But even the useful blade work input is compromised by various losses that occur in the stage components. Losses in the impeller reduce the blade work input to produce a head coefficient μ_2 at the impeller discharge; losses in the diffuser further reduce the head coefficient to μ_4 at the diffuser exit. The discharge (volute or return system) losses reduce it further to the final overall stage head coefficient.

Each of the stage components (impeller, diffuser, etc.) will have an optimum operating condition where its losses are a minimum. The designer normally seeks to "match" the various components to produce their best performance at the stage's design flow coefficient. Hence, it is normal for losses to increase as the stage operates further from its design flow (both higher and lower flows). This imposes some definite limits on the operating range of the stage. At flow coefficients sufficiently below the design value, increasing losses will produce a maximum in the head curve. At flows lower than this, the head curve will show a positive slope, which is theoretically unstable. The onset of an unstable operation is commonly called "surge." Compressor surge is a very complex phenomenon that is highly dependent on the complete system involved—not just on the compressor. Therefore, it is a useful simplification to associate this surge with the maximum head point. Similarly, at flow coefficients sufficiently above the design value, the increasing losses will reduce the stage head coefficient to zero. This is commonly referred to as "choke," "stonewall" or "overload." In some cases, this is caused by fluid dynamic choking in one of the stage components, but it may also just relate to high losses in components operating far from their optimum conditions.

1.4 Similitude

The dimensionless parameters introduced in Section 1.2 lead quite naturally to a very useful concept, referred to as similitude or similarity. Two compressor stages are completely similar if the ratios of all their corresponding length dimensions, velocities and forces are constant (Sheppard, 1956). If the two stages are completely similar, it follows that their performance, as expressed by the dimensionless parameters, will also be similar. Complete similitude is rarely (if ever) achieved in practice, but often it is closely enough approximated to make this an extremely valuable concept. For example, similitude regularly permits the aerodynamicist to conduct affordable performance tests on a small prototype unit and apply the results to a large production unit.

The simplest application of similitude is to geometrically scale a known stage to a different size and adjust its speed to produce the same rotational Mach number, M_U. Then, if the working fluid and inlet thermodynamic conditions are identical for the two stages, they will be completely similar except for effects due to the Reynolds number. For modest differences in the Reynolds number, these effects can usually be ignored. Alternatively, specific corrections for Reynolds number effects can be used, as discussed in Section 1.2. This case becomes more complicated if the working fluids for the two stages are different. A different working fluid may produce different ratios of gas density across the impeller, often referred to as a volume ratio effect. This means the ratios of velocity between the two stages is not constant everywhere, so complete similitude is not achieved. The volume ratio effect may be small enough to be neglected, or suitable correction procedures can be used to permit exploiting the concept of similitude.

Similitude is regularly employed to create multiple geometrical sizes of a stage or to use a stage in different applications, all based on a single prototype stage test. Indeed, with a little care, the concept can be used to modify the stage design somewhat to produce different flow capacity, head levels, etc. But it should be recognized that use of similitude always involves a judgment by the aerodynamicist as to the impact any lack of perfect similitude may have on the forecasted performance based on the known prototype stage.

1.5 Units and Conventions

This book is relatively independent of the type of units preferred by the reader. To accomplish this, it is necessary to use fully consistent units for all phases of design and analysis. In some cases, this may require care and adjustment by readers who do not follow that practice. Quite often, individual organizations use different types of units for different disciplines, largely for historical reasons. Thermodynamics and aerodynamics are typical examples, where it is not unusual for energy terms, equation of state data and gas velocity to be specified by inconsistent units. It is up to the reader to recognize the need for conversion factors, etc. The flow and blade angle convention used is to measure angles with respect to the tangential direction, and absolute temperatures are used in all cases. The wide range of topics covered makes it very difficult to employ a common nomenclature throughout this book—there are not enough different symbols available for each one to be unique. For this reason, each chapter will contain a nomenclature after its Introduction to ensure clarity; to the extent possible, a consistent nomenclature has been used for the various parameters that are common to different chapters.

EXERCISES

1.1   Eye seal clearances for covered impellers are typically no larger than blade clearances for open impellers. In light of this observation, give at least two reasons why the covered impeller style might be preferred for a multistage compressor, where clearances are difficult to control.

1.2   A compressor stage is to be applied to a variety of working fluids but with no other modifications. Performance maps are to be provided in the form of head (H) versus volume flow (Q_0) for a series of rotation speeds in rpm (N).

Derive suitable dimensionless forms of H, Q_0 and N based on similitude. Since geometrical parameters will not vary in this case, do not include them in your dimensionless parameters.

1.3 A compressor stage has an impeller tip diameter of 50 cm, a rotation speed of 9,000 rpm and an inlet volume flow of 0.925 m^3/s. Using Fig. 1-9, estimate the head and efficiency to be expected from this stage.

1.4 You are told to redesign the compressor stage in exercise 1.3 to increase its efficiency by 5 percentage points. Your new design must pass the same inlet volume flow and produce the same head as the original stage, while changing the rotation speed no more than necessary to improve efficiency. Using Fig. 1-9, choose values for the impeller tip diameter and rotation speed that can be expected to meet the requirements.

1.5 A large compressor stage is producing 5% more head than required, thus consuming excess power. Cost and time constraints prohibit replacing the stage with a new design, but simple modifications to the existing stage are acceptable. How could you modify the existing stage to correct this problem?

1.6 Due to a change in requirements, an existing compressor stage with a flow coefficient of 0.02 needs to be replaced. The replacement compressor must pass approximately 10% less flow than the current stage while producing at least as much head. There is insufficient time to conduct a new stage development program. It has been suggested that the existing compressor design be used with all flow passage widths reduced by 10%. Using Fig. 1-9 as a guide, comment on the probability of success. Can similitude be used in this case? Should you consider any other design modifications to meet the new requirements? Can any changes in operation of the new stage be beneficial?

Chapter 2

THERMODYNAMICS

Thermodynamics is fundamental to all aspects of aerodynamic design and analysis of centrifugal compressors. This chapter introduces the key thermodynamic principles, with special emphasis on this application. No attempt is made to provide complete derivations or "proofs" of the concepts presented. Readers who desire more detailed development of any of these concepts can refer to any basic thermodynamics textbook.

NOMENCLATURE

A = Helmholtz energy, $dA = -pdv$ and parameters defined in Eqs. (2-58), (2-59) and (2-62)

A_0 = BWR constant in Eq. (2-29)

a = sound speed, gas constant defined in Eq. (2-25), BWR constant in Eq. (2-29)

B = a parameter defined in Eq. (2-63)

B_0 = BWR constant in Eq. (2-29)

b = gas constant defined in Eq. (2-25), BWR constant in Eq. (2-29)

C = gas velocity and a parameter defined in Eq. (2-64)

C_p = specific heat at constant pressure, pressure recovery coefficient.

C_v = specific heat at constant volume

C_0 = BWR constant in Eq. (2-29)

c = gas constant defined in Eq. (2-27), BWR constant in Eq. (2-29)

f = fugacity

H = head

h = specific enthalpy

K = phase equilibrium constant

L = liquid mole fraction

M = molecular weight

$\dot{m}$ = mass flow rate

n = exponent in the modified Redlich-Kwong equation, Eq. (2-27)

p = pressure

q = specific heat transfer

$\dot{q}$ = rate of heat transfer

R = gas constant, R_U/M

R_U = universal gas constant

s = specific entropy

T = absolute temperature
u = specific internal energy
v = specific volume
w = specific work input
x = mole fraction
$\dot{w}$ = power input
z = compressibility factor, $pv/(RT)$
α = BWR constant in Eq. (2-29)
γ = c_p/c_v, BWR constant in Eq. (2-29)
ρ = gas density = $1/v$
ω = acentric factor, Eq. (2-26)
$\overline{\omega}$ = total pressure loss coefficient

Subscripts

ad = adiabatic-reversible (isentropic) process
c = value at the critical point
d = stage or component discharge parameter
i = stage or component inlet condition
L = liquid phase property
p = polytropic process
R = reduced parameter (normalized by its critical point value)
r = a reference condition
t = total condition
v = condition or vapor saturation curve or a vapor phase parameter

Superscripts

L = liquid phase value
v = vapor phase value
0 = condition at a reference pressure where the ideal gas model applies, $P^0 v^0 = RT^0$

2.1 Fundamental Laws of Thermodynamics

The first law of thermodynamics is the basic principle of conservation of energy. If we restrict attention to steady flow, it can be applied to a compressor. Since the compressor is an open system, steady flow is the only case for which the first law is applicable. It requires the change in fluid energy across the compressor to balance the sum of the rate of heat transfer, $\dot{q}$, and the power input, $\dot{w}$. Hence, the appropriate energy equation is

$$\dot{q} + \dot{w} = \dot{m}\Delta\left[u + \frac{C^2}{2} + pv\right] \qquad (2\text{-}1)$$

where $\dot{m}$ = mass flow rate, and changes in potential energy due to gravitational forces are assumed to be negligible. The terms in brackets represent the energy in the flow, including the internal energy, kinetic energy and the flow work (pv). Flow work is the work required to move the fluid across the systems boundaries, a concept developed in most basic thermodynamics textbooks (e.g., Zemansky, 1957). While the term pv can be evaluated for any fluid, it represents a work term only where steady flow is crossing a system boundary (in our case, the compressor inlet and discharge). The definition of specific enthalpy is

$$h = u + pv \qquad (2\text{-}2)$$

A total (or stagnation) thermodynamic condition is defined as the value of a parameter that would exist if a fluid were brought to zero velocity without any heat or external work transfer. Hence total enthalpy is defined by

$$h_t = h + \frac{C^2}{2} \qquad (2\text{-}3)$$

In the absence of coolers, it is usually reasonable to neglect heat transfer, i.e., the flow is assumed to be adiabatic. Hence, the energy equation for a compressor can be expressed as

$$\dot{w} = \dot{m}\Delta h_t = \dot{m}[h_{td} - h_{ti}] \qquad (2\text{-}4)$$

Specific entropy, s, is defined as

$$ds = \frac{dq_{rev}}{T} \qquad (2\text{-}5)$$

where the subscript, rev, identifies a reversible process. A process is reversible if the system and its surroundings can be completely returned to their initial states after the process has occurred. Any processes not satisfying this requirement are said to be irreversible. For example, processes involving friction and heat transfer are irreversible. For any adiabatic flow process, the second law of thermodynamics requires

$$\Delta s \geq 0 \qquad (2\text{-}6)$$

It follows that an adiabatic, reversible process is also a constant entropy or isentropic process. If we apply the first law of thermodynamics to a fluid element as a closed system,

$$(dq)_{rev} = T ds = du + dw = du + p dv \qquad (2\text{-}7)$$

we obtain a fundamental thermodynamic relation for entropy, which is valid for any process

$$T ds = dh - v dp \qquad (2\text{-}8)$$

2.2 Head and Efficiency

Figure 2-1 shows an enthalpy-entropy diagram for a compressor operating between inlet total conditions (p_{ti}, T_{ti}) and discharge total conditions (p_{td}, T_{td}). The work input required to produce the pressure rise is denoted by Δh_t. But, if the compressor were to be operated as an ideal isentropic machine, this pressure rise would be produced by the smaller work input denoted as H_{ad} in Fig. 2-1. One common definition of compressor efficiency is based on the ratio of these two values. It is generally referred to as the adiabatic efficiency, since it compares the adiabatic-reversible work to the actual work (sometimes it is called the isentropic efficiency, which is a more precise term). H_{ad} is usually referred to as the adiabatic head. It is nothing more than the work that would be required if the process were isentropic. H_{ad} is often expressed in units of length, in analogy with pump practice, where head represents the height of a column of fluid that could be supported by the machine's pressure increase. For compressible flow,

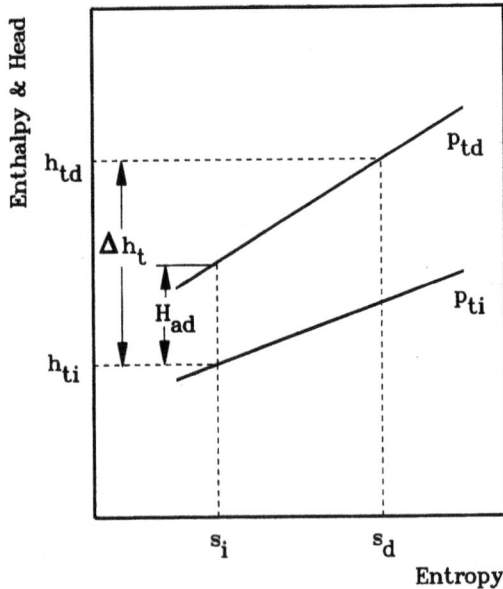

FIGURE 2-1. A Stage h-s Diagram

that interpretation is not relevant, so the present development will assume head and enthalpy are expressed in the same units. H_{ad} is given by

$$H_{ad} = \int_i^d vdp \ (s = \text{const}) \tag{2-9}$$

where the integral is carried out along a path of constant entropy. Hence, calculation of the adiabatic head requires selection of an equation of state. The adiabatic efficiency is defined as

$$\eta_{ad} = \frac{H_{ad}}{\Delta h_t} \tag{2-10}$$

The foregoing development expresses the total-to-total adiabatic efficiency, i.e., total thermodynamic conditions define the end points of the process. If the discharge state is defined by static thermodynamic conditions, Eqs. (2-9) and (2-10) yield the total-to-static adiabatic efficiency. By using static conditions at both end points, the static-to-static adiabatic efficiency is obtained. When interpreting efficiency information, it is important to be sure which definition is being used. For multistage compressors, total-to-total efficiencies are normally used. But for many single-stage compressors, the discharge kinetic energy is of no value to the application, prompting the use of total-to-static efficiency as the relevant means of evaluation.

Adiabatic head and efficiency have the advantage of being well defined, but they include an undesirable thermodynamic effect. As illustrated in Fig. 2-1, lines of constant pressure diverge on an h-s diagram. Consequently, two compressor stages designed for different pressure ratios and with equivalent aerodynamic design quality will not have the same adiabatic efficiency. The higher pressure ratio stage will have a lower efficiency due to a thermodynamic effect. Consequently, adiabatic head and efficiency do not evaluate the true aerodynamic design quality of a compressor stage. A more serious weakness can be seen in a multistage machine. If a series of stages—with all having the same adiabatic head and efficiency—are run in a multistage machine, the overall compressor efficiency will be less than the stage efficiencies. In addition, the overall adiabatic head will be less than the sum of the stage adiabatic heads. Again, this is a result of the thermodynamic effect inherent in the definition of adiabatic head and efficiency.

These weaknesses are eliminated by using polytropic head and efficiency (sometimes called the "small-stage" or "true aerodynamic" efficiency). Conceptually, polytropic performance evaluation is quite straightforward, but for many years its application was quite difficult except in the case of ideal gases. Recent development of the concept as well as availability of accurate real gas equations of state have largely eliminated that difficulty today, making polytropic evaluation suitable for real gases, too. Instead of choosing an isentrope as the reversible path, polytropic evaluation chooses a path of constant efficiency, defined by

$$\eta_p = v \, \frac{dp}{dh} \tag{2-11}$$

where η_p is defined such that the path passes through the end points of the process, e.g., (p_{ti}, h_{ti}) and (p_{td}, h_{td}). Early application was accomplished by assuming the path could be approximated by pv^e = constant, which is suitable for ideal gases but often leads to large errors for real gases. A classic paper by Schultz (1962) first addressed this weakness by developing a head correction factor to obtain more consistent results for real gases. While it was fairly approximate and somewhat difficult to apply, Schultz succeeded in extending the concept to real gases. Mallen and Saville (1977) developed a simpler method that provided comparable results. Huntington (1985) explored the problem in some detail by numerically solving Eq. (2-11) for many test cases, to provide a basis to evaluate the various approaches. This confirmed that the simpler Mallen-Saville method was somewhat more accurate than the Schultz method, making it an obvious preferred choice. Huntington also developed an approximate method, offering somewhat better accuracy than either of the aforementioned methods by including an intermediate point as well as the end points in his path integration process. His method is certainly very practical but somewhat more difficult to apply in a routine fashion, basically requiring numerical solution. Where extreme accuracy is required (e.g., compressor application work), Huntington's method is certainly the best available, but for the vast majority of aerodynamic design and analysis activity, the much simpler Mallen-Saville model is quite adequate.

The Mallen-Saville model employs an empirical path equation defined by

$$T \, \frac{ds}{dT} = \text{constant} \tag{2-12}$$

which, when combined with Eq. (2-8) yields the following relation for the total-to-total polytropic head.

$$H_p = h_{td} - h_{ti} - \frac{(s_d - s_i)(T_{td} - T_{ti})}{\ln(T_{td}/T_{ti})} \tag{2-13}$$

and total-to-static or static-to-static values are computed in the same fashion as previously described for adiabatic head. Note that this model requires only thermodynamic state data for the two end points of the path, making it no more difficult to compute than the adiabatic head. Polytropic efficiency is then given by

$$\eta_p = \frac{H_p}{h_{td} - h_{ti}} \tag{2-14}$$

Consequently, the aerodynamicist can now enjoy all the advantages of polytropic performance evaluation with no more effort than that required for adiabatic evaluation. It is only necessary to specify a suitable equation of state to directly apply this model.

2.3 The Gas Equation of State

Centrifugal compressors are applied to a wide range of gases and gas mixtures and over a very broad range of temperatures and pressures. In many cases, these gases show very nonideal behavior, requiring use of real gas models. Figure 2-2 illustrates a p-h diagram for a typical fluid. Under various conditions the fluid may be liquid, vapor, or both. Under a phase change, there is a corresponding enthalpy change due to the heat of vaporization, Δh_{vap}. Centrifugal compressors are rarely (intentionally) applied to two-phase flows; thus, gas-phase models cover most of the problems of interest.

Any aerodynamic analysis will require thermodynamic state equations in addition to the fundamental equations of fluid dynamics and thermodynamics. These equations characterize the behavior of the working fluid. Equations of state are typically derived from kinetic theory or statistical mechanics, or developed from experimental data. At least two equations of state are required to characterize the fluid. The thermal equation of state relates state variables to one another, typically in the functional form $p = p(T, \rho)$. The caloric equation of state relates the energy content of the fluid to state variables. The specific internal energy, $u = u(T, p)$, is often used, but the specific enthalpy, $h = h(T, p)$, is a more common choice for turbomachinery. These two equations are sufficient for single-phase fluid dynamics applications.

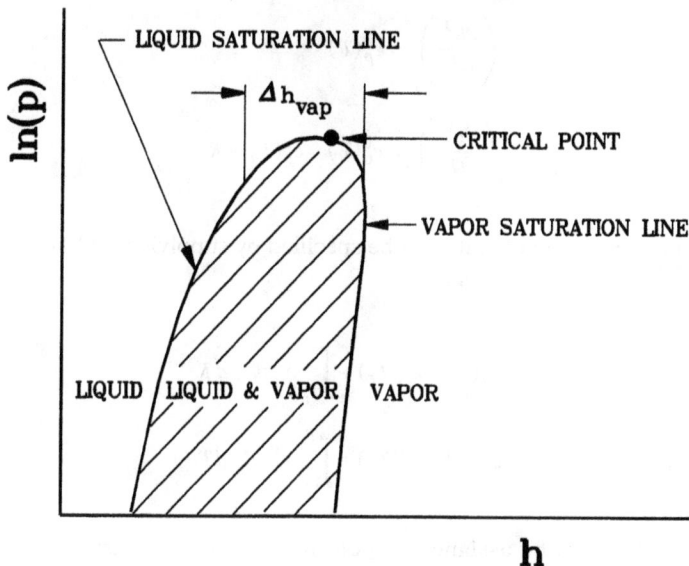

FIGURE 2-2. Fluid Pressure-Enthalpy Diagram

2.4 Thermally Perfect Gases: The Caloric Equation of State

The simplest thermal equation of state is commonly called the perfect gas equation of state,

$$p = \rho R T \tag{2-15}$$

where R = gas constant for the specific fluid. R is calculated from the universal gas constant, R_U, and the molecular weight, M, where $R = R_U/M$. Equation (2-15) is a reasonable approximation for many fluid dynamics problems. But all real fluids can be liquified, as illustrated in Fig. 2-2. The highest temperature at which liquid and vapor can coexist defines the critical point for the fluid. Measured critical point properties (T_c, p_c, ρ_c) clearly show that all fluids are far from thermally perfect at this point. As a general rule, the thermally perfect equation of state can be used when temperatures are much greater than T_c and pressures are much less than p_c. Indeed, all fluids obey the thermally perfect gas equation at sufficiently low values of ρ. Hence, this model is quite useful even when a real gas model is required. For example, the caloric equation of state for any fluid can be greatly simplified if it is specified for conditions at which the fluid is thermally perfect. Here, a superscript 0 will be used to designate parameters for thermally perfect gases. Basic thermodynamics shows that the energy content of a thermally perfect fluid is a function of only temperature, i.e., $h^0 = h^0(T)$ and $u^0 = u^0(T)$. The specific heat at constant pressure, c_p^0, and that at constant specific volume, c_v^0, are defined by

$$\left(\frac{\partial h^0}{\partial T} \right)_p = c_p^0(T) \tag{2-16}$$

$$\left(\frac{\partial u^0}{\partial T} \right)_v = c_v^0(T) = c_p^0(T) - R \tag{2-17}$$

The caloric equation of state can be specified by supplying $c_p^0(T)$ or $c_v^0(T)$ and using Eq. (2-16) or Eq. (2-17), i.e.

$$h^0(T) = h^0(T_r) + \int_{T_r}^{T} c_p^0(T)\, dT \tag{2-18}$$

$$u^0(T) = u^0(T_r) + \int_{T_r}^{T} c_v^0(T)\, dT \tag{2-19}$$

where u^0 and h^0 can be assigned any desired values at the arbitrary reference point (T_r, p_r). Since data for $c_p^0(T)$ are readily available for almost any fluid (e.g., Ried et al., 1977, Appendix A), the specification of a caloric equation of state is relatively simple. The specific entropy, s, is given by Eq. (2-8). Hence, for any pressure, p^0, where the gas is thermally perfect, s^0 is given by

$$s^0(T) = s^0(T_r) + \int_{T_r}^{T} c_p^0(T) \frac{dT}{T} - R \ln(p^0/p_r) \qquad (2\text{-}20)$$

An entropy equation is always required for fluid dynamics analysis: Even for isentropic flows, this equation is required for a conversion between total and static thermodynamic properties, which are related by an isentropic change in enthalpy. If c_p^0 is constant, properties at any two state points having the same entropy can be related analytically by direct integration of Eq. (2-20). Otherwise, a numerical solution is needed to satisfy the requirement that the two state points have the same entropy. This is always the case for real gases, as will be seen later in this chapter. An entropy equation is also required for nonisentropic processes, such as a total pressure loss mechanism. This type of process usually occurs at constant enthalpy, requiring a new value of entropy for the total-to-static property conversion. An entropy equation is also required for the adiabatic and polytropic head calculations discussed earlier in this chapter. From fundamental thermodynamics, the speed of sound for a thermally perfect gas is given by

$$(a^0)^2 = \gamma \left(\frac{\partial p}{\partial \rho} \right)_T = \gamma RT \qquad (2\text{-}21)$$

$$\gamma = c_p/c_v \qquad (2\text{-}22)$$

2.5 The Thermal Equation of State for Real Gases

The thermal equation of state for a real gas can be written

$$p/(\rho RT) = z(T, p) \qquad (2\text{-}23)$$

The gas compressibility factor, $z(T, p)$ must be obtained from either tabular data (e.g., Nelson and Obert, 1954; Pitzer et al., 1955) or from an appropriate real gas equation of state. A real gas equation of state is the best choice for fluid dynamics analysis: It results in better computational speed and provides direct calculation of the many thermodynamic parameters required. There are many special real gas equations for specific classes of fluids (e.g., refrigerants), but these will not be covered in this chapter; however, some useful general-purpose real gas thermal equations of state will be discussed.

The simple two-parameter real gas equations of state are adequate for most centrifugal compressor fluid dynamics analyses and offer very good computational speed. The Redlich-Kwong equation (Redlich and Kwong, 1949) is generally considered to be one of the most accurate of the two-parameter equations. It is given by

$$p = \frac{RT}{v-b} - \frac{a}{v(v+b)T_R^{0.5}} \qquad (2\text{-}24)$$

where $v = 1/\rho$ = specific volume; $T_R = T/T_c$ = reduced temperature, and

$$a = 0.42747 R^2 T_c^2 / p_c; \qquad b = 0.08664 R T_c / p_c \qquad (2\text{-}25)$$

Equation (2-24) yields rather good accuracy except for points close to the critical point. Several investigators have attempted to improve the accuracy of the original Redlich-Kwong equation by including the acentric factor, ω, as an additional correlating factor (Aungier, 1994, 1995; Barnes, 1973; Soave, 1972; Wilson, 1966). The definition of ω (Pitzer et al., 1955) is

$$\omega = -\log_{10}(p_v/p_c) - 1; \qquad T_R = 0.7 \qquad (2\text{-}26)$$

where p_v = pressure on the vapor saturation line (Fig. 2-2). Aungier (1994, 1995) presents an extensive evaluation of the original Redlich-Kwong equation as well as these modified forms. It was concluded that only the original equation and Aungier's modified form are applicable to all of the compounds and thermodynamic conditions considered in that evaluation. Aungier's modified Redlich-Kwong equation is given by Eq. (2-25) and

$$p = \frac{RT}{v-b+c} - \frac{a}{v(v+b)T_R^{n}} \qquad (2\text{-}27)$$

$$n = 0.4986 + 1.1735\omega + 0.4754\omega^2 \qquad (2\text{-}28)$$

The constant, c, is included to provide accurate predictions near the critical point, where the Redlich-Kwong equation becomes very inaccurate. It is computed from Eq. (2-27) using specified critical point properties p_c, T_c and ρ_c. Equation (2-28) is an empirical curve fit of the values of n required to minimize prediction errors for a series of different fluids (Fig. 2-3). Aungier's (1994, 1995) model requires specification of two more parameters (ρ_c and ω) than does the Redlich-Kwong model, but was shown to achieve a reduction in the root mean square (RMS) error levels in predicted pressure of about 50%. Figures 2-4 and 2-5 each show a summary of the errors in predicted pressure obtained from the two equations for about one third of the experimental data points evaluated in Aungier (1994, 1995). Both equations yield good accuracy, with Aungier's modified form showing significant improvement. Methods to extend application of Eqs. (2-24) and (2-27) to gas mixtures are included in the original references and in Section 2.11. Gas property data required for both models are readily available for almost any compound of interest (e.g., Ried et al., 1977; Ried and Sherwood, 1966).

The eight-parameter Benedict-Webb-Rubin (BWR) equation (Benedict et al., 1951) is a highly accurate real gas equation of state

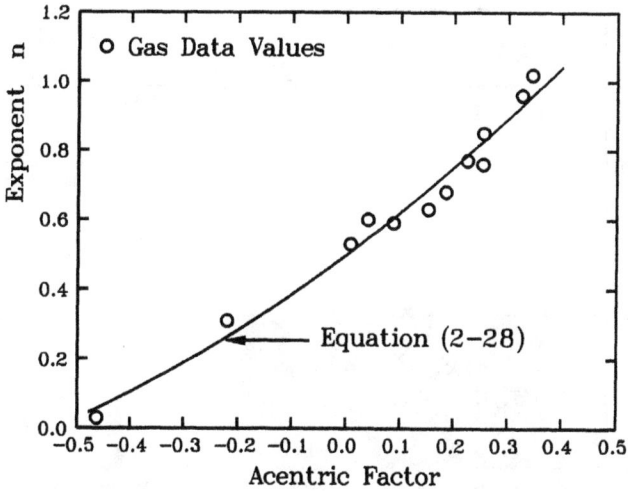

FIGURE 2-3. Empirical Correlation for *n*

$$p = \rho RT + (B_0 RT - A_0 - C_0/T^2)\rho^2 + (bRT - a)\rho^3 + a\alpha\rho^6$$
$$+ (1 + \gamma\rho^2)\exp(-\gamma\rho^2)c\rho^3/T^2 \qquad (2\text{-}29)$$

where A_0, B_0, C_0, a, b, c, α and γ = constants for the gas or gas mixture. Its major drawbacks for centrifugal compressor analysis is that the necessary data are less

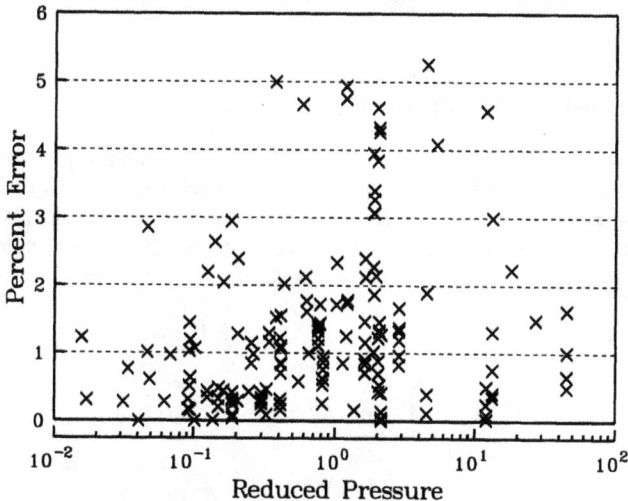

FIGURE 2-4. Classical Redlich-Kwong Model Accuracy

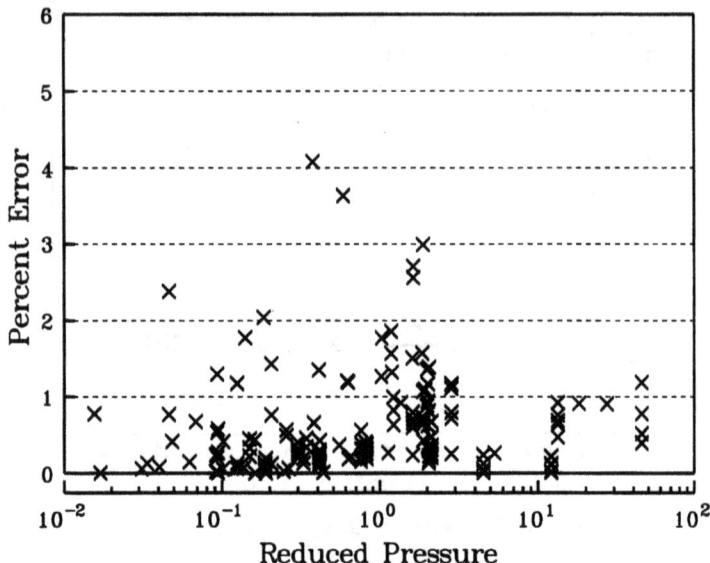

FIGURE 2-5. Modified Redlich-Kwong Model Accuracy

readily available than for the previous models and it requires more computer time for complex analyses. Tabulations of the constants for many commonly used compounds are available (Cooper and Goldfrank, 1967), and generalized constants for other compounds have been reported (Bishnoi et al., 1974). This model is useful when very high accuracy is required, e.g., for test data reduction or compressor application work.

2.6 Thermodynamic Properties of Real Gases

For real gases, h and u are functions of pressure as well as temperature. Following Ried et al. (1977), the differences in h and u with respect to the thermally perfect gas state (h^0 and u^0) are defined by departure functions. The pressure, p^0, at which the gas is thermally perfect must be specified, from which the corresponding specific volume, $v^0 = RT/p^0$, is defined for any temperature, T. The relevant departure functions are (Ried et al., 1977)

$$A - A^0 = -\int_{\infty}^{v} (p - RT/v)dv - RT\ln(v/v^0) \tag{2-30}$$

$$s - s^0 = -\left(\frac{\partial(A - A^0)}{\partial T}\right)_v \tag{2-31}$$

$$h - h^0 = (A - A^0) + T(s - s^0) + RT(z - 1) \tag{2-32}$$

$$u - u^0 = (A - A^0) + T(s - s^0) \tag{2-33}$$

where A = Helmholtz energy. For the Redlich-Kwong models, Eqs. (2-30) through (2-33) yield

$$h - h^0 = pv - RT - \frac{a}{b}(n + 1)T_R^{-n} \ln\left[\frac{v + b}{v}\right] \tag{2-34}$$

$$\frac{s - s^0}{R} = \ln\left[\frac{v}{v^0} \frac{v - b + c}{v}\right] - \frac{na}{RbT} T_R^{-n} \ln\left[\frac{v + b}{v}\right] \tag{2-35}$$

where $n = 0.5$ for the original Redlich-Kwong equation. Benedict et al. (1951) provide departure functions for the BWR equation. Real gas equations for the specific heats and the sound speed follow from basic thermodynamics, i.e.

$$c_p = \left(\frac{\partial h}{\partial T}\right)_p \tag{2-36}$$

$$c_v = \left(\frac{\partial u}{\partial T}\right)_v \tag{2-37}$$

$$a^2 = \gamma\left(\frac{\partial p}{\partial \rho}\right)_T = \frac{\gamma z RT}{1 - \frac{p}{z}\left(\frac{\partial z}{\partial p}\right)_T} \tag{2-38}$$

2.7 Thermally and Calorically Perfect Gases

Thermodynamic relations are greatly simplified when the gas can be considered thermally perfect ($z = 1.0$) and calorically perfect (c_p, c_v and γ are constants). Equation (2-20) can then be integrated directly to obtain

$$s - s_r = c_p \ln(T/T_r) - R\ln(p/p_r) \tag{2-39}$$

Hence, for isentropic (or adiabatic-reversible) calculations, we have

$$T/T_r = (p/p_r)^{(\gamma - 1)/\gamma} = (\rho/\rho_r)^{\gamma - 1} \tag{2-40}$$

From Eqs. (2-3) and (2-18), the total and static temperatures are related by

$$T_t = T + \frac{C^2}{2C_p} \tag{2-41}$$

Hence Eqs. (2-40) and (2-41) provide simple expressions to relate total and static thermodynamic conditions for all fluid dynamics calculations. Similarly, our efficiency calculations simplify to

$$\eta_{ad} = \frac{(p_{td}/p_{ti})^{(\gamma - 1)/\gamma} - 1}{T_{td}/T_{ti} - 1} \tag{2-42}$$

$$\eta_p = \frac{\gamma - 1}{\gamma} \frac{\ln(p_{td}/p_{ti})}{\ln(T_{td}/T_{ti})} \tag{2-43}$$

The more general cases are far more complex. If the gas is calorically imperfect, Eqs. (2-18) through (2-20) must normally be integrated numerically. When real gas effects are considered, the departure functions in Eqs. (2-30) through (2-33) also must be applied.

2.8 Perfect Gas Models Applied to Real Gases

The thermally and calorically perfect gas models from the previous section can dramatically reduce the computation time over any real gas model for fluid dynamics analysis. Most centrifugal compressor aerodynamic analysis applications can benefit from a similar simplification by using the pseudo-perfect gas approximation described in Aungier (1998). This involves employing a real gas equation of state to compute fictitious gas constants $(\bar{R}, \bar{c}_p, \bar{c}_v, \bar{\gamma})$ to approximate the real gas using perfect gas relations. This approximation is almost always adequate for centrifugal compressor analysis, yet almost never exploited by developers of those analyses. Some care in formulating the fluid dynamics analysis is required. All perfect gas relations that employ the relationships between c_p, c_v, γ and R expressed in Eqs. (2-17) and (2-22) must be avoided; this does not complicate formulation of the analysis. But, if it is not done, the dramatic simplification offered by the pseudo-perfect gas approximation cannot be exploited. It should be emphasized that use of fictitious gas constants in an existing analysis that assumes the interrelation of R, c_p and c_v is extremely risky and should be avoided. Equations (2-39) through (2-43) properly formulate the thermally and calorically perfect gas model for this approximation.

To employ this model, the isentropic limits of the range of thermodynamic conditions to be approximated must be selected. For centrifugal compressor analyses, the fluid total thermodynamic conditions and the static conditions corresponding to sonic flow can be used as these limits, since they normally include all thermodynamic conditions encountered. If subscripts 1 and 2 designate the two isentropic end points of this range, the real gas model can be used to compute the fictitious gas constants using the following relations

$$\bar{R} = R(z_1 z_2)^{0.5} \tag{2-44}$$

$$\bar{c}_p = (h_2 - h_1)/(T_2 - T_1) \tag{2-45}$$

$$\bar{c}_v = (u_2 - u_1)/(T_2 - T_1) \tag{2-46}$$

$$\bar{\gamma} = \ln(p_2/p_1)/\ln(v_2/v_1) \tag{2-47}$$

Here $\bar{\gamma}$ is an isentropic exponent rather than the ratio of specific heats. As long as z_1 and z_2 are not too different, this approximation yields good accuracy.

The author employs a modular computerized equation of state package, based on Aungier (1994, 1995), which is common to all aerodynamic design and analysis software. The pseudo-perfect gas model is included in this package as an option. When that option is invoked, the full real gas model is used to compute the fictitious gas constants, but the actual analysis is conducted with the much simpler thermally and calorically perfect gas model. For most analyses, this results in a very dramatic reduction in computational time and excellent prediction accuracy.

2.9 Component Performance and Losses

The stage performance can be evaluated with the adiabatic or polytropic efficiency definitions described earlier. They also serve the same purpose for the impeller. But these definitions are meaningless for a stationary component, since, in the absence of heat transfer, the total enthalpy is constant. It is often useful to evaluate the performance of stationary components against some ideal, reversible standard. Figure 2-6 illustrates an h-s diagram for a stationary component that is diffusing the flow. The flow passes from the inlet to the discharge, undergoing an entropy increase, but, in the process, increasing the static pressure from p_i to p_d. If the purpose of the component is to diffuse the flow, it is useful to compare the actual diffusion to what would be achieved if the process were adiabatic-reversible or isentropic. As seen in Fig. 2-6, the actual change in static enthalpy is Δh. But if the process were isentropic with the same inlet and discharge static pressures, a change of Δh_{ad} would be required. Hence, a diffuser efficiency can be defined as

$$\eta_{\text{diff}} = \frac{\Delta h_{ad}}{\Delta h} \tag{2-48}$$

Another popular parameter used for diffusing components is the pressure recover coefficient, c_p, which simply quantifies the fraction of the inlet dynamic head, $p_{ti} - p_i$, that is recovered in the form of static pressure

$$c_p = \frac{p_d - p_i}{p_{ti} - p_i} \tag{2-49}$$

Although a compressor is basically a diffusing device, some components do have the purpose of accelerating the flow—for example, inlets and inlet guide vanes. Such a process is illustrated in Fig. 2-7. It is reasonable to compare the increase in kinetic energy achieved relative to what would be achieved by an isentropic

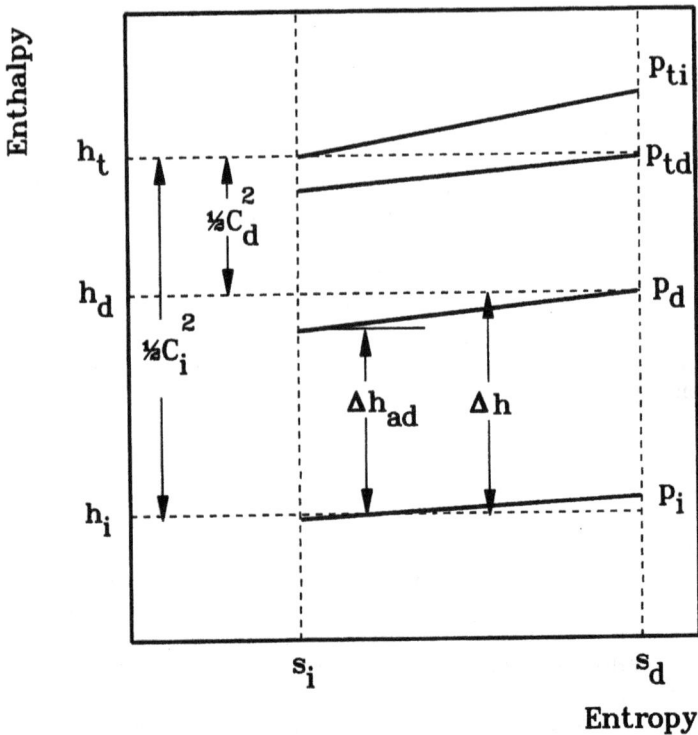

FIGURE 2-6. A Diffuser *h-s* Diagram

process, again assuming the component operates with the same inlet and discharge static pressures. This is the well-known nozzle efficiency, given by

$$\eta_{noz} = \frac{C_d^2 - C_i^2}{C_{ad}^2 - C_i^2} \tag{2-50}$$

A direct measure of irreversibility is the loss coefficient, a concept fundamental to all centrifugal compressor performance analysis and evaluation. Since performance analysis continues to depend on empirical loss models to a large degree, the usual practice is to develop empirical models for loss coefficients, typically qualified by observed loss coefficients from compressor testing. The nature of flow processes in a centrifugal compressor leads to some subtle features requiring careful consideration when defining relevant loss coefficients. The direct measure of irreversibility in any flow process is the entropy rise as defined in Eq. (2-8). Since centrifugal compressor technology had its roots in incompressible machines (pumps and blowers), early practice was to define an adiabatic head loss by

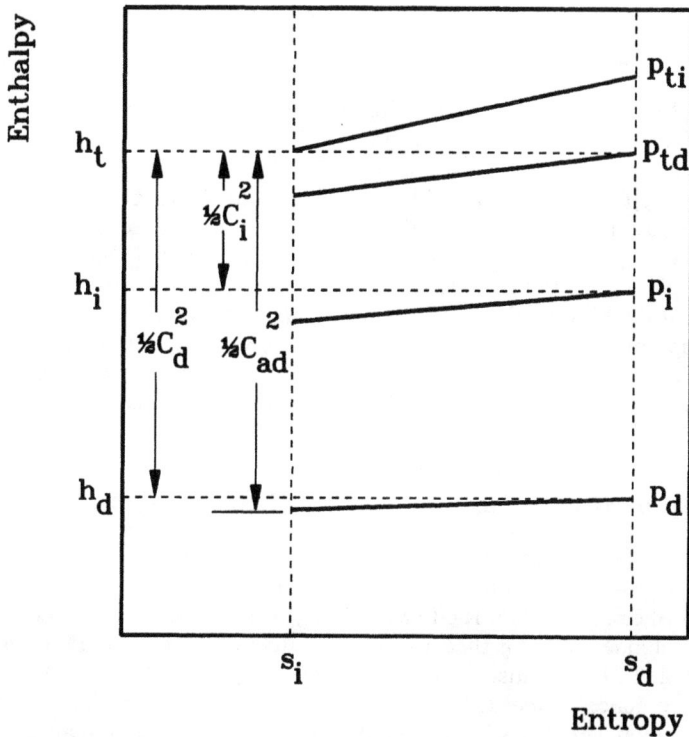

FIGURE 2-7. A Nozzle *h-s* Diagram

$$\Delta H_{ad} = \int T ds = - \int v dp \qquad (2\text{-}51)$$

This concept suffers from the same weakness as the adiabatic efficiency, as discussed in Section 2.2, namely the thermodynamic effect caused by the divergence of constant pressure lines on an *h-s* diagram. Hence, if one imposes a head loss in the analysis of a compressor, it will produce different effects depending upon the pressure ratios involved. Thus, use of the head loss concept should be limited to incompressible flow machines. Entropy rise can be used as the measure of loss but it is difficult to interpret, having little physical meaning to most people. Consequently, total pressure loss is the more popular choice. If total enthalpy is constant, Eq. (2-8) yields

$$v \Delta p_t = v(p_{ti} - p_{td}) = T \Delta s \qquad (2\text{-}52)$$

Since loss is usually proportional to kinetic energy, a logical definition for loss coefficient is obtained by dividing Eq. (2-52) by $(1/2)C^2$. But a more appropriate definition of a total pressure loss coefficient is

$$\bar{\omega} = \frac{\Delta p_t}{p_t - p} \tag{2-53}$$

which is found to better correlate losses over a wide range of Mach numbers (Aungier, 1993b, 1995). This definition works quite well for stationary components, where total enthalpy is constant. But Aungier (1993b, 1995) points out that when applied to impellers, where total enthalpy is not constant (even in the impeller's rotating frame of reference), additional care is required. If a loss is considered to occur at one location (e.g., the impeller inlet) but is applied at a different point (e.g., the impeller discharge), then Eq. (2-8) must be employed to account for the effect of the total enthalpy change on total pressure loss, such that the entropy increase remains the same. Therefore, even with a valid state-point loss parameter such as total pressure, there is potential for a thermodynamic effect that is unrelated to the true aerodynamic performance.

2.10 Approximate Liquid and Two-Phase Flow Models

In principal, the real gas equations discussed previously apply to liquid phase calculations if gas-liquid equilibrium models are incorporated (e.g., Ried et al., 1977; Ried and Sherwood, 1966; Ried et al., 1987). But in the case of fluid dynamics analyses intended only for the gas-phase, excursions into the liquid or gas-liquid region (Fig. 2-2) can cause serious convergence problems. If the solution is far from convergence, numerical errors can occasionally cause such excursions into the "wet region." Analyses intended only for the gas phase should always prevent these excursions to avoid problems. The vapor saturation pressure is known to vary linearly with $1/T$ (Ried and Sherwood, 1966). Thus, from the specified acentric factor and critical point data, a good estimate for vapor saturation line data is given by

$$\log_{10}(p_v/p_c) = 7(1 + \omega)(1 - T_c/T)/3 \tag{2-54}$$

Equation (2-54) is easily inverted to predict the saturation temperature as a function of pressure. Other vapor saturation line data can be computed from the equation of state. For gas-phase flow analyses, it is prudent to use these data to limit the values of state variables acceptable for the real gas calculations.

There are occasions when the centrifugal compressor aerodynamicist must consider liquid or two-phase fluid properties. First, it is often necessary to confirm that the application being analyzed is really in the gas-phase region. For this purpose, Eq. (2-54) can be used for single component fluids, or if a fluid mixture is approximated as a pure substance. In rare cases, fluid dynamics analysis of two-phase flows may be required. Except for the equilibrium flash calculation discussed in the next section of this chapter, rigorous liquid-vapor equilibrium models can easily lead to prohibitively long computation times. Aungier (1998) recommends approximate models for those cases. Equation (2-54) and the equation of state can be used to compute the vapor saturation line data. With suitable approximations for the liquid saturation line properties, two-phase flows can be reasonably well approximated. The heat of vaporization, Δh_{vap} (Fig. 2-2), has been approximated by Chen (1965) as

$$\Delta h_V = T(7.9T_R - 7.82 - 7.11 \log_{10}(P_{Rv})/(1.07 - T_R)/M \qquad (2\text{-}55)$$

Gunn and Yamada (1971) estimate the liquid specific volume, v_L, as

$$v_L = A(1 - \omega\Gamma)(0.292 - 0.0967\omega)RT_c/P_c \qquad (2\text{-}56)$$

where

$$\Gamma = 0.29607 - 0.09045T_R - 0.04842T_R^2 \qquad (2\text{-}57)$$

$$A = 0.33593(1 - T_R) + 1.51941T_R^2 - 2.02512T_R^3 + 1.11422T_R^4; T_R \le 0.8 \qquad (2\text{-}58)$$

$$A = 1 + 1.3\sqrt{1 - T_R}\log_{10}(1 - T_R)$$
$$- 0.50879(1 - T_R) - 0.91534(1 - T_R)^2; T_R > 0.8 \qquad (2\text{-}59)$$

Ried et al. (1977) review and evaluate several models for the liquid specific heat at constant pressure, c_{pL}. The method attributed to Sternling and Brown (no reference given) appears to be quite accurate, i.e.,

$$c_{pL} = c_p^0 + R(0.5 + 2.2\omega)[3.67 + 11.64(1 - T_R)^4 + 0.634/(1 - T_R)] \qquad (2\text{-}60)$$

where c_p^0 = ideal gas-specific heat. These approximations strictly apply to a single component fluid (i.e., a pure substance); their application to mixtures is only an approximation, since different components liquify under different conditions. The more general phase equilibrium problem is discussed in the following section.

2.11   Equilibrium Flash or Liquid Knockout Calculations

There are some situations where the approximate two-phase treatment, described in Section 2.10, cannot be employed. The most common case in centrifugal compressor aerodynamics occurs when intercoolers are present. In the process of cooling the working fluid, the intercooler will often cause certain gas mixture components to liquify (e.g., water in an air-water vapor mixture), thereby changing the composition of the compressor's working fluid and its mass flow. The process of computing the compositions and quantity of vapor and liquid phases in these cases is commonly referred to as an equilibrium flash or liquid knockout calculation. This involves computing the amount of liquid formed and the composition of the remaining vapor at the local temperature and pressure. Here, the calculation procedure for Aungier's (1994, 1995) modified version of the Redlich-Kwong equation of state will be described. After some manipulation, this equation of state can be reduced to the form

$$z^3 + (pC - 1)z^2 + (p^2CB - p^2B^2 - pB + pA)z + pA(pC - pB) = 0 \qquad (2\text{-}61)$$

where

$$A = a/[R^2 T^{(2+n)}] \qquad (2\text{-}62)$$
$$B = b/(RT) \qquad (2\text{-}63)$$
$$C = c/(RT) \qquad (2\text{-}64)$$

The process of calculating the roots of a cubic equation of this type is well-known and listed in many standard mathematics references (e.g., Selby, 1965). When two phases of the fluid exist, Eq. (2-61) will have three roots: The largest root is the vapor value, the smallest is the liquid value. Phase equilibrium is governed by a parameter called fugacity, f. For a pure substance, f is defined by

$$\ln(f/p) = \int_0^p \left(\frac{v}{RT} - \frac{1}{p} \right) dp \qquad (2\text{-}65)$$

The liquid and vapor phases will be in equilibrium when the fugacity of the two phases are equal, i.e., $f^L = f^v$. The extension to gas mixtures is accomplished by requiring the fugacities of the two phases to be equal for each mixture component, i.e., $f_i^L = f_i^v$. Ried et al. (1987) have carried out complex calculations for the mixture component fugacities for general cubic equations of state, such as the Redlich-Kwong equation and its modified forms. The result is

$$\ln\left[\frac{f_i}{px_i} \right] = \frac{b_i}{b}(z - 1) - \ln(z - B) - \frac{A}{B}\left[2\sqrt{\frac{a_i}{a}} - \frac{b_i}{b} \right] \ln\left(1 + \frac{B}{z} \right) \qquad (2\text{-}66)$$

where x_i = component mole fraction and the mixture "mixing rules" are

$$\sqrt{a} = \sum x_i \sqrt{a_i} \qquad (2\text{-}67)$$

$$b = \sum x_i b_i = \sum x_i T_{ci}/p_{ci} \qquad (2\text{-}68)$$

$$z_c = \sum x_i z_{ci} \qquad (2\text{-}69)$$

$$\omega = \sum x_i \omega_i \qquad (2\text{-}70)$$

$$M = \sum x_i M_i \qquad (2\text{-}71)$$

Let us use the notation x_i for the complete mixture mole fractions, x_{vi} for the vapor phase mole fractions, and x_{Li} for the liquid phase mole fractions. Equations (2-66) through (2-71) can be applied to both phases by substituting x_{vi} or x_{Li} for x_i. The phase equilibrium constant for a mixture component is defined as

$$K_i = x_{vi}/x_{Li} \qquad (2\text{-}72)$$

To perform a flash calculation at a temperature, T, and pressure, p, a first estimate of the equilibrium constants is

$$K_i = p_{vi}/p \qquad (2\text{-}73)$$

which can be evaluated using Eq. (2-54) for each gas component. If $K_i > 1$, it indicates that the component is entirely in the vapor phase. As a second check, solve Eq. (2-61) for each component. If this does not yield three roots, the equation of state indicates the component cannot be in the liquid phase. If these checks indicate that no liquid exists for any component, the flash calculation is complete. Otherwise, assume a value for L, the mass fraction of fluid in the liquid phase, and estimate x_{vi} and x_{Li} from the initial K_i values. Then, start an iterative calculation to establish the phase equilibrium condition. This process is as follows:

1. Apply Eqs. (2-66) through (2-71) to each component of the fluid in both the liquid and vapor phases to obtain the fugacity coefficients.
2. Calculate a new estimate of the equilibrium constants as

$$K_i = (f_i^L/x_{Li})/(f_i^v/x_{vi}) \qquad (2\text{-}74)$$

3. Calculate the new estimates of the liquid and vapor phase concentrations from

$$x_{Li} = x_i/[L + K_i(1 - L)]$$
$$x_{vi} = K_i x_{Li} \qquad (2\text{-}75)$$

4. Compute a new estimate of the liquid phase mole fraction, L, from King (1980)

$$L = L - \sum x_i \phi_i \Big/ \sum x_i \phi_i^2$$
$$\phi_i = [K_i - 1]/[K_i + (1 - K_i)L] \qquad (2\text{-}76)$$

5. Check for convergence on the quantity $\sum(x_{vi} - x_{Li}) < 0.0001$ (or a tolerance of your choice).
6. If there is no convergence, normalize x_{vi} and x_{Li} so that each adds up to 1.0 for all components, and then return to Step 1.

When converged, the mole fraction of liquid present, L, and the composition of each phase, x_{vi} and x_{Li}, are known. In the case of liquid knockout for a compressor (e.g., in an intercooler), the vapor composition is used to redefine the gas mixture and equation of state, and the compressor mass flow is reduced by the amount of liquid formed.

EXERCISES

2.1 Two compressors operate on a thermally and calorically perfect gas with a ratio of specific heats of 1.4. Both have a polytropic efficiency of 80%. One compressor has a pressure ratio of 1.5, while the other's pressure ratio is 6.0. Compute the adiabatic efficiencies of both compressors.

2.2 A thermally and calorically perfect gas with a molecular weight of 28 and a ratio of specific heats of 1.4 is flowing through a diffuser with insulated walls. The diffuser inlet total temperature is 300°K, the inlet total pressure is 200 kPa and the inlet static pressure is 170 kPa. The diffuser's static pressure recovery coefficient is 0.65 and its loss coefficient is 0.1, based on inlet dynamic head ($p_t - p$). Compute the diffuser's discharge conditions and the diffuser efficiency [Note: $R_U = 8314$ Pa-m^3/(kmole-°K)].

2.3 A compressor is to be tested with refrigerant R134a as the working fluid with an inlet pressure of 100 kPa (Note: $M = 102.031$, $T_c = 374.3$°K, $p_c = 4{,}065$ kPa, $\omega = 0.3254$). It is imperative that liquid phase flow be avoided, both to obtain valid test data and to protect the compressor from liquid erosion. Estimate the lowest inlet temperature that can be used.

2.4 A thermally and calorically perfect gas with a molecular weight of 28 and a ratio of specific heats of 1.4 is flowing through a nozzle with insulated walls. The inlet total temperature is 300°K, the inlet pressure is 100 kPa, the inlet static pressure is 85 kPa, the discharge static pressure is 70 kPa and the nozzle efficiency is 95%. Assuming the flow is uniform across the nozzle passage (i.e., one-dimensional flow), compute the inlet and discharge velocities. Compute the nozzle loss coefficient [Note: $R_U = 8314$ Pa-m^3/(kmole-°K)].

Chapter 3

FLUID MECHANICS

Fluid dynamics and thermodynamics are the fundamental sciences used in centrifugal compressor design and analysis. This chapter develops the fundamental fluid dynamics concepts and governing equations used throughout the book. Fluid dynamics analysis is applied in several forms for centrifugal compressors. One-dimensional analysis with empirical work input and loss models is the basis for most aerodynamic performance analysis. Detailed aerodynamic design is normally based on two- and three-dimensional inviscid flow analysis, often supported by boundary layer analysis. Computational methods based on this technology are commonly referred to as Euler codes to distinguish them from viscous flow codes; these codes are based on a mature technology. They can and should be available to any centrifugal compressor designer. Representative methods for all of these analysis techniques are covered in detail here.

The use of viscous flow computational fluid dynamics (CFD) codes is rapidly becoming fairly standard practice. Most centrifugal compressor design groups now employ one or more of the several commercially available CFD codes. The time, resources and expertise needed for CFD code development are substantial, making it impractical for the average compressor design organization. In particular, CFD technology is changing so rapidly that only investigators dedicated to it can hope to stay abreast of the latest developments. Consequently, CFD technology is not covered in this book, beyond recognizing its emerging role as an important analysis tool.

Currently, CFD is used for the final evaluation of specific centrifugal compressor component designs. Computer running time for CFD analyses is simply too long to employ them in the basic design phase, where numerous alternatives and refinements must be investigated. Also, right now CFD is employed in a somewhat qualitative fashion. The flow fields in centrifugal compressors are too complex to be completely modeled by current CFD technology. Both fundamental numerical methods and turbulence modeling require considerable advancement before CFD can supply definitive quantitative results. Thus, for example, basic one-dimensional performance analysis continues to be more reliable than performance estimates from CFD codes. Nevertheless, CFD adds an important new dimension to the aerodynamic analysis of centrifugal compressors. It is now possible to visualize the critical viscous effects that really govern the performance level of a compressor component.

NOMENCLATURE

a = sound speed
b = stream sheet thickness
C = absolute velocity
E = boundary layer entrainment function
$\vec{e}$ = unit vector
f = body force
H = boundary layer shape factor
h = enthalpy
M = Mach number
m = meridional coordinate
$\dot{m}$ = mass flow
n = normal coordinate
p = pressure
R = rothalpy
r = radius
$\vec{r}$ = general position vector of a point in space
s = entropy
T = temperature
t = time
W = relative velocity
$\dot{w}$ = power
y = distance normal to a wall
z = axial coordinate
α_C = streamline slope angle with axis
δ = boundary layer thickness
δ^* = boundary layer displacement thickness
θ = tangential coordinate (polar angle) and boundary layer momentum thickness
κ = curvature
ρ = gas density
τ = torque and boundary layer shear stress
ϕ = general function
ω = rotation speed

Subscripts

e = boundary layer edge parameter
h = parameter on the hub contour
m = meridional component
n = normal component
r = radial component
s = parameter on the shroud contour
t = total thermodynamic condition
U = tangential component

w = parameter at a wall
0 = impeller eye condition
1 = impeller blade inlet condition
2 = impeller tip condition
θ = θ component

Superscripts

* = sonic flow condition
′ = value relative to rotating frame of reference

3.1 Flow in a Rotating Coordinate System

The analysis of the flow in impellers is best accomplished in a coordinate system that rotates with the impeller. A general curvilinear coordinate system (θ, m, n) will be used, where m is measured along a stream surface, θ is the usual polar angle of cylindrical coordinates and n is normal to the stream surface. A stream surface is defined as having no fluid velocity component normal to it, i.e., $W_n = 0$. This coordinate system is illustrated in Figs. 3-1 and 3-2. If we designate velocities in the stationary or absolute frame of reference as C and relative velocities in the rotating coordinates as W, they can be related by

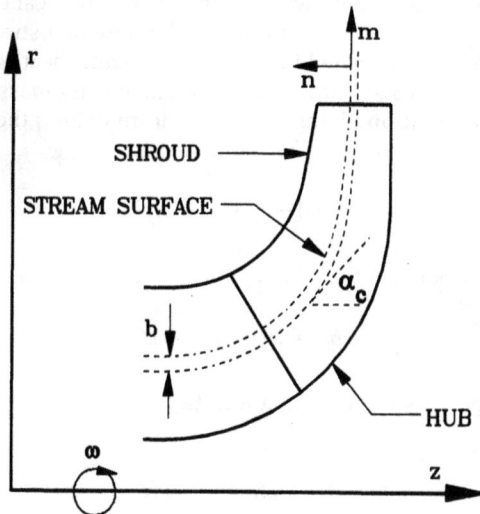

FIGURE. 3-1. Curvilinear Coordinate System

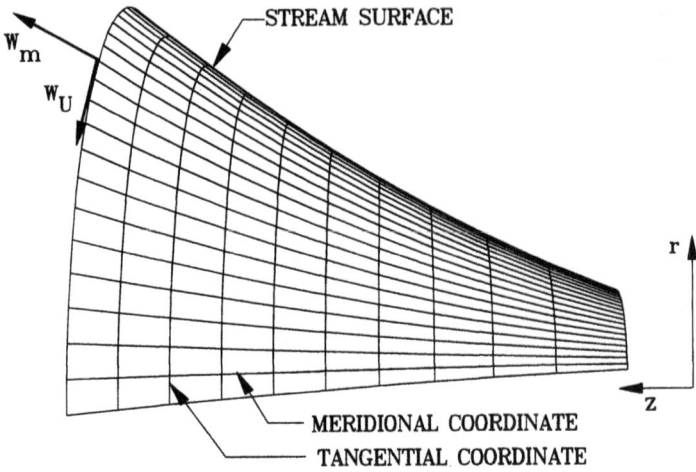

FIGURE. 3-2. Coordinates in a Stream Surface

$$W_U = C_U - \omega r$$
$$W_m = C_m$$
$$W = \sqrt{W_m^2 + W_U^2} \tag{3-1}$$

where the subscripts m and U designate the meridional and tangential velocity components, respectively.

The impeller is the component which transfers mechanical energy to the fluid to increase its energy and pressure. Consider the flow in a stream sheet passing through the impeller, as illustrated in Fig. 3-1. A stream sheet is just a thin annular passage bound by two stream surfaces such that its mass flow is constant for steady flow. Conservation of angular momentum defines the torque, τ, on the fluid in the stream sheet, supplied by the impeller.

$$\tau = \dot{m}[r_2 C_{U2} - r_1 C_{U1}] \tag{3-2}$$

This torque must balance the power input, i.e.,

$$\dot{w} = \omega\tau = \dot{m}\omega[r_2 C_{U2} - r_1 C_{U1}] \tag{3-3}$$

The basic energy conservation equation, Eq. (2-4), can be combined with Eq. (3-3) to yield

$$h_{t2} - h_{t1} = \omega(r_2 C_{U2} - r_1 C_{U1}) \tag{3-4}$$

which is the well-known Euler turbine equation. If we define rothalpy, R, by

$$R = h_t - \omega r C_U \tag{3-5}$$

it can be seen from Eq. (3-4) that rothalpy is conserved on streamlines through the impeller. Rothalpy is the basic parameter expressing energy conservation for rotating blade rows such as impellers. Indeed, it can be used for any component since it is identical to h_t if $\omega = 0$. In the absence of work input to the fluid, Eq. (2-4) requires that h_t be conserved. The fluid dynamics analysis of centrifugal compressors requires systematic relationships for the flow in the stationary and rotating frames of reference. Equations (3-1) supply the relationship for velocity. If we note that static thermodynamic conditions are identical for both frames, the relative total enthalpy in rotating coordinates, h_t', can be related to h_t by

$$h = h_t' - \frac{W^2}{2} = h_t - \frac{C^2}{2} \tag{3-6}$$

From Eqs. (3-1) and (3-5), this yields

$$h_t' = h_t - \omega r C_U + \frac{(\omega r)^2}{2} = R + \frac{(\omega r)^2}{2} \tag{3-7}$$

Since entropy is identical in the two frames of reference, all other relative total thermodynamic parameters can be calculated from the equation of state as a function of (h_t', s). This typically requires computation of the isentropic change in the parameter of interest from its static value with a change in enthalpy from h to h_t'. Hence, all fluid dynamic and thermodynamic parameters in either frame of reference can be computed directly from those in the other frame of reference.

3.2 Governing Equations for Adiabatic Inviscid Compressible Flow

Adiabatic inviscid flow analysis involves solving the basic conservation equations for mass, momentum and energy under the assumption that the gas viscosity and thermal conductivity are both zero. These equations will be developed in the rotating coordinate system for generality, noting that if $\omega = 0$, the equations reduce to the stationary frame of reference form. In the vector form, the momentum equation for compressible, inviscid flow (see, e.g., Novak, 1967; Wu, 1952; Vavra, 1960) is

$$\frac{d\vec{C}}{dt} = -\frac{1}{\rho} \vec{\nabla} p = \frac{d\vec{W}}{dt} + 2(\vec{\omega} \times \vec{W}) + \vec{\omega} \times (\vec{\omega} \times \vec{r}) \tag{3-8}$$

where time derivative is the substantial time derivative and the last two terms in Eq. (3-8) are the Coriolis and centrifugal accelerations imposed by the rotating coordinates. Using the definition of the substantial derivative

$$\frac{\partial \vec{W}}{\partial t} + (\vec{W} \cdot \vec{\nabla})\vec{W} + 2(\vec{\omega} \times \vec{W}) + \vec{\omega} \times (\vec{\omega} \times \vec{r}) = -\frac{1}{\rho} \vec{\nabla} p \qquad (3-9)$$

Using standard vector identities, Eq. (3-9) can be written

$$\frac{\partial \vec{W}}{\partial t} - \vec{W} \times (\vec{\nabla} \times \vec{W} + 2\ \vec{\omega}) - \omega^2 r \vec{e}_r + \frac{1}{2} \vec{\nabla} W^2 = -\frac{1}{\rho} \vec{\nabla} p \qquad (3-10)$$

Introducing Eqs. (2-3), (2-8), (3-5) and (3-6), an alternative form is obtained

$$\frac{\partial \vec{W}}{\partial t} - \vec{W} \times (\vec{\nabla} \times \vec{W} + 2\ \vec{\omega}) = -\vec{\nabla} R + T\ \vec{\nabla} s \qquad (3-11)$$

The continuity and energy equations are

$$\frac{\partial \rho}{\partial t} + \vec{\nabla} \cdot (\rho \vec{W}) = 0 \qquad (3-12)$$

$$\frac{\partial R}{\partial t} - \frac{1}{\rho} \frac{\partial p}{\partial t} + (\vec{W} \cdot \vec{\nabla})R = 0 \qquad (3-13)$$

The vector equations, Eqs. (3-10) through (3-13), are general to any coordinate system. To express these governing equations in the (θ, m, n) coordinate system, standard curvilinear coordinate transformations are used, which can be found in most advanced calculus textbooks that cover vector field theory (e.g., Sokolnikoff and Redheffer, 1958). Appendix A in Vavra (1960) provides a detailed derivation of the relevant vector operators and the governing equations in the (θ, m, n) coordinate system. For general reference, the primary vector operators are listed at the end of this chapter without derivation. The resulting equations are

$$\frac{\partial \rho}{\partial t} + \frac{1}{r} \left[\frac{\partial r \rho W_m}{\partial m} + \frac{\partial \rho W_U}{\partial \theta} \right] + \kappa_n \rho W_m = 0 \qquad (3-14)$$

$$\frac{\partial W_m}{\partial t} + W_m \frac{\partial W_m}{\partial m} + \frac{W_U}{r} \frac{\partial W_m}{\partial \theta} - \frac{\sin \alpha_C}{r} [W_U + \omega r]^2 = -\frac{1}{\rho} \frac{\partial p}{\partial m} \qquad (3-15)$$

$$\frac{\partial W_U}{\partial t} + W_m \frac{\partial W_U}{\partial m} + \frac{W_U}{r} \frac{\partial W_U}{\partial \theta} + \frac{W_m \sin \alpha_C}{r} [W_U + 2\omega r] = -\frac{1}{r\rho} \frac{\partial p}{\partial \theta} \qquad (3-16)$$

$$\kappa_m W_m^2 + \frac{\cos \alpha_C}{r} [W_U + \omega r]^2 = \frac{1}{\rho} \frac{\partial p}{\partial n} \qquad (3-17)$$

$$\frac{\partial R}{\partial t} - \frac{1}{\rho} \frac{\partial p}{\partial t} + W_m \frac{\partial R}{\partial m} + \frac{W_U}{r} \frac{\partial R}{\partial \theta} = 0 \qquad (3-18)$$

where we have noted that $W_n = 0$. The stream sheet curvature, κ_m, and n-surface curvature, κ_n, are computed from the stream surface angle, α_C, and stream sheet thickness, b (Fig. 3-1).

$$\kappa_m = -\frac{\partial \alpha_C}{\partial m}$$

$$\kappa_n = \frac{\partial \alpha_C}{\partial n} = \frac{1}{b}\frac{\partial b}{\partial m} \tag{3-19}$$

Equations (3-15) through (3-17) can be written in the alternate form from vector equations, Eq. (3-11)

$$\frac{\partial W_m}{\partial t} + \frac{W_U}{r}\left[\frac{\partial W_m}{\partial \theta} - \frac{\partial (rW_U + \omega r^2)}{\partial m}\right] = -\frac{\partial R}{\partial m} + T\frac{\partial s}{\partial m} \tag{3-20}$$

$$\frac{\partial W_U}{\partial t} - \frac{W_m}{r}\left[\frac{\partial W_m}{\partial \theta} - \frac{\partial (rW_U + \omega r^2)}{\partial m}\right] = -\frac{\partial R}{\partial \theta} + T\frac{\partial s}{\partial \theta} \tag{3-21}$$

$$\kappa_m W_m^2 + \frac{W_U}{r}\frac{\partial (rW_U + \omega r^2)}{\partial n} + W_m\frac{\partial W_m}{\partial n} = \frac{\partial R}{\partial n} - T\frac{\partial s}{\partial n} \tag{3-22}$$

It should be noted that one of the momentum equations is redundant in this (θ, m, n) coordinate system. This follows from the fact that there are only two velocity components, i.e.,

$$\vec{W} = W_m \vec{e}_m + W_U \vec{e}_U \tag{3-23}$$

In effect, one of the momentum equations has been replaced by the assumption that stream surface geometries are known. In fact, they are not, so the problem hasn't truly been simplified. If specific "stream surface" geometries were specified in advance, they would undoubtedly no longer be true stream surfaces—rather, they would be general curvilinear coordinate surfaces. In that case, our governing equations would have to include a velocity component, W_n, since it no longer needs to be zero. The various fluid dynamic analyses described in this book do not require this more general form. Nevertheless, in the interest of completeness, the general form of Eqs. (3-14) through (3-18) in curvilinear coordinates is given here.

$$\frac{\partial \rho}{\partial t} + \frac{1}{r}\left[\frac{\partial r\rho W_m}{\partial m} + \frac{\partial \rho W_U}{\partial \theta} + \frac{\partial r\rho W_n}{\partial n}\right] + \kappa_n \rho W_m + \kappa_m W_n = 0 \qquad (3\text{-}24)$$

$$\frac{\partial W_m}{\partial t} + W_m\frac{\partial W_m}{\partial m} + W_n\frac{\partial W_m}{\partial n} + \frac{W_U}{r}\frac{\partial W_m}{\partial \theta} - \frac{\sin \alpha_C}{r}[W_U + \omega r]^2$$

$$- \kappa_n W_n^2 + \kappa_m W_n W_m = -\frac{1}{\rho}\frac{\partial p}{\partial m} \qquad (3\text{-}25)$$

$$\frac{\partial W_U}{\partial t} + W_m\frac{\partial W_U}{\partial m} + W_n\frac{\partial W_U}{\partial n} + \frac{W_U}{r}\frac{\partial W_U}{\partial \theta} + \frac{1}{r}[W_m \sin \alpha_C$$

$$+ W_n \cos \alpha_C][W_U + 2\omega r] = -\frac{1}{r\rho}\frac{\partial p}{\partial \theta} \qquad (3\text{-}26)$$

$$\frac{\partial W_n}{\partial t} + W_m\frac{\partial W_n}{\partial m} + W_n\frac{\partial W_n}{\partial n} + \frac{W_U}{r}\frac{\partial W_n}{\partial \theta} - \frac{\cos \alpha_C}{r}[W_U + \omega r]^2$$

$$+ \kappa_n W_n W_m - \kappa_m W_m^2 = -\frac{1}{\rho}\frac{\partial p}{\partial n} \qquad (3\text{-}27)$$

$$\frac{\partial R}{\partial t} - \frac{1}{\rho}\frac{\partial p}{\partial t} + W_m\frac{\partial R}{\partial m} + w_n\frac{\partial R}{\partial n} + \frac{W_\theta}{r}\frac{\partial R}{\partial \theta} = 0 \qquad (3\text{-}28)$$

3.3 Adiabatic Inviscid Compressible Flow Analysis

The governing equations developed in the previous section can be employed in several forms to support centrifugal compressor design and analysis activity. The most general application would be a three-dimensional, time-unsteady solution using Eqs. (3-24) through (3-28). But such an analysis would involve computer running times comparable to a CFD code, without the benefit of treating viscous effects. With the emergence of viable CFD technology, a full three-dimensional, time-unsteady Euler code has little merit. Rather, Euler codes need to provide computationally fast analyses suitable for the rapid iteration process inherent to the centrifugal compressor design process—a capability that CFD currently does not have.

A common application is solution of the flow in a stream sheet, commonly referred to as a blade-to-blade flow analysis. This requires that the geometry of the stream sheet be specified, either by an estimate or from some other analysis. It is common to assume that the stream sheet is an axisymmetric surface, although that is not a necessary assumption. The classical approach is to solve the time-steady form of the equations and to assume the flow at the inlet to the solution domain is axisymmetric. Under these assumptions it is necessary that the gradients of R and s be zero. This implies that R and s are constant, although variations can be treated. Variations in R and s can be specified as being supplied by some external mechanism. The only requirement is that their local gradients be zero. Thus, a distributed entropy increase could be imposed to match discharge conditions obtained from experiment or performance analyses. Under these conditions, the time-steady form of Eqs. (3-14) and (3-21) are typically solved. One difficulty regularly encountered is that the basic mathematical character of the governing equations depends on the Mach number. When $M < 1$, the equations are elliptic in form, while when $M > 1$ they are hyperbolic. These two types of equations require different numerical methods for solution. Elliptic equations represent a boundary value problem, where the solution is determined by the conditions imposed at all boundaries. Hyperbolic equations represent an initial-value problem, where the solution involves a marching technique (e.g., the method of characteristics). For centrifugal compressors, the steady, blade-to-blade flow analysis is normally used for subsonic flow. When treating problems involving mixed subsonic-supersonic flows, a common approach is to employ the time-unsteady form of the equations of motion, e.g., Eqs. (3-14), (3-15), (3-16) and (3-18). These equations are hyperbolic in time, regardless of the Mach number level. The procedure is to advance the solution in time until the flow asymptotically approaches its steady-state solution. This technique is commonly referred to as the "time-marching" or "time-dependent" method.

Similarly, a two-dimensional solution can be computed for a stream sheet extending from the hub to the shroud, commonly called a hub-to-shroud solution. Again, the stream sheet geometry must be supplied, either as an estimate or from another analysis. A variant on this is to solve the flow as an axisymmetric one, viewed as the average flow in the hub-to-shroud direction. In this case, W_U or flow angle data must be supplied. By solving Eqs. (3-14), (3-22) and (3-18) in their time-steady, axisymmetric form, this type of solution is fairly direct.

One can solve both the hub-to-shroud and blade-to-blade solutions iteratively, obtaining the required specified data for each solution from the other solution, until two analyses are consistent with each other. This is referred to as a quasi-three-dimensional flow analysis. Originally suggested by Wu (1952), it is a rather common type of Euler code used in centrifugal compressor design.

3.4 Boundary Layer Analysis

It is often useful to employ boundary layer analysis to evaluate viscous effects not considered by inviscid flow analyses. The basic assumption of boundary layer theory is that viscous effects are confined to a thin layer close to the physical surfaces. This is far from true for a centrifugal compressor, but there are enough situations where the boundary layer approximations are reasonable to warrant including a boundary layer calculation to support the inviscid flow analyses. Boundary layers in centrifugal compressors are always highly three-dimensional in nature. Except for the axisymmetric swirling flow case, the three-dimensional boundary layer problem is very complex, leading to excessive computer running times. Like the general three-dimensional inviscid flow analysis, attempts to solve the general three-dimensional boundary layer problem has little merit today. It is normally more logical to employ a viscous CFD code. Today, the principal use of boundary layer analysis in centrifugal compressor aerodynamics is for qualitative evaluation of viscous effects, employing either two-dimensional or axisymmetric, three-dimensional models.

The governing equations for two-dimensional boundary layer flow over an adiabatic wall in (x, y) coordinates with corresponding velocity components (u, v) are

$$\frac{\partial \rho b u}{\partial x} + \frac{\partial \rho b v}{\partial y} = 0 \tag{3-29}$$

$$u \frac{\partial u}{\partial x} + v \frac{\partial u}{\partial y} + \frac{1}{\rho} \frac{\partial p}{\partial x} = \frac{1}{\rho} \frac{\partial \tau}{\partial y} \tag{3-30}$$

where τ = shear stress. The stream sheet thickness term, b, is included in Eq. (3-29) in anticipation of its use for cases where b is a function of x, i.e., where the wall surface streamlines are converging or diverging. The usual boundary layer assumption is that the pressure is constant across the boundary layer. The usual approach for turbomachinery applications is to integrate Eqs. (3-29) and (3-30) across the boundary layer to obtain the well-known momentum integral equation. Integration of Eq. (3-29) across the boundary layer yields

$$\frac{\partial}{\partial x} \int_0^\delta b \rho u \, dy = b \rho_e u_e \frac{\partial \delta}{\partial x} - b \rho_e v_e = \frac{\partial}{\partial x} [b \rho_e u_e (\delta - \delta^*)] \tag{3-31}$$

where the Liebnitz rule is used to interchange the order of integration and differentiation; the subscript e denotes boundary layer edge (or inviscid flow) conditions, and the displacement thickness, δ^*, is defined by

$$\rho_e u_e \delta^* = \int_0^\delta [\rho_e u_e - \rho u] \, dy \tag{3-32}$$

The displacement thickness has a simple interpretation illustrated by Eq. (3-31). A fictitious thickness will conserve mass in the boundary layer if the flow is assumed to be identical to the free stream flow except for the thickness δ^*, for which the mass flow is assumed to be zero. it is sometimes called the mass-defect thickness. Combining Eqs. (3-29) and (3-30), the momentum equation becomes

$$\frac{1}{b} \frac{\partial b \rho u^2}{\partial x} + \frac{\partial \rho u v}{\partial y} + \frac{\partial p}{\partial x} = \frac{\partial \tau}{\partial y} \tag{3-33}$$

Defining the momentum thickness, θ, by

$$\rho_e u_e^2 \theta = \int_0^\delta \rho u [u_e - u] \, dy \tag{3-34}$$

which can be interpreted similar to δ^*, but for momentum; Eq. (3-33) becomes

$$\frac{1}{b}\frac{\partial}{\partial x}\int_0^\delta b\rho u^2\,dy - \rho_e u_e^2\frac{\partial \delta}{\partial x} + \rho_e u_e v_e + \delta\frac{\partial p}{\partial x} = -\tau_w \tag{3-35}$$

where τ_w is the wall shear stress. Combining Eqs. (3-31) and (3-35)

$$\frac{1}{b}\frac{\partial}{\partial x}[b\rho_e u_e^2(\delta - \delta^* - \theta)] - \frac{u_e}{b}\frac{\partial}{\partial x}[b\rho_e u_e(\delta - \delta^*)] + \delta\frac{\partial p}{\partial x} = -\tau_w \tag{3-36}$$

Imposing the usual boundary layer approximation that pressure is constant across the boundary layer, Eq. (3-36) can be rearranged to yield

$$\frac{1}{b}\frac{\partial b\rho_e u_e^2\theta}{\partial x} + \delta^*\rho_e u_e\frac{\partial u_e}{\partial x} - \tau_w = \delta\left[\frac{\partial p}{\partial x} + \rho_e u_e\frac{\partial u_e}{\partial x}\right] \tag{3-37}$$

The right-hand-side of Eq. (3-37) is zero, as can be seen by evaluating Eq. (3-30) at the boundary layer edge, where the derivatives of u and τ with respect to y are zero. Hence, we obtain the well-known momentum integral equation for two-dimensional boundary layers.

$$\frac{1}{b}\frac{\partial b\rho_e u_e^2\theta}{\partial x} + \rho_e u_e\delta^*\frac{\partial u_e}{\partial x} = \tau_w \tag{3-38}$$

Noting that the stream sheet thickness, b, for two-dimensional, axisymmetric flow (i.e., $C_U = 0$) is proportional to r, the momentum integral equation for that case is

$$\frac{1}{r}\frac{\partial r\rho_e u_e^2\theta}{\partial x} + \rho_e u_e\delta^*\frac{\partial u_e}{\partial x} = \tau_w \tag{3-39}$$

For laminar boundary layers, the solution procedure typically involves assuming specific flow profile shapes to permit integration of the momentum integral equation. In the case of turbulent boundary layers, several empirical models are required for solution, and may include boundary layer flow profile assumptions. Usually turbulent boundary layer analysis involves solving a second conservation equation, e.g., mass, moment of momentum or energy (see Rotta, 1996, for a discussion of various alternatives). For example, mass conservation, known as the entrainment equation, is developed from Eq. (3-31)

$$\frac{d}{dx}[b\rho_e u_e(\delta - \delta^*)] = b\rho_e u_e E \tag{3-40}$$

where E is an empirical entrainment function governing the rate at which fluid is entrained from the inviscid free stream into the boundary layer. The premise is that this entrainment rate can be predicted with sufficient accuracy using an empirical model derived from experiment.

Two-dimensional boundary layer analyses can be conducted along the inviscid flow streamlines at the blade or end-wall surfaces to provide a qualitative evaluation of expected viscous effects. However, this neglects the secondary flows or cross-flows (flow normal to the x-direction), which are always substantial in centrifugal compressors. These calculations are computationally very fast and yield more insight than an inviscid flow analysis alone. When a CFD code is used in a final evaluation, the designer soon develops a "calibration" on the interpretation of the simple boundary layer results that can significantly reduce the need to modify designs to obtain satisfactory CFD results. Hence, these simple boundary layer analyses are well worth doing.

Several investigators have employed axisymmetric three-dimensional boundary layer analyses for annular vaneless passages, such as inlet passages, vaneless diffusers and crossover bends (e.g., Davis, 1976; Senoo et al., 1977; Aungier, 1988). They are also used within blade passages to provide an evaluation of the average or mean boundary layer behavior, as discussed by Horlock (1970). This application is similar to the approach used for hub-to-shroud flow inviscid analysis, as shown in the previous section. The boundary layer equations for this case can be written as

$$\frac{1}{r}\frac{\partial r\rho W_m}{\partial m} + \frac{\partial \rho W_y}{\partial y} = 0 \tag{3-41}$$

$$W_m\frac{\partial W_m}{\partial m} + W_y\frac{\partial W_m}{\partial y} - \frac{\sin \alpha_C}{r}(W_U + \omega r)^2 = \frac{1}{\rho}\left[f_m - \frac{\partial p_e}{\partial m} - \frac{\partial \tau_m}{\partial y}\right] \tag{3-42}$$

$$W_m\frac{\partial W_U}{\partial m} + W_y\frac{\partial W_U}{\partial y} + \frac{\sin \alpha_C}{r}W_m(W_U + 2\omega r) = \frac{1}{\rho}\left[f_U - \frac{\partial \tau_U}{\partial y}\right] \tag{3-43}$$

Since a primary boundary condition for boundary layer analysis is that the velocity is zero at the wall, it is most convenient to perform the analysis in a coordinate system fixed to the wall. For turbomachinery, the end wall may be either rotating or stationary. Consequently, the foregoing equations are written for the rotating coordinate system but apply to a stationary wall if ω is set to zero. The terms f_m and f_U are body force terms, included for cases where blade forces act on the fluid. These blade forces can be evaluated from the boundary layer edge flow data, directly from the momentum equations, i.e.

$$f_{me} = \rho_e W_{me}\frac{\partial W_{me}}{\partial m} + \frac{\partial p_e}{\partial m} - \rho_e(W_{Ue} + \omega r)^2\frac{\sin \alpha_C}{r} \tag{3-44}$$

$$f_{Ue} = \rho_e W_{me}\left[\frac{\partial W_{Ue}}{\partial m} + \frac{\sin \alpha_C}{r}(W_{Ue} + 2\omega r)\right] = \frac{\rho_e W_{me}}{r}\frac{\partial r C_{Ue}}{\partial m} \tag{3-45}$$

The momentum integral equations can be developed in the manner similar to that for the two-dimensional case, but with somewhat more tedious algebra. The result is

$$\frac{\partial}{\partial m}[r\rho_e W_{me}(\delta - \delta^*)] = r\rho_e W_e E \tag{3-46}$$

$$\frac{\partial}{\partial m}[r\rho_e W_{me}^2 \theta_{11}] + \delta_1^* r\rho_e W_{me}\frac{\partial W_{me}}{\partial m} - \sin\alpha_C\rho_e W_{Ue}[W_{Ue}(\delta_2^* + \theta_{22}) + 2\omega r\delta_2^*]$$

$$= r\tau_{mw} + r\delta(f_{me} - f_m) \tag{3-47}$$

$$\frac{\partial}{\partial m}[r^2\rho_e W_{me}W_{Ue}\theta_{12}] + r\delta_1^*\rho_e W_{me}\left[r\frac{\partial W_{Ue}}{\partial m} + \sin\alpha_C(W_{Ue} + 2\omega r)\right]$$

$$= r^2\tau_{Uw} + r^2\delta(f_{Ue} - f_U) \tag{3-48}$$

where the various defect thicknesses are defined by

$$\rho_e W_{me}\delta_1^* = \int_0^\delta (\rho_e W_{me} - \rho W_m)\,dy \tag{3-49}$$

$$\rho_e W_{me}^2 \theta_{11} = \int_0^\delta \rho W_m(W_{me} - W_m)\,dy \tag{3-50}$$

$$\rho_e W_{me}W_{Ue}\theta_{12} = \int_0^\delta \rho W_m(W_{Ue} - W_U)\,dy \tag{3-51}$$

$$\rho_e W_{Ue}\delta_2^* = \int_0^\delta (\rho_e W_{Ue} - \rho W_U)\,dy \tag{3-52}$$

$$\rho_e W_{Ue}^2 \theta_{22} = \int_0^\delta \rho W_U(W_{Ue} - W_U)\,dy \tag{3-53}$$

If it is assumed that the blade force is constant through the boundary layer when analyzing end-wall boundary layers, the last term in both Eqs. (3-47) and (3-48) are zero. They are retained here for cases where body force defects are to be included. For example, Aungier (1988) employs this type of terms to handle the special cases of merged or separated boundary layers. Smith, Jr. (1970) and Hunter and Cumpsty (1982) report experimental measurements that show the existence of tangential force defects in axial flow compressors, i.e., $f_U < f_{Ue}$. There is little doubt that similar effects are present in centrifugal compressors. And since blade forces are approximately normal to the flow streamlines, it is clear that the meridional blade force must show a similar defect. When conducting end-wall boundary layer analyses within blade rows, consideration of the blade force defects would appear essential to realistically represent the physics of the problem.

3.5   Vector Operators

The basic vector operators—gradient, divergence, curl and Laplacian—expressed in an axisymmetric (θ, m, n) coordinate system are listed here for general reference. Readers interested in complete derivations should consult Vavra (1960, Appendix A).

$$\vec{\nabla}\phi = \frac{\partial\phi}{\partial m}\,\vec{e}_m + \frac{\partial\phi}{\partial n}\,\vec{e}_n + \frac{1}{r}\frac{\partial\phi}{\partial\theta}\,\vec{e}_\theta \tag{3-54}$$

$$\vec{\nabla}\cdot\vec{V} = \frac{1}{r}\frac{\partial r V_m}{\partial m} + \frac{1}{r}\frac{\partial r V_n}{\partial n} + \frac{1}{r}\frac{\partial V_\theta}{\partial\theta} + \kappa_n V_m + \kappa_m V_n \tag{3-55}$$

$$\vec{\nabla}\times\vec{V} = \left[\frac{\partial V_n}{\partial m} + \kappa_n V_n - \frac{\partial V_m}{\partial n} - \kappa_m V_m\right]\vec{e}_\theta + \frac{1}{r}\left[\frac{\partial r V_\theta}{\partial n} - \frac{\partial V_n}{\partial\theta}\right]\vec{e}_m$$

$$+ \frac{1}{r}\left[\frac{\partial V_m}{\partial\theta} - \frac{\partial r V_\theta}{\partial m}\right]\vec{e}_n \tag{3-56}$$

$$\nabla^2\phi = \frac{1}{r^2}\frac{\partial^2\phi}{\partial\theta^2} + \frac{\partial^2\phi}{\partial m^2} + \frac{\partial^2\phi}{\partial n^2} + \left[\frac{1}{r}\frac{\partial r}{\partial m} + \kappa_n\right]\frac{\partial\phi}{\partial m} + \left[\frac{1}{r}\frac{\partial r}{\partial n} + \kappa_m\right]\frac{\partial\phi}{\partial n} \tag{3-57}$$

The convective derivative can be evaluated using the above operators with the vector identity

$$(\vec{V}\cdot\vec{\nabla})\,\vec{V} = \tfrac{1}{2}\vec{\nabla}V^2 - \vec{V}\times(\vec{\nabla}\times\vec{V}) \tag{3-58}$$

where $\vec{\nabla}V^2$ follows directly from the gradient operator, i.e.,

$$\tfrac{1}{2}\vec{\nabla}V^2 = \tfrac{1}{2}\vec{\nabla}(\vec{V}\cdot\vec{V}) = V\frac{\partial V}{\partial m}\,\vec{e}_m + V\frac{\partial V}{\partial n}\,\vec{e}_n + \frac{V}{r}\frac{\partial V}{\partial\theta}\,\vec{e}_\theta \tag{3-59}$$

$$V = \sqrt{V_m^2 + V_n^2 + V_\theta^2} \tag{3-60}$$

EXERCISES

3.1   Equations (3-14) through (3-18) govern general inviscid, adiabatic flow with swirl. These equations can be simplified to a one-dimensional flow approximation by neglecting variations in the flow parameters across the passage. Derive the governing equations for one-dimensional, time-steady flow in a (nonrotating) vaneless passage, where the flow is circumferentially uniform. The passage width, b, can vary with the meridional distance, m.

3.2   For the same assumptions as in Exercise 3.1, use Eq. (3-21) to derive an equation governing the variation in C_U along a stream surface. Compare the result to the tangential momentum equation derived in Exercise 3.1.

3.3 For a two-dimensional boundary layer, express the integrated mass (ρu) and momentum (ρu^2) fluxes in the boundary layer in terms of the boundary layer edge flow parameters, δ, δ^* and θ.

3.4 A two-dimensional boundary layer analysis is conducted for a simple two-dimensional channel with a constant passage height, b, having two identical boundary layers. For each boundary layer, the data at the end of the passage are ρ_e, p_e, u_e, δ, δ^* and θ. A useful approximation to convert boundary layer data to a total pressure loss is based on "instant mixing" of the boundary layer fluid with the free stream fluid at constant static pressure to form a (fictitious) uniform flow. For simplicity, assume the fluid is incompressible (i.e., ρ = constant and $p_t = p + \frac{1}{2}\rho u^2$). By requiring conservation of mass and momentum between the actual flow at the end of the passage and the "mixed-out" uniform flow, show that the total prssure loss for this case is given by $\Delta p_t = \rho u_e^2 [2\theta/b + 2(\delta^*/b)^2]$.

3.5 A one-dimensional performance analysis for impellers provides all flow data and stream surface geometry on the mean stream surface. An important factor it does not consider is the matching between the flow angles on the hub and shroud at the blade's leading edge, α_{1h} and α_{1s} (where $\tan\alpha = W_m/W_U$), and the corresponding leading edge blade angles. Develop an expression to evaluate these flow angles in terms of mean stream surface data (r_1, W_{m1}, κ_{m1}, etc.) rotation speed, ω, and the hub and shroud radii, r_{1h} and r_{1s}, for a full-inducer impeller such as that shown in Fig. 1-3. Assume that total enthalpy, h_t, entropy, s, and angular momentum, rC_U, are all constant upstream of the impeller blade's leading edge.

Chapter 4

The Impeller Work Input

The total enthalpy rise imparted to the fluid by the impeller is termed the impeller work input. An accurate prediction of the impeller work input is fundamental to all aspects of centrifugal compressor aerodynamic design and analysis. As discussed in Section 1.3, the work input supplied by the impeller blades is the useful work, which supplies the pressure rise. There are also various sources of parasitic work, which can be considered wasted energy. The fluid in the clearance gaps between the impeller and the housing exert frictional forces on the impeller, consuming part of the power supplied to it. This is a parasitic work commonly called the windage and disk friction loss. For covered impellers, fluid leakage from the impeller discharge back to the inlet through the impeller eye seal (see Fig. 1-4) is another source of parasitic work, known as the leakage loss. Since the pressure rise imparted to this fluid is dissipated in the clearance gap and seal, the impeller works on this portion of the fluid twice, with no addition contribution to the pressure rise. These two sources of parasitic work are easily recognized physical phenomena. From experience, it is known that other parasitic work models are needed to explain the observed work input curves from compressor test results. These take the form of models postulated from fluid dynamics reasoning and are validated by comparing predictions with experimental data. Here, the models of Aungier (1993b, 1995) will be reviewed. It is reasonable to also expect a leakage loss in open impellers, where the fluid leaks through the blade clearance gaps, dissipating a portion of the pressure rise, to be re-energized by the impeller in the next blade passage. In highly diffusing impellers, another parasitic work, the recirculation loss, is observed. It is believed to result from a portion of the fluid at the impeller tip reversing to re-enter the impeller at the tip, again dissipating part of its pressure rise and requiring the impeller to re-energize it. It is convenient to express the impeller work in dimensionless form as the work input coefficient defined in Eq. (1-6). Expanding this equation to include all sources of work input, the total work input coefficient on the mean stream surface is written as

$$I = \Delta h_t / U_2^2 = I_B + I_{DF} + I_L + I_R \qquad (4\text{-}1)$$

The blade work input coefficient follows directly from Eq. (3-4)

$$I_B = C_{U2}/U_2 - U_1 C_{U1}/U_2^2 \qquad (4\text{-}2)$$

If the fluid were perfectly guided by the blades, e.g., if there were an infinite number of infinitesimally thin blades, the first term on the right-hand side of Eq. (4-2) could be written as

$$C_{U2\infty}/U_2 = 1 - \lambda \dot{m} \cot \beta_2/(\rho_2 A_2 U_2) = 1 - \lambda \phi_2 \cot \beta_2 \qquad (4\text{-}3)$$

where $\phi_2 = \dot{m}/(\rho_2 A_2 U_2)$ is the tip flow coefficient and λ is the tip distortion factor, which is related to the tip area blockage due to viscous effects or flow profile distortion, i.e.

$$\lambda = 1/(1 - B_2) \qquad (4\text{-}4)$$

But observation shows that the actual blade work input coefficient is somewhat less than this ideal value. Hence, the blade work input coefficient is written as

$$I_B = \sigma(1 - \lambda \phi_2 \cot \beta_2) - U_1 C_{U1}/U_2^2 \qquad (4\text{-}5)$$

where σ is called the slip factor. If $C_{U1} = 0$, it is the ratio of the actual blade work input to the value that would exist if the flow received perfect guidance from the blades. As illustrated in Fig. 4-1, it can be viewed simply as the blade work input coefficient that would exist if the impeller could operate with zero mass flow, i.e., with $\phi_2 = 0$. If the distortion factor is constant, the blade work input coefficient is linear with ϕ_2 as illustrated in Fig. 4-1. But, normally λ is found to vary with ϕ_2, so generally this is not the case. If we add the various parasitic work input coefficients to the blade work input coefficient, the total work input coefficient is obtained as illustrated in Fig. 4-1. It can be seen that the prediction of the work

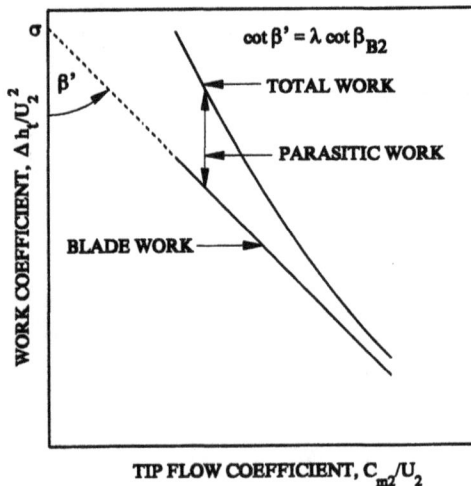

FIGURE 4-1. Impeller Work Characteristics

input requires modeling of σ, λ, I_{DF}, I_L and I_R. This chapter describes useful models for predicting each of these parameters.

NOMENCLATURE

A = an area inside the blade passage

a = sound speed

B = fractional area blockage

b = hub-to-shroud passage width

C = absolute velocity

C_c = seal carryover coefficient

C_M = disk torque coefficient

C_m = absolute meridional velocity

C_r = seal contraction coefficient

C_t = seal throttling coefficient

C_U = absolute tangential velocity

D_{eq} = equivalent diffusion factor

d = diameter

d_H = hydraulic diameter

h = enthalpy

I = work input coefficient

K = clearance gap swirl parameter, $C_U/(\omega r)$

K_F = impeller tip swirl parameter, C_{U2}/U_2

L = blade mean streamline meridional length

L_B = length of blade mean camberline

M = Mach number

M_U = rotational Mach number, U_2/a_{0t}

m = meridional coordinate

$\dot{m}$ = mass flow

N = number of seal fins

P = seal pitch

p = pressure

p_R = seal pressure ratio

P_v = velocity pressure, $p_t - p$

R = gas constant

Re = Reynolds number

r = radius

s = clearance gap width

t = seal fin thickness

U = blade speed, ωr, and leakage tangential velocity

W = relative velocity

W_U = relative tangential velocity

z_{FB} = number of full length blades

z_{SB} = number of splitter blades

z = effective number of blades, $z_{FB} + z_{SB}L_{SB}/L$

α = flow angle with respect to tangent

α_c = streamline slope angle with axis
β = blade angle with respect to tangent
δ = seal clearance
ϵ = impeller meanline radius ratio, r_1/r_2
λ = impeller tip distortion factor
μ = gas viscosity and head coefficient
ρ = gas density
ξ = distance along the mean camberline
σ = slip factor, $C_{U2}/C_{U2\infty}$
τ = torque due to windage and disk friction
ϕ = stage flow coefficient, $\dot{m}/(\pi\rho_t{}_0 r_2^2 U_2)$
ϕ_2 = tip flow coefficient, $\dot{m}/(\rho_2 A_2 U_2)$
ω = rotation speed
$\overline{\omega}_{SF}$ = skin friction total pressure loss coefficient

Subscripts

B = blade parameter
C = cover parameter
CL = clearance gap parameter
D = disk parameter
DF = disk friction parameter
FB = full blade parameter
h = parameter on the hub contour
L = leakage parameter
R = recirculation parameter
r = value for a rough disk
SB = splitter blade parameter
s = parameter on the shroud contour, value for a smooth disk
t = total thermodynamic condition
0 = impeller eye condition
1 = impeller blade inlet condition
2 = impeller tip condition
∞ = condition for an infinite number of blades

4.1 The Slip Factor

Figure 4-2 illustrates the basic origin of the slip factor, or slip. The fluid entering the impeller can be considered to be irrotational. In the rotating frame of reference, a relative eddy rotating in a direction opposite to the impeller is required to maintain an irrotational flow in the absolute frame. It is quite apparent that the flow will not be perfectly guided by the blades due to the presence of this relative eddy. Indeed, the simple kinematic arguments of Fig. 4-2 are the basis of the slip factor model by Stodola (1927)

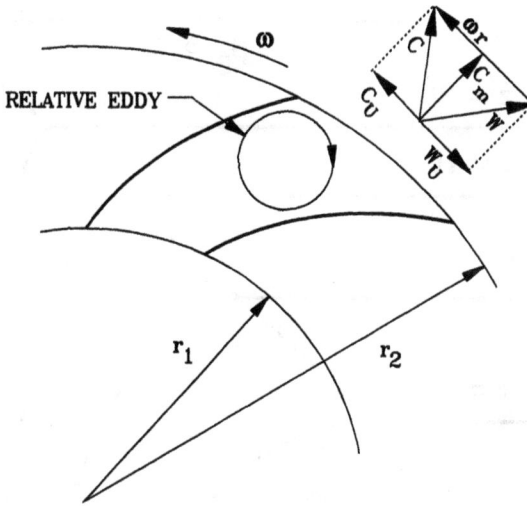

FIGURE 4-2. The Concept of Slip Factor

$$\sigma = 1 - \pi \sin\beta_2 \sin\alpha_{C2}/z \qquad (4\text{-}6)$$

where α_{C2} is the angle between the streamline slope and the axial direction. This equation is purely kinematic in nature. Busemann (1928), reviewed in Wislicenus (1947) analyzed the flow in impellers with logarithmic spiral vanes to obtain a somewhat more accurate model for the slip factor, but one that requires a family of curves for calculation. Only a portion of Busemann's results are needed to satisfy requirements for the present work input formulation—the slip factor at zero flow. Wiesner (1967) successfully approximated the Busemann slip factor at zero flow by the empirical relation

$$\sigma = 1 - \sqrt{\sin\beta_2}\,\sin\alpha_{C2}/z^{0.7} \qquad (4\text{-}7)$$

From Busemann's results, the slip factor can be treated as constant up to a limiting radius ratio, ϵ_{Lim}, beyond which it falls off rapidly with ϵ. Wiesner presents a correction procedure for larger values of ϵ, using a formula given by Sheets (1950) for the limiting radius ratio

$$\epsilon_{LIM} = \exp(-8.16\sin\beta_2/z) \qquad (4\text{-}8)$$

A more accurate approximation to Busemann's results is obtained in Aungier (1993b, 1995). Defining

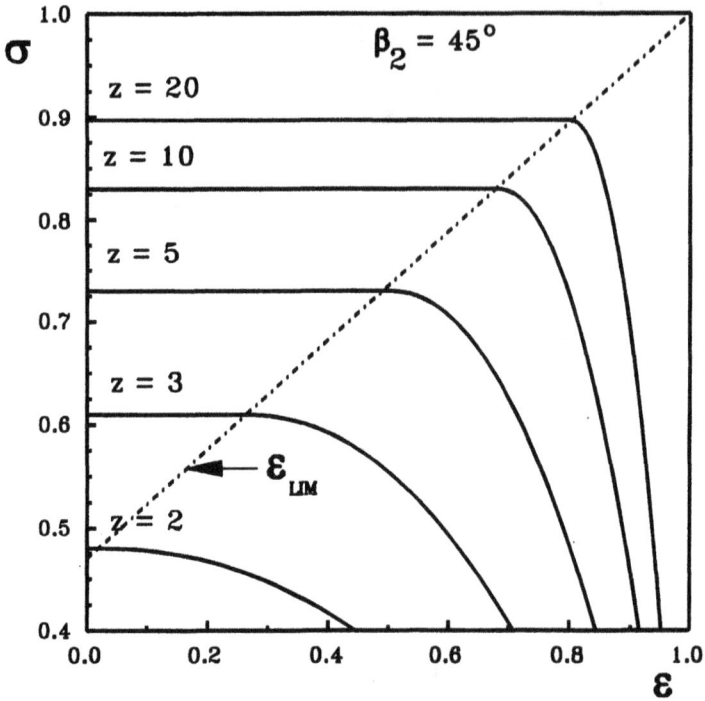

FIGURE 4-3. Wiesner's Slip Factor Model

$$\epsilon_{LIM} = \frac{\sigma - \sigma*}{1 - \sigma*}$$

$$\sigma* = \sin(19^0 + 0.2\beta_2) \tag{4-9}$$

a corrected slip factor, σ_{COR}, for $\epsilon > \epsilon_{Lim}$ is given by

$$\sigma_{COR} = \sigma \left[1 - \left(\frac{\epsilon - \epsilon_{LIM}}{1 - \epsilon_{LIM}} \right)^{\sqrt{\beta_2/10}} \right] \tag{4-10}$$

where σ is given by Eq. (4-7). Figure 4-3 illustrates typical results obtained from Eqs. (4-7) and (4-10) for a specific blade tip angle.

When centrifugal compressors are designed for high rotational Mach numbers, it is fairly common practice to employ splitter blades. As illustrated in Fig. 4-4, these are partial-length blades between adjacent full blades. Their purpose is to maintain an acceptable blade solidity while reducing the blade metal blockage at the minimum passage area or throat. This provides a larger throat area

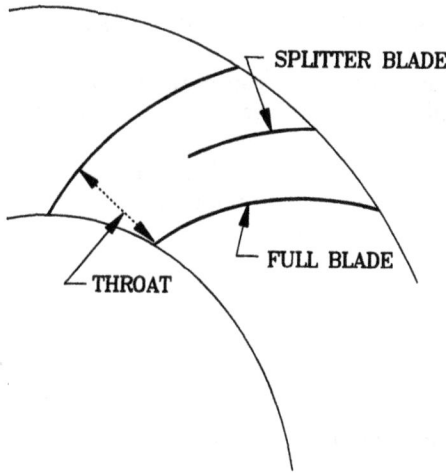

FIGURE 4-4. Splitter Blades

for the impeller to pass more mass flow before impeller choke occurs. Special procedures are necessary when applying the slip factor models to splitter blades. Aungier (1993b, 1995) recommends use of an effective number of blades, i.e.

$$z = z_{FB} + z_{SB}L_{SB}/L_{FB} \tag{4-11}$$

where L = meridional length of the blade. In addition, the limiting radius ratio should be checked for both the full blades and the splitter blades, since extremely short splitter blades may be relatively ineffective.

4.2 The Impeller Distortion Factor

The impeller distortion factor, λ, or tip blockage, B_2, is a key parameter in the blade work input equation. The literature offers little guidance with respect to estimating this parameter, with the exception of Aungier (1993b, 1995), which provides a specific empirical equation. That empirical equation was derived by inverting the work input equation to calculate the tip blockage from experimental work input curves. The weakness in this approach lies in the necessity to assume that the slip factor and the parasitic work models are all "correct." For this reason, the empirical tip blockage model most likely includes corrections to compensate for any weaknesses in those other models. Based on the impeller tip blockage levels needed to match a wide range of experimental work input curves, the following observations were made:

- For very low flow coefficients, blockage varies directly with the skin friction loss.

- Blockage increases with velocity head (ρW^2) diffusion from the blade passage throat to the passage discharge.
- Blockage increases with blade aspect ratio (b_2/L_B).
- For open impellers, blockage increases with blade clearance.

From these observations, the empirical tip blockage equation of Aungier (1993b, 1995) was formulated. Some minor refinements were subsequently incorporated. The present model used is

$$B_2 = \overline{\omega}_{SF} \frac{p_{v1}}{p_{v2}} \sqrt{\frac{W_1 d_H}{W_2 b_2}} + \left[0.3 + \frac{b_2^2}{L_B^2} \right] \frac{A_R^2 \rho_2 b_2}{\rho_1 L_B} + \frac{\delta_{CL}}{2 b_2} \qquad (4\text{-}12)$$

where p_v = velocity pressure; $p_t - p$ and A_R = a passage area ratio defined by

$$A_R = A_2 \sin \beta_2 / (A_1 \sin \beta_{th}) \qquad (4\text{-}13)$$

The angle β_{th} is the blade suction surface angle at the throat. In effect, A_R is the ratio of the impeller tip flow area to the throat area. It is formulated in this fashion to permit analysts to make somewhat arbitrary adjustments to the throat area when they desire to match performance predictions to an experimentally observed impeller choke limit without having an undesired effect on the work input prediction. The first term on the right-hand side of Eq. (4-12) contains the impeller skin friction total pressure loss coefficient, $\overline{\omega}_{SF}$, which will be presented in Chapter 5. The last term in Eq. (4-12) applies only to open impellers, basically assuming that just half of the clearance gap area is available for through-flow. This empirical equation models a wide range of blockage producing mechanisms, including skin friction, hub-to-shroud profile distortion, density ratio, area ratio (flow diffusion), blade aspect ratio and clearance effects. These various mechanisms assume vastly different significances for different impeller types. For example, the terms dominant for low flow coefficient impellers will be negligible for high flow coefficient impellers and vice versa.

4.3 Clearance Gap Flows

The parasitic work contributions due to leakage, windage and disk friction are all dependent upon the flow in the clearance gaps. The disk/housing clearance gap flows for the disk and shroud sides for covered impellers, as well as blade/housing clearance gap flows for open impellers all require consideration. Aungier (1993b, 1995) presents an approximate analysis for the disk/housing gap flows using a forced vortex flow model. The governing equation is

$$\frac{\partial p}{\partial r} = K^2 \omega^2 \rho r \qquad (4\text{-}14)$$

where $K = C_U/(\omega r)$ is assumed constant in the gap. Empirical models for K were developed using an internal flow analysis for clearance gap flows (Moussa, 1978). The internal flow analysis showed that seal leakage has a significant influence due to the influx of angular momentum into the gap. The analysis of Aungier (1993b, 1995) starts with an empirical expression for K for no seal leakage, derived from the data in Daily and Nece (1960a)

$$K_0 = 0.46/(1 + 2s/d) \tag{4-15}$$

Designating $K_F = C_U/U_2$ as the swirl parameter for seal leakage flow entering the gap, the value of K in the gap is adjusted by

$$K = K_0 + C_q(1.75K_F - 0.316)r_2/s$$

$$C_q = \frac{\dot{m}_L(\rho r_2 U_2/\mu)^{1/5}}{2\pi\rho r_2^2 U_2} \tag{4-16}$$

For seal leakage from the impeller tip, $K_F = C_{U2}/U_2$, while for seal leakage toward the tip, $K_F = 0$. As reported in Aungier (1993b, 1995), the validity of these approximations was established by comparing clearance gap radial static pressure profiles predicted by Eqs. (4-14) through (4-16) with experimental data for several compressor stages. To use this model, the seal leakage flow rate must be known. But the pressure distributions in the clearance gap must be known to define the seal pressure ratio, which governs the seal leakage. Consequently, the clearance gap flow analysis requires an iterative solution procedure.

For open impellers, flow leakage occurs in the blade/housing clearance gap due to the pressure difference on the two sides of the blade. The velocity of this clearance gap leakage flow is estimated by

$$U_{CL} = 0.816\sqrt{2\Delta p_{CL}/\rho_2} \tag{4-17}$$

where gas density is considered constant, equal to the tip value. The throttling coefficient of 0.816 was computed by assuming an abrupt contraction loss when the flow enters the gap, followed by an abrupt expansion loss as the flow leaves the gap. The pressure difference across the blade surfaces creates the force that must balance the impeller torque given in Eq. (3-2). Consequently, the average pressure difference across the gap can be estimated from the change in fluid angular momentum through the impeller.

$$\Delta p_{CL} = \frac{\dot{m}(r_2 C_{U2} - r_1 C_{U1})}{z\bar{r}\bar{b}L}$$

$$\bar{r} = (r_1 + r_2)/2$$

$$\bar{b} = (b_1 + b_2)/2 \tag{4-18}$$

Then, the blade clearance gap leakage flow is given by

$$\dot{m}_{CL} = \rho_2 z s L U_{CL} \tag{4-19}$$

4.4 Windage and Disk Friction Work

The classic papers of Daily and Nece (1960a, 1960b) are the best available sources for windage and disk friction losses. They considered rotating disks in a housing for smooth and rough disks. Defining the disk torque coefficient by

$$C_M = \frac{2\tau}{\rho \omega^2 r^5} \tag{4-20}$$

where r = disk radius; and ω = rotation speed. Daily and Nece consider four different flow regimes:

1. Laminar, merged boundary layers
2. Laminar, separate boundary layers
3. Turbulent, merged boundary layers
4. Turbulent, separate boundary layers

If s is the clearance between the disk and the housing and Re is a Reynolds number defined by

$$Re = \frac{\rho \omega r^2}{\mu} \tag{4-21}$$

the torque coefficients for these four regimes are

$$C_{M1} = \frac{2\pi}{(s/r)Re} \tag{4-22}$$

$$C_{M2} = \frac{3.7(s/r)^{0.1}}{\sqrt{Re}} \tag{4-23}$$

$$C_{M3} = \frac{0.08}{(s/r)^{1/6}Re^{1/4}} \tag{4-24}$$

$$C_{M4} = \frac{0.102(s/r)^{0.1}}{Re^{0.2}} \tag{4-25}$$

To determine which flow regime exists, one has to compute all four torque coefficients. The correct flow regime will be the one that yields the largest value of C_M. At sufficiently large values of Re, roughness effects will increase the torque coefficient. As Re continues to increase, the disk eventually becomes "fully rough" and the torque coefficient ceases to vary with Re. If e represents the peak-to-valley surface roughness height, the fully rough disk torque coefficient is given by

$$\frac{1}{\sqrt{C_{Mr}}} = 3.8 \log_{10}(r/e) - 2.4(s/r)^{0.25} \qquad (4\text{-}26)$$

Daily and Nece (1980b) provide the following estimate for the Reynolds number, Re_s, where roughness effects first appear:

$$Re_s\sqrt{C_M} = 1,100(e/r)^{-0.4} \qquad (4\text{-}27)$$

To compute Re_s, Eq. (4-27) can be combined with Eq. (4-24) or (4-25), as appropriate to the turbulent flow regime relevant to the problem. Daily and Nece (1960b) provide a similar method to compute the Reynolds number, Re_r, where the disk becomes fully rough, which is questionable. Using Fig. 6 from Daily and Nece (1960b), this author formulated the following empirical equation, which better matches experimental results:

$$Re_r = 1,100 r/e - 6 \cdot 10^6 \qquad (4\text{-}28)$$

Between the limits of Re_s and Re_r, C_M varies linearly with $\log(Re)$, i.e., in this transition zone

$$C_M = C_{Ms} + (C_{Mr} - C_{Ms}) \log(Re/Re_s)/\log(Re_r/Re_s)] \qquad (4\text{-}29)$$

Figure 4-5 shows typical results predicted with this formulation of the Daily and Nece model.

Aungier (1993b, 1995) applies some empirical corrections to these ideal disk torque coefficients for centrifugal compressor impellers. First, a correction for the clearance gap leakage flow is imposed. Denoting the Daily and Nece torque coefficient by C_{M0}, the corrected value is given by

$$C_M = C_{M0}(1 - K)^2/(1 - K_0)^2 \qquad (4\text{-}30)$$

Equation (30) is simply an intuitive correction. Its validity is supported by the accurate work input predictions it provides for ultralow flow coefficient stages, where leakage, windage and disk friction are substantial portions of the stage work input. The impeller disk friction torque coefficient is computed independently for the disk and the cover (if one exists) and the results are adjusted by

$$C_{MD} = 0.75 C_M$$
$$C_{MC} = 0.75 L C_M [1 - (d_{1s}/d_2)^5]/(r_2 - r_1) \qquad (4\text{-}31)$$

where the constant 0.75 is an empirical "experience" factor developed from correlation studies of predictions with experimental stage work input curves. Note that the Daily and Nece C_M supplies the torque from both sides of the disk. Thus, the factor of 0.75 imposes a torque coefficient about 50% higher than the Daily

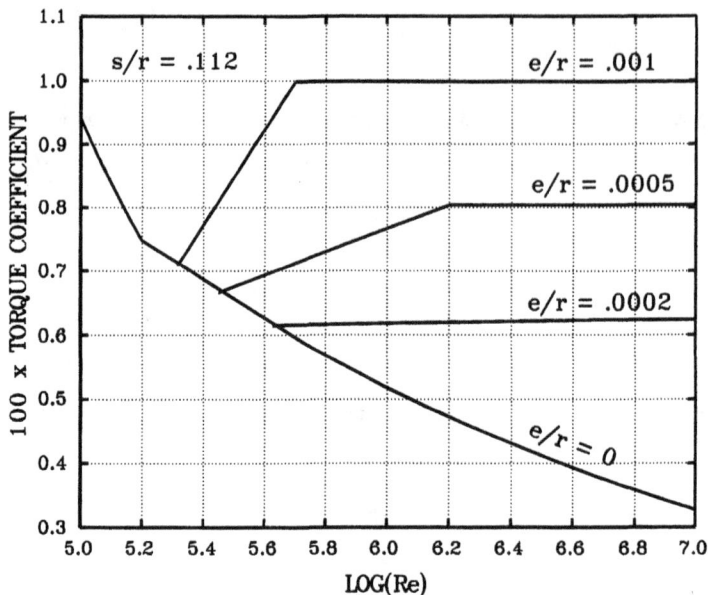

FIGURE 4-5. Daily-Nece Torque Coefficient Model

and Nece correlation. C_{MC} includes a correction for the different impeller cover surface area relative to that of a flat disk. The power consumed by windage and disk friction is $\omega\tau$. Hence, Eqs. (1-6), (2-4), (4-20), (4-30) and (4-31) yield

$$I_{DF} = (C_{MD} + C_{MC})\rho_2 U_2 r_2^2/(2\dot{m})$$ (4-32)

4.5 Leakage Work

Prediction of the work input for covered impellers requires consideration of eye seal leakage. Typically, a simple straight-through labyrinth eye seal is employed to reduce the leakage. Figure 4-6 illustrates the seal geometrical parameters. A number of seal teeth or fins form a series of annular orifices with the seal land to restrict the fluid flow. To compute the seal leakage, the pressure ratio across the seal must be known. Equation (4-14) can be integrated to predict the seal pressure in the clearance gap. Since leakage affects this pressure distribution, an iterative solution procedure is required. The classic paper by Egli (1935) provides an acceptable method for estimating seal leakage. This defines the leakage mass flow by

$$\dot{m}_L = \pi d\delta C_t C_c C_r \rho \sqrt{RT}$$ (4-33)

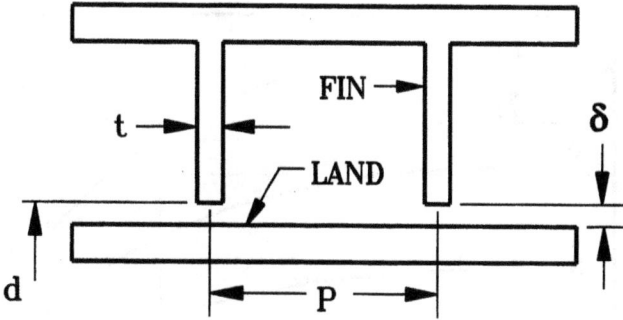

FIGURE 4-6. Labyrinth Seal Geometry

where C_t, C_c and C_r = empirical coefficients; and ρ and T are evaluated on the upstream (higher pressure) side of the seal. C_r is the contraction ratio, which is primarily a function of δ/t. Egli also shows a dependence on t, but the use of a dimensional parameter to compute a dimensionless coefficient must be considered very dubious. Indeed, other sources available to this author suggest that the lower of the two curves presented by Egli should be used for all values of t. An empirical equation to fit this graphical data is

$$C_r = 1 - \cfrac{1}{3 + \left[\cfrac{54.3}{1 + 100\delta/t} \right]^{3.45}} \qquad (4\text{-}34)$$

Figure 4-7 shows the variation of C_r with δ/t obtained from this equation. The seal throttling coefficient, C_t, is also supplied graphically by Egli (1935). It is a function of the seal pressure ratio, p_R, and the number of fins, N, in the seal. An empirical equation that correlates Egli's graphical data is

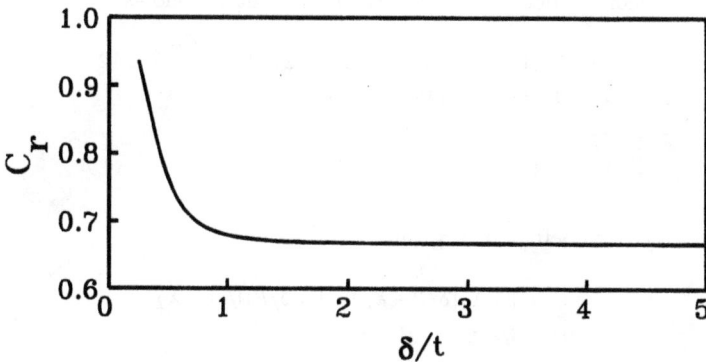

FIGURE 4-7. Seal Contraction Ratio

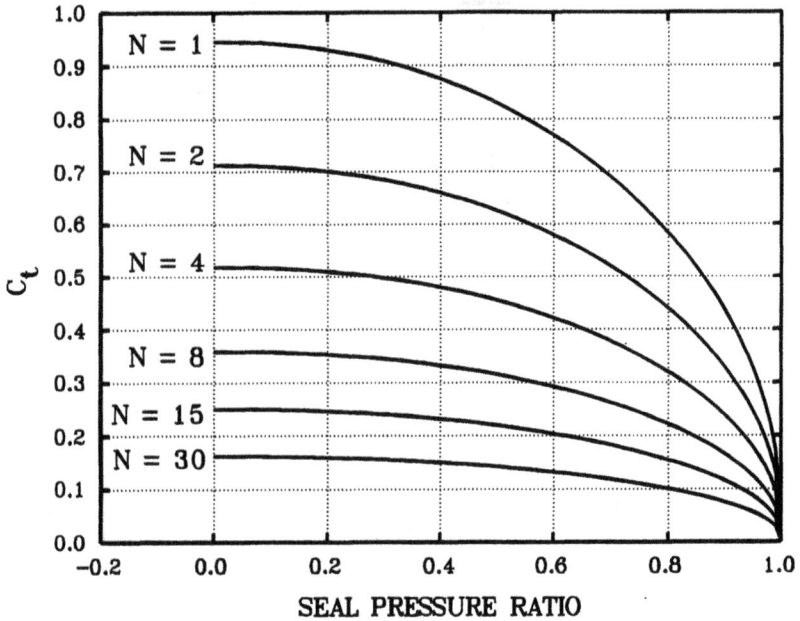

FIGURE 4-8. Seal Throttling Coefficient

$$C_t = \frac{2.143[\ln(N) - 1.464]}{N - 4.322} [1 - p_R]^{(0.375 p_R)} \tag{4-35}$$

Figure 4-8 illustrates the functional form produced by Eq. (4-35). The carry-over coefficient, C_c, accounts for the residual kinetic energy carried through from one restriction to the next. It is a function of the number of fins and δ/P, where P is the pitch shown in Fig. (4-6). Again, Egli presents this coefficient in a graphical form; it has been reduced to an empirical curve fit as follows:

$$X_1 = 15.1 - 0.05255 \exp[.507(12 - N)]; \qquad N \leq 12$$
$$X_1 = 13.15 + 0.1625N; \qquad N > 12 \tag{4-36}$$
$$X_2 = 1.058 + 0.0218N; \qquad N \leq 12$$
$$X_2 = 1.32; \qquad N > 12 \tag{4-37}$$

and the carryover coefficient is given by

$$C_c = 1 + X_1[\delta/P - X_2 \ln(1 + \delta/P)]/(1 - X_2);$$
$$\delta/P \leq X_2 - 1 \tag{4-38}$$

where Eq. (4-38) notes an upper limit on δ/P where the model is valid. This cor-

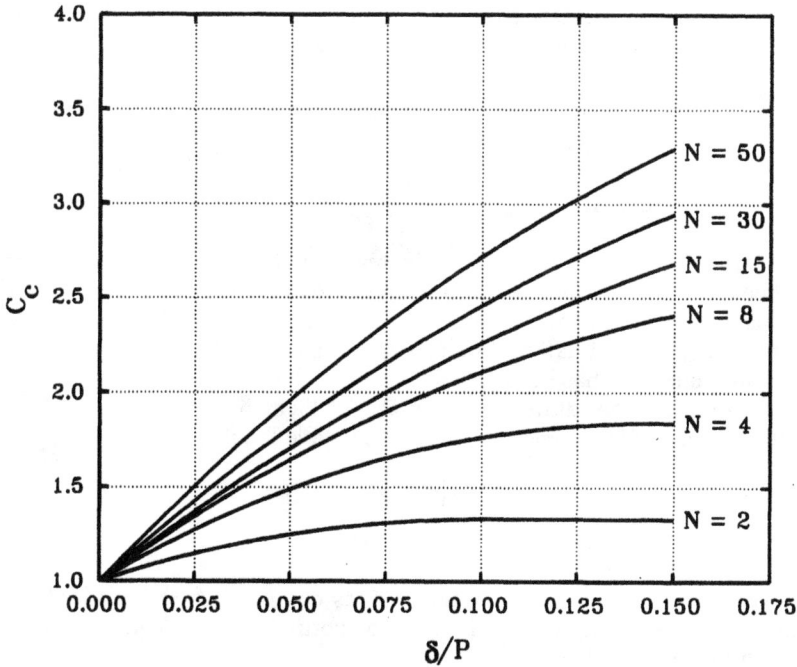

Figure 4-9. Seal Carryover Coefficient

responds to a maximum in the curves, which exceeds the range of Egli's data and is rarely encountered—except in poor seal designs. This author's practice is to use the maximum value when this occurs. Figure 4-9 illustrates results from this model.

This seal leakage model is also applicable to shaft seals, which control leakage in the clearance gap between the impeller disk and back wall. Note that this also has an effect on the work input prediction since disk friction depends on the leakage flow. For better control over the leakage, successive seal fins, called "stepped seals," are sometimes applied at different seal diameters. Although rare in centrifugal compressors, another seal style is the "staggered seal," in which successive fins are attached to opposite walls. Both these styles can be treated with the above models simply by setting $C_c = 1$.

The leakage loss for covered impellers follows directly from the eye seal leakage flow, i.e.,

$$I_L = \dot{m}_L I_B / \dot{m} \qquad (4-39)$$

which simply accounts for the fact that the eye seal leakage flow is worked on by the impeller a second time. For open impellers, Aungier (1993b, 1995) employs the assumption that half of the blade clearance leakage flow is reentrained into the blade passage flow and reenergized by the impeller. This assumption is empirical, and is derived from matching predicted and experimental work input curves for open impellers. It can be expressed as

$$I_L = \dot{m}_{CL} U_{CL}/(2U_2\dot{m}) \tag{4-40}$$

4.6 Recirculation Work

Some impellers show a pronounced increase in work input at low mass flow rates, believed to be associated with recirculation of flow back into the impeller tip. This is common for very high head coefficient impellers with excessive blade loading and low tip relative flow angles. Aungier (1993b, 1995) provides an empirical model that has been found effective in handling these cases. The axial compressor equivalent diffusion factor, D_{eq}, of Lieblein (1959) is generalized to radial and mixed flow blades to evaluate the blade loading.

$$W_{max} = (W_1 + W_2 + \Delta W)/2$$
$$D_{eq} = W_{max}/W_2 \tag{4-41}$$

The average blade velocity difference, ΔW, is computed from standard irrotational flow relations, assuming the ideal or optimum blade loading style illustrated in Fig. 4-10. This yields

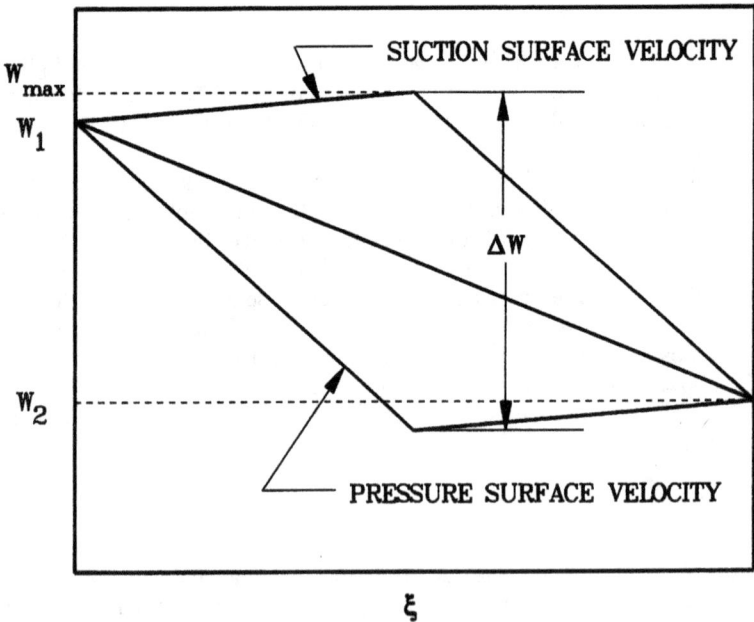

FIGURE 4-10. Simple Blade Loading Model

$$\int_0^{L_B} (W_s - W_p)d\xi = 2\pi(r_2 C_{U2} - r_1 C_{U1})/z$$

$$W_s - W_p = \Delta W[1 - |2\xi/L_B - 1|]$$

$$\Delta W = 2\pi d_2 U_2 I_B/(zL_B) \tag{4-42}$$

Lieblein found that blade stall can be expected when $D_{eq} > 2$. This was found to be an appropriate blade stall limit for impellers also. Hence, when $D_{eq} > 2$, a recirculation work input coefficient is computed from

$$I_R = (D_{eq}/2 - 1)[W_{U2}/C_{m2} - 2\cot\beta_2]$$

$$I_R \geq 0 \tag{4-43}$$

For most impellers, the limit constraint will result in $I_R = 0$ even when $D_{eq} > 2$. Only specific types of tip velocity triangles will result in a recirculation loss.

EXERCISES

4.1 A straight-through labyrinth seal is designed with $\delta/t = 2$, $\delta/p = 0.15$ and operates at a seal pressure ratio of 0.6. Currently, the seal has four fins. If the manufacturing cost is directly proportional to the number of fins, comment on the cost-effectiveness of doubling the number of fins to further reduce seal leakage.

4.2 Figure 4-2 shows the velocity vectors at the impeller tip in a very useful form known as a "velocity diagram." Construct the velocity diagram at the impeller tip for a radially bladed impeller ($\beta_2 = 90°$), where $\sigma = 0.88$, $\phi_2 = 0.3$ and $C_{U1} = 0$, normalizing all velocity components by the tip speed, U_2.

4.3 A smooth disk rotating in a housing has $s/r = 0.02$ and $Re = 2,000,000$. Compute the disk torque coefficient. What type of flow regime is present in this case?

4.4 If the disk in Exercise 4.3 has a surface roughness of $e/r = 0.001$, should you authorize an increase in manufacturing cost to polish this disk to reduce the roughness by 50%?

4.5 An impeller is to be designed for an adiabatic head coefficient of 0.52 and an adiabatic efficiency of 0.85. If the sum of all parasitic loss coefficients is 0.02, $\lambda = 1.1$ and $\phi_2 = 0.3$, what impeller tip blade angle, β_2, is required?

4.6 An impeller is designed with $\beta_2 = 90°$, $\lambda = 1.1$, $\sigma = 0.88$ and $C_{U1} = 0$. Sketch the blade work input coefficient as a function of the tip flow coefficient for $0.2 \leq \phi_2 \leq 0.3$. Assume the sum of all head coefficient losses is given by $I_B - \mu = 0.2 + (0.25 - \phi_2)^2$. Add the stage head coefficient curve over the same tip flow coefficient range. The ratio μ/I_B is referred to as the "hydraulic" efficiency or "internal" efficiency, since it neglects parasitic losses. Add the hydraulic efficiency curve to your sketch. What general statement can be made about the location of the maximum hydraulic efficiency for a radial bladed impeller? Can you expect to operate this stage consistently at its peak hydraulic efficiency?

4.7 Repeat Exercise 4.6 for an impeller with a 40° backward lean ($\beta_2 = 50°$). What are your prospects for operating this stage at its peak hydraulic efficiency? A coworker sees your evaluation of the two stages, notes the reduced work coefficient and hydraulic efficiency of the backward-leaning impeller, and comments that the radial bladed impeller will always achieve higher work input and efficiency. What is wrong with this conclusion?

Chapter 5

ONE-DIMENSIONAL AERODYNAMIC PERFORMANCE ANALYSIS

Due to the highly complex flow in centrifugal compressors, sophisticated two- and three-dimensional inviscid and viscous flow analysis codes are routinely employed in modern aerodynamic design and analysis. Indeed, these sophisticated analyses are leading to levels of efficiency believed to be unattainable a few years ago. Yet, the very complexity of these flows results in one of the many contradictions encountered by the centrifugal compressor aerodynamicist—namely, that the most accurate and reliable performance analysis techniques continue to be based on one-dimensional flow models. This situation is certainly not unique to centrifugal compressor analysis. Whenever fundamental analysis techniques prove inadequate for a problem, investigators routinely employ a combination of fundamental analysis and empirical models to address the problem. In fact, the entire field of boundary layer analysis is based on this approach.

One-dimensional centrifugal compressor performance analyses function in this manner. The flow is analyzed along a mean stream surface through the various stage components, using numerous empirical fluid dynamics and loss correlations to supply the information not obtainable through the basic theoretical methods. The quality of results obtained directly depends on the validity of the empirical models employed. The losses in typical stage components are produced by many different fluid dynamics mechanisms. It is almost never possible to experimentally measure a specific loss. At best, the investigator formulating the analysis will have experimental data to define the overall component losses. Given a sufficient number of such experimental results for widely varying stage types, it may be possible to validate an empirical loss system for a component. This involves refining and "tuning" the postulated loss mechanisms until contradictions to the experimental data are removed and satisfactory performance predictions are achieved. The process of postulating the loss mechanisms plays an important role in this process. Purely empirical loss models (e.g., curve fits of experimental results) are almost always doomed to failure for centrifugal compressors. A reasonable foundation of fluid dynamics principles is critical to success. The result of this complex process is a loss system for a component, validated as a whole against experimental data. It is not possible to search the literature for the most appealing loss models, incorporate them into an analysis and expect them to work well. Since the developers of various models have no way to validate their models individually, one simply cannot separate the parts from the whole. The literature can supply candidate loss mechanisms that have been successful, but validation and tuning of the complete loss system against experimental data is still a necessary step.

It should be apparent from the foregoing discussion that a performance analysis based on one-dimensional flow models has inherent limitations, since the definition of the stage geometry used for a one-dimensional analysis is quite limited. There are an infinite number of ways a component can be designed to produce the geometry specific to the performance analysis. The best a one-dimensional performance analysis can do is define the performance achievable if the detailed stage geometry is based on "good design practice." Obviously, good design practice is also a relative term. As design technology improves, periodic upgrading of the performance analysis becomes necessary. One of the most challenging features of developing and maintaining a one-dimensional performance analysis is that of obtaining reasonable performance prediction accuracy for older, existing designs, which may incorporate a poor design practice by modern standards, and for new designs supported by modern, multidimensional internal flow analysis. This might appear to be an impossible goal, but it can be achieved with reasonable success. Lacking modern design methods, designers of older designs had little chance of "guessing" the optimum configuration, even in the sense of one-dimensional performance analysis. Yet, it is certainly true that the weakest aspect of a well-formulated, modern performance analysis is its accuracy when applied to obsolete stage designs. Normally, this type of design is correctly identified as a poor design, but it may, in fact, perform worse than predicted.

Descriptions of most highly-developed one-dimensional performance analyses are simply not available to the aerodynamicist. Normally, such descriptions are available to members of the specific design groups or perhaps to purchasers, in the case of commercially available codes. One such analysis has been published in the literature (Aungier, 1990, 1993, 1995; Weber and Koronowski, 1986) and will be described in detail here. The present description differs somewhat from the original references. The principal change is that the analysis has been revised to use a total pressure loss formulation rather than the obsolete head loss formulation still in use when the original references were published. The initial reference to this analysis was made by Wiesner (1979), who formulated the first version in 1972. Since 1972, this author has continued development of the analysis, which included six major revisions as feedback was obtained from new stage development programs. Due to continued availability of a very large collection of experimental performance data, this analysis has benefited from extensive validation studies. It has been qualified in detail against experimental data for more than a hundred different stages. These validation studies include stage flow coefficients ranging from 0.009 to 0.16, covering most of the range of interest. But like all one-dimensional methods, its range of validation has limits, and its application has been limited to industrial compressors. The detailed validation studies extend to pressure ratios up to about 3.5 and its actual application range has been limited to pressure ratios up to about 4.2. The fact that its validity beyond the range for which it has been qualified is not established is an important aspect potential users should recognize.

The advantages of describing a complete performance prediction should be apparent from the above discussion. But it omits the more general discussion of various alternative approaches used for this purpose. Readers interested in alternative approaches are referred to the excellent summary provided by Whitfield and Baines (1990).

NOMENCLATURE

A = an area inside the blade passage
A_R = area ratio
a = sound speed and location of vane maximum camber measured along chord
B = fractional area blockage
b = hub-to-shroud passage width
C = absolute velocity
C_m = absolute meridional velocity
C_U = absolute tangential velocity
c = blade chord
c_f = skin friction coefficient
c_r = throat contraction ratio
D = passage divergence parameter, Eqs. (5-48) and (5-50)
D_{eq} = equivalent diffusion factor
D_m = parameter defined in Eqs. (5-49) and (5-51)
d = diameter
d_H = hydraulic diameter
E = diffusion efficiency
f_c = head loss correction factor
h = enthalpy
h_{th} = blade-to-blade throat width
I = work input coefficient
I_C = vaneless passage curvature loss term
I_D = vaneless passage diffusion loss term
i = incidence angle, $\beta - \alpha$
L = mean streamline meridional length
L_{SB} = splitter blade mean streamline meridional length
L_B = length of blade mean camberline
M = Mach number
M_U = rotational Mach number, U_2/a_{0t}
m = meridional coordinate
$\dot{m}$ = mass flow
n = coordinate normal to the mean stream surface
p = pressure
R = rothalpy
Re_d = Reynolds number based on pipe diameter
Re_e = Reynolds number based on surface roughness
r = radius
S = entropy

s = clearance gap width
U = blade speed, ωr, and leakage velocity
u = general gas velocity
v = gas specific volume = $1/\rho$
W = relative velocity
W_U = relative tangential velocity
w = blade-to-blade width = $(2\pi r \sin\beta)/z$
z = effective number of blades, $z_{FB} + z_{SB}L_{SB}/L$
z_{FB} = number of full-length blades
z_{SB} = number of splitter blades
α = flow angle with respect to tangent
α_C = streamline slope angle with axis
β = blade angle with respect to tangent
γ = blade stagger angle
δ = deviation angle
η = adiabatic efficiency
θ = polar angle, vane camber angle
θ_C = diffuser divergence angle
κ = blade angle with respect to meridional direction
κ_m = streamline curvature
λ = impeller tip distortion factor
μ = adiabatic head coefficient and gas viscosity
ξ = distance along blade mean camberline
ρ = gas density
σ = blade row solidity, $zc/(2\pi r)$
ϕ = stage flow coefficient, $\dot{m}/(\pi\rho_{0t}r_2^2 U_2)$
ϕ_2 = tip flow coefficient, $\dot{m}/(\rho_2 A_2 U_2)$
ω = rotation speed
$\overline{\omega}$ = total pressure loss coefficient, $\Delta p_t/(p_t - p)_{in}$

Subscripts

B = a blade parameter
C = cover parameter
CL = clearance gap parameter
D = disk parameter
h = parameter on the hub contour
id = ideal, isentropic (no loss) condition
L = leakage parameter
l = laminar flow value
m = meridional component or maximum value
max = a maximum condition
p = blade pressure surface parameter
R = recirculation parameter
r = fully rough surface condition
S = value at onset of blade stall

s = shroud contour or blade suction surface parameter
t = total thermodynamic condition
th = throat parameter
U = tangential component
-1 = inlet guide vane leading edge condition
0 = impelle eye condition
1 = impeller blade inlet condition
2 = impeller tip condition
3 = vaned diffuser inlet condition
4 = diffuser exit condition
5 = crossover or volute inlet condition
6 = crossover or volute exit condition
7 = return channel vane or exit cone exit condition
8 = return channel exit condition

Superscripts

$*$ = sonic flow condition, condition at minimum loss incidence angle
$'$ = value relative to rotating frame of reference

5.1 One-Dimensional Flow Analysis

One-dimensional flow analysis involves many features common to the different types of components to be analyzed. It is convenient to cover these common features first. The basic solution is accomplished on a mean stream surface, subject to appropriate boundary conditions and conservation equations, and supported by suitable empirical models. Normally, the mean stream surface is defined such that equal passage flow areas reside on either side of it.

For vaned components (inlet guide vanes, impellers, vaned diffusers and return channels), fluid dynamics data are generated at the inlet and discharge stations, with an additional calculation at the passage throat to identify possible choking limits. The inlet boundary conditions will include the total thermodynamic conditions and angular momentum obtained from analysis of the previous component or specified as initial conditions. The discharge boundary conditions are supplied by empirical methods, normally consisting of the discharge flow angle or angular momentum and the total pressure loss in the passage. The solution process consists of satisfying conservation of mass and energy subject to these boundary conditions. Because some of the empirical models will depend on the discharge flow conditions, the solution procedure is iterative in nature, seeking convergence on the discharge mass flow and thermodynamic conditions.

For vaneless components (inlet passages, vaneless diffusers, crossovers, etc.) the approach is slightly different. Here, the basic one-dimensional mass, momentum and energy equations are solved normally, subject to wall skin friction forces and other empirical loss models. The inlet boundary conditions are the same as for vaned components. The problem is generally formulated as a "marching" one, meaning no boundary conditions are needed at the discharge. This is possible by virtue of the fact that the mean stream surface is predefined.

At higher Mach number levels, choking may occur in one of the components analyzed. As flow velocity increases, gas density decreases, resulting in a maximum mass flow that can be passed through a given passage area. For fully one-dimensional flow (i.e., vaned components) this condition is known to occur when the local Mach number is unity. Choking in a vaneless component is also possible, subject to a different constraint—that the tangential velocity is fixed due to conservation of angular momentum. In this case choke can be shown to occur when the meridional Mach number is sonic, i.e., when $C_m = a$.

One loss mechanism common to all stage components is wall friction. Due to the wide range of operating conditions imposed on centrifugal compressors, a very general formulation for the skin friction coefficient is required, including laminar and turbulent flow and the influence of surface finish. All component analyses described in this section employ a common skin friction model based on generalized pipe friction data. Skin friction coefficients are correlated as a function of the Reynolds number based on pipe diameter, $Re_d = \rho V d / \mu$ and peak-to-valley surface roughness, e, where V represents the relevant velocity. Three well-established models are used for this purpose (e.g., see Schlichting, 1979). For $Re_d < 2{,}000$, the flow is laminar and

$$c_{fl} = 16/Re_d \tag{5-1}$$

If $Re_d > 2{,}000$, the flow is turbulent. If the wall is smooth

$$\frac{1}{\sqrt{4c_{fts}}} = -2\log_{10}\left[\frac{2.51}{Re_d\sqrt{4c_{fts}}}\right] \tag{5-2}$$

and for turbulent flow over a fully rough surface

$$\frac{1}{\sqrt{4c_{ftr}}} = -2\log_{10}\left[\frac{e}{3.71d}\right] \tag{5-3}$$

For $Re_d < 2{,}000$, Eq. (5-1) is the relevant form. Transition between laminar and turbulent or smooth and rough zones are modeled as weighted averages of the above values. The surface roughness becomes significant when

$$Re_e = (Re_d - 2{,}000)e/d > 60 \tag{5-4}$$

Hence, set the turbulent skin friction coefficient as

$$c_{ft} = c_{fts}; \quad Re_e < 60$$
$$c_{ft} = c_{fts} + (c_{ftr} - c_{fts})(1 - 60/Re_e); \quad Re_e \geq 60 \tag{5-5}$$

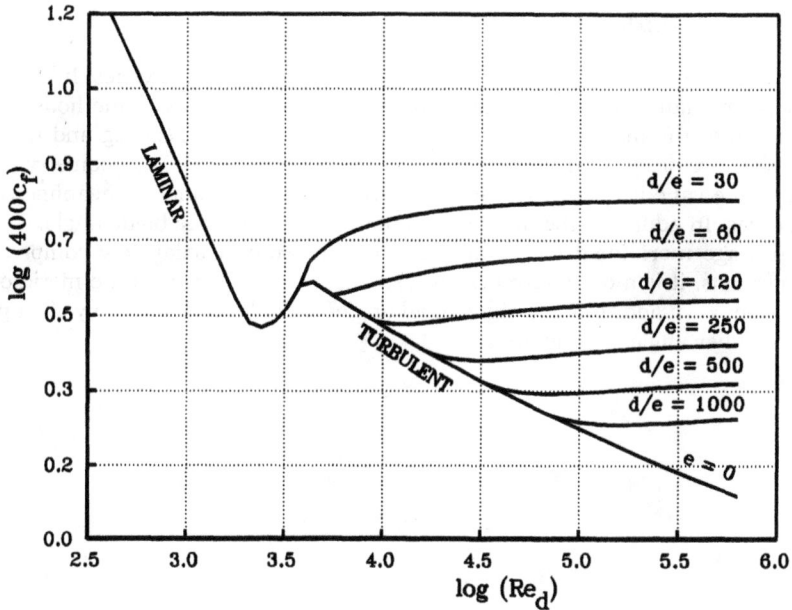

FIGURE 5-1. The Skin Friction Correlation

which is the relevant value when $Re_d > 4{,}000$. When Re_d lies between 2,000 and 4,000, the skin friction coefficient is given by

$$c_f = c_{fl} + (c_{ft} - c_{fl})(Re_d/2{,}000 - 1) \qquad (5\text{-}6)$$

Figure 5-1 shows the typical results obtained from these approximations. These results are in excellent agreement with the experimental pipe skin friction coefficients of Nikuradse (1930). In applying this generalized skin friction model to general passages, the conventional practice is to replace d with the hydraulic diameter, d_H, defined as

$$d_H = 4(\text{cross-sectional area})/(\text{wetted perimeter}) \qquad (5\text{-}7)$$

For simple annular passages, d can be replaced by the passage width, b. Application of Eq. (5-7) to annular passages yields $d_H = 2b$, which is often used but is less consistent with the basic concept of a universal velocity profile for turbulent flow. This model can also be used for boundary layer flows with a boundary layer thickness, δ, by setting $d_H = 2\delta$. In many cases, the surface finish is measured in terms of the root-mean-square value. A reasonable method of converting these to peak-to-valley values is by assuming a sine wave form, i.e.,

$$e = 2e_{rms}/0.707 = e_{rms}/0.3535 \qquad (5\text{-}8)$$

5.2   Inlet Guide Vane Performance

Figures 5-2 and 5-3 illustrate the basic geometry of inlet guide vanes. Inlet guide vane performance is computed using basic axial flow compressor methods transformed into the more general meridional stream surface. Leading and trailing edge data are designated by subscripts -1 and 0, respectively. The geometry specified at the vane leading and trailing edges are the blade angles, meanline radii and areas. In addition, the number of vanes, the midpassage blade angle, $\bar{\beta}$, and the passage meridional length are specified. To employ axial flow compressor correlations, it is more convenient to use angles with respect to the meridional direction, κ. In addition, the signs of all angles must be corrected to yield a positive camber angle, i.e., define

$$K = \frac{\beta_0 - \beta_{-1}}{|\beta_0 - \beta_{-1}|} \tag{5-9}$$

$$\kappa_{-1} = K(90^0 - \beta_{-1}) \tag{5-10}$$

$$\kappa_0 = K(90^0 - \beta_0) \tag{5-11}$$

$$\bar{\kappa} = K(90^0 - \bar{\beta}) \tag{5-12}$$

Then, the blade camber angle, θ, stagger angle, γ, and solidity, σ, are estimated by

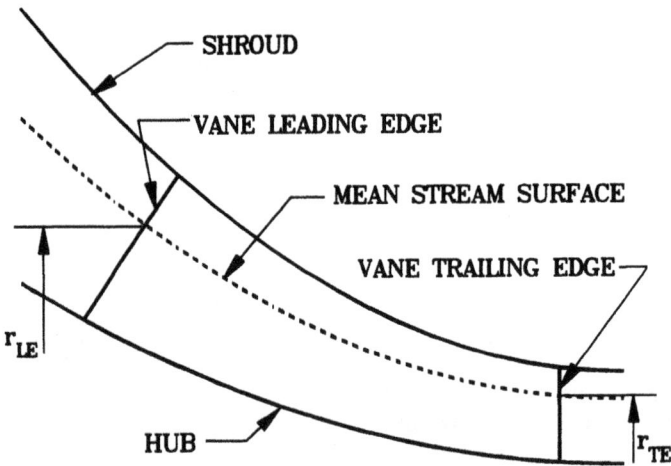

FIGURE 5-2. Inlet Guide Vane Passage

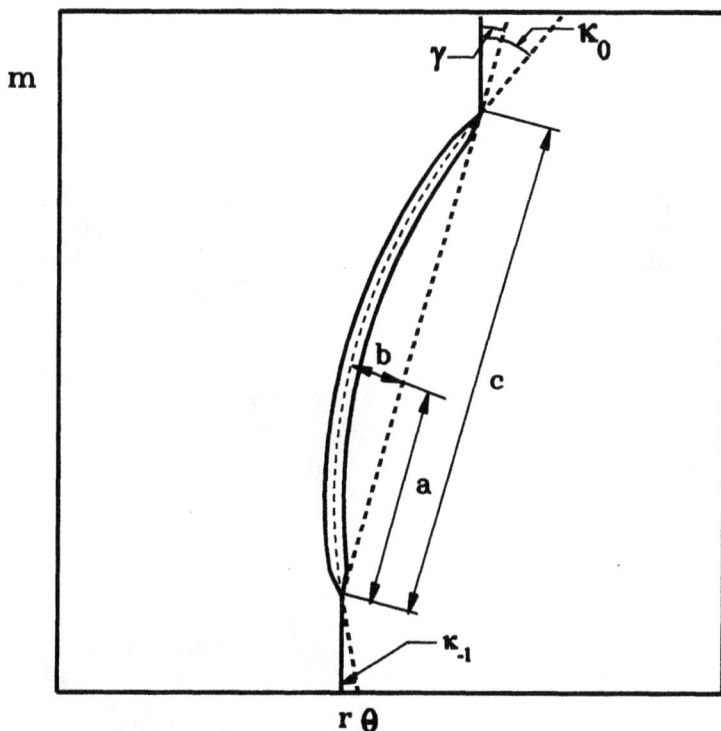

FIGURE 5-3. Inlet Guide Vane Geometry

$$\theta = (\kappa_{-1} - \kappa_0)$$

$$\gamma = (\kappa_{-1} + \bar{\kappa} + \kappa_0)/3$$

$$\sigma = \frac{z(m_0 - m_{-1})}{2\pi r_0 \cos(\gamma)} \tag{5-13}$$

The location of the point of maximum camber is estimated from

$$a/c = (2 - |\bar{\kappa} - \kappa_0|/\theta)/3 \tag{5-14}$$

The discharge flow angle is computed using Howell's (1947) deviation angle correlation

$$\delta^* = \frac{\theta[0.92(a/c)^2 + 0.02\kappa_0]}{\sqrt{\sigma} - 0.02\theta} \qquad (5\text{-}15)$$

$$\alpha_0 = 90^0 - K(\kappa_0 + \delta^*) \qquad (5\text{-}16)$$

The minimum loss, i^*, positive stall, i_S, and negative stall, i_C, incidence angles are computed from unpublished correlations of National Advisory Committee for Aeronautics (NACA) cascade data (Herrig et al., 1957) developed by this author for axial flow compressor performance analysis.

$$i^* = (3.6 + \theta/3.42)\sqrt{\sigma} + \gamma - \kappa_{-1} \qquad (5\text{-}17)$$
$$i_S - i^* = 4(a/c)^2(10^0 + 0.36\theta) \qquad (5\text{-}18)$$
$$i^* - i_C = 2(a/c)(9^0 - 0.06\theta) \qquad (5\text{-}19)$$

The equivalent diffusion factor (Lieblein, 1959) is defined by

$$D_{eq} = \frac{[1.12 + 0.61\Gamma \cos^2(\kappa_{-1} + i)]A_0 \cos(\kappa_0 + \delta^*)}{A_{-1} \cos(\kappa_{-1} + i)}$$

$$\Gamma = [\tan(\kappa_{-1} + i) - r_0 A_0 \tan(\kappa_0 + \delta^*)/(r_{-1}A_{-1})]/\sigma$$

$$i = K(90^0 - \alpha_{-1}) - \kappa_{-1} \qquad (5\text{-}20)$$

The total pressure loss coefficient is given by

$$\overline{\omega} = \frac{0.018\sigma(1 + X^3)}{\cos(\kappa_0 + \delta^*)} \left[\frac{A_{-1} \cos(\kappa_{-1} + i^*)}{A_0 \cos(\kappa_0 + \delta^*)} \right]^2 D_{eq}^2$$

$$X = (i - i^*)/(i_S - i^*); \qquad i \geq i^*$$

$$X = (i^* - i)/(i^* - i_C); \qquad i < i^* \qquad (5\text{-}21)$$

This formulation uses the simplifying assumption that $C_{m-1}A_{-1}/(C_{m0}A_0)$ is constant, which eliminates the need for an iterative solution. This is valid if gas density is about constant across the blade row, which is normally a reasonable assumption for inlet guide vanes. Since the discharge total pressure and flow angle are given by these empirical models and total enthalpy is constant through the vane row, simple mass conservation at the discharge yields all other discharge flow conditions.

5.3 Impeller Performance

Prediction of the impeller work input has been described in Chapter 4. Here, the remaining aspects of the impeller analysis are described, based on the understanding that work input prediction is an integral part of predicting impeller performance. Figures 5-4 and 5-5 illustrate some of the key geometrical features in the meridional and mean stream surfaces, respectively.

The impeller analysis must be conducted in the rotating frame of reference. From the known absolute inlet flow conditions, Eqs. (3-1), (3-5) and (3-6) define the relative velocities and total enthalpy at the inlet

$$h'_{t1} = h_{t1} - U_1 C_{U1} + U_1^2/2 \tag{5-22}$$

where the prime notation designates a relative condition in rotating coordinates. Since entropy is the same in both frames of reference, p'_t, T'_t, etc., can be computed using the appropriate equation of state. Since rothalpy, R, is conserved in the rotating frame of reference, the relative total enthalpy at the blade passage exit is also known from Eq. (3-7)

$$h'_{t2} = h'_{t1} + (U_2^2 - U_1^2)/2 \tag{5-23}$$

From which the ideal or isentropic discharge relative total conditions, p'_{t2id}, T'_{t2id}, etc., can be computed using the appropriate equation of state. Then if the total pressure loss is known, these ideal conditions can be corrected to actual discharge relative conditions noting that h'_t, is constant. But a subtle problem is

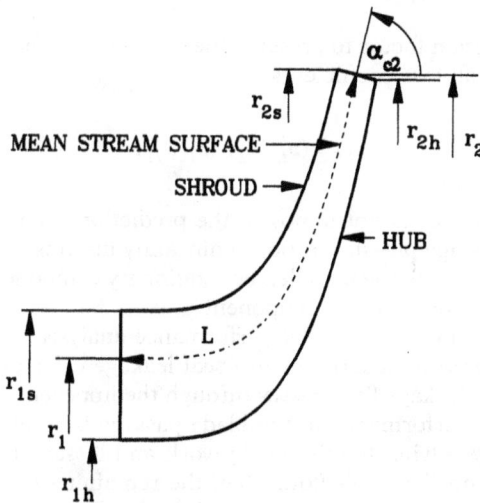

FIGURE 5-4. Impeller Passage Geometry

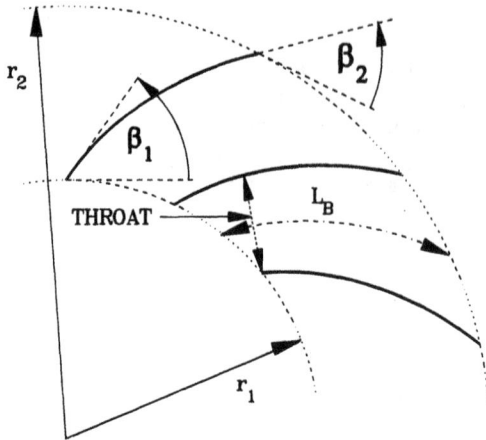

FIGURE 5-5. Impeller Blade Geometry

created by our use of total pressure loss coefficients based on the inlet velocity pressure, Eq. (2-53). It must be recognized that the measure of an irreversible process is the increase in entropy that occurs. Since we base our loss coefficients on the inlet conditions, but apply them at the discharge, they must be corrected to ensure the rise in entropy is unchanged. From Eq. (2-8) it is easily shown that the discharge relative total pressure is given by

$$p'_{t2} = p'_{t2id} - f_c(p'_{t1} - p_1) \sum_i \overline{\omega}_i \qquad (5\text{-}24)$$

where f_c = a correction factor to preserve the entropy rise unchanged, applied to the summation of all loss coefficients.

$$f_c = (\rho'_{t2} T'_{t2})/(\rho'_{t1} T'_{t1}) \qquad (5\text{-}25)$$

If this correction factor is not imposed, the prediction accuracy level achieved will vary with the stage pressure ratio, significantly limiting its range of validity. This problem is unique to the impeller. For stationary components, total pressure loss is invariant throughout the component.

A basic flow chart of the impeller performance analysis is shown in Fig. 5-6. Note that for covered impellers, the eye seal leakage calculation is part of the analysis, since that leakage flow passes through the impeller. In the analysis process, first the flow performance in the blade passage is computed. The process involves iteratively solving for the blade work and losses until convergence is obtained on the impeller mass flow. Then the remaining parasitic work terms and the shaft seal leakage are computed. Now the flow conditions outside the blade passage (i.e., at the diffuser inlet) can be computed from the passage per-

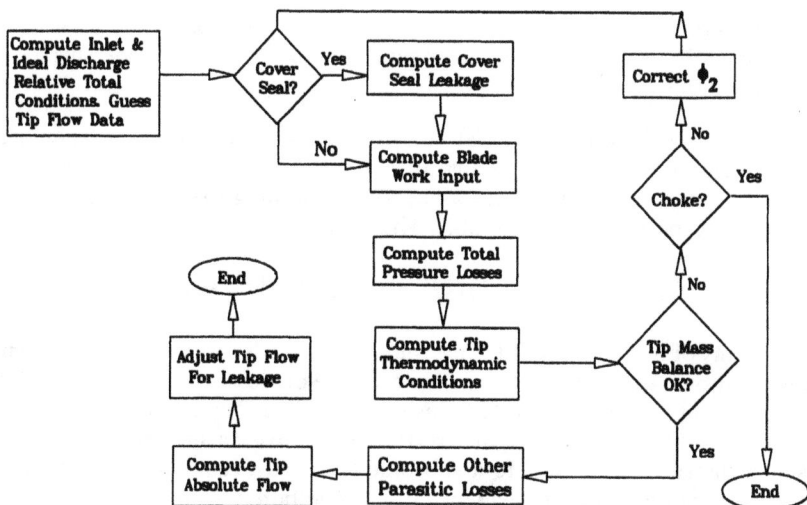

FIGURE 5-6. Flow Chart of the Impeller Analysis

formance by including the parasitic work and using isentropic equation of state calculations. The tip flow must be adjusted for all seal leakage flows such that the proper mass flow into the diffuser is specified. The precise corrections depend on the direction of leakage flows. The analysis may be terminated by sensing a choke condition. The "choke" check shown in the flow chart is more general than true aerodynamic choke. The analysis is terminated when the impeller efficiency is below a prescribed limit, say, e.g., $\eta_{2t} < 10\%$. The loss models of Aungier (1993b, 1995) are used, but are expressed as total pressure loss rather than head loss. The following models are used:

1. Normal shock wave loss if $M_1' > 1$ to account for bow shocks.
2. Incidence loss to account for flow adjustment from the actual inlet flow angle to the blade angle.
3. Entrance diffusion loss to account for excessive flow diffusion from the blade inlet to the throat.
4. Choking loss to account for losses as the throat Mach number approaches unity.
5. Blade loading loss to account for blade-to-blade pressure gradients, which produce secondary flows and may lead to blade stall.
6. Hub-to-shroud loading loss, analogous to the blade loading loss but in the hub-to-shroud direction.
7. Skin friction loss to account for the loss resulting from wall friction.
8. Distortion loss to account for mixing of distorted C_m profiles.
9. Blade clearance loss for open impellers to account for losses due to clearance gap leakage and pressure drop.
10. Mixing loss to account for losses due to mixing of blade wakes with the free stream flow.

11. Supercritical Mach number loss to account for shock wave losses or shock induced, boundary layer separation losses when blade surfaces velocities are supersonic.

The entrance flow is defined at the hub, mean and shroud surfaces using an approximate stream surface curvature correction procedure. This is accomplished by integrating Eqs. (3-19) and (3-22), assuming the gradients of rC_U, R and S with respect to n are negligible. This yields

$$C_{mh1} = C_{m1}[1 + \kappa_{m1}b_1/2]; \qquad C_{ms1} = C_{m1}[1 - \kappa_{m1}b_1/2] \qquad (5\text{-}26)$$

When any of the resulting entrance velocities exceeds sonic conditions, normal shock relations are imposed to reduce these velocities to subsonic conditions. Since nonideal gases are permitted, this is accomplished by conserving mass, ρW, momentum, ρW^2, and energy h_t', across the shock wave via an iteration procedure to shock the supersonic flow to a subsonic level. This requires a total pressure loss, $\overline{\omega}_{sh}$, to be imposed. An incidence loss is computed from

$$\overline{\omega}_{inc} = 0.8[1 - C_{m1}/(W_1 \sin \beta_1)]^2 + [z_{FB}t_{b1}/(2\pi r_1 \sin \beta_1)]^2 \qquad (5\text{-}27)$$

where Eq. (5-27) is applied at the hub, mean and shroud stream surfaces. The overall inlet shock and incidence loss coefficients are defined as weighted averages of their hub, mean and shroud values, where the mean values are weighted 10 times as heavy as the hub and shroud values. The second term on the right-hand side of Eq. (5-27) accounts for the effect of the abrupt flow area contraction at the blade leading edge due to the latter's thickness; it is normally negligible, unless an unusually thick blade leading edge is used. For some impellers, the diffusion of flow between the leading edge and the throat has been found to be more significant than the leading edge adjustment of the flow due to incidence. An entrance diffusion loss, $\overline{\omega}_{DIF}$, is computed to account for these cases

$$\overline{\omega}_{DIF} = 0.8[1 - W_{th}/W_1]^2 - \overline{\omega}_{inc}; \qquad \overline{\omega}_{DIF} \geq 0 \text{ required} \qquad (5\text{-}28)$$

where W_{th} = throat velocity obtained from conservation of mass and rothalpy on the mean stream surface at the blade passage throat, including the throat aerodynamic blockage (discussed subsequently).

Some impellers are observed to exhibit a rather pronounced inducer stall, which is often sufficient to preclude operation of the compressor at a lower flow due to either flow instability or stage surge. Flow diffusion between the inlet and the throat has been found to be a significant indicator of inducer stall. The specific inducer stall criterion used is

$$W_{1s}/W_{th} \geq 1.75 \qquad (5\text{-}29)$$

Subsequent to development of this stall criterion, it was noted that Kosuge et al. (1982) employed a similar parameter to estimate the onset of rotating stall in impellers. When the inducer stall is predicted, the diffusion loss is limited by

$$\bar{\omega}_{DIF} \geq [(W_{1s} - 1.75W_{th})/W_1]^2 - \bar{\omega}_{inc} \tag{5-30}$$

The aerodynamic blockage in the impeller throat is modeled by a contraction ratio correlation

$$C_r = \sqrt{A_1 \sin \beta_1 / A_{th}}$$
$$C_r \leq 1 - (A_1 \sin \beta_1 / A_{th} - 1)^2 \tag{5-31}$$

Denoting the area for which the assigned mass flow will yield a sonic velocity as A^*, the contracted throat area and A^* are used to impose a choking loss

$$X = 11 - 10C_r A_{th}/A^*$$
$$\bar{\omega}_{CH} = 0; \qquad X \leq 0$$
$$\bar{\omega}_{CH} = \tfrac{1}{2}(0.05X + X^7); \qquad X > 0 \tag{5-32}$$

The loss due to wall skin friction is computed from

$$\bar{\omega}_{SF} = 4c_f(\overline{W}/W_1)^2 L_B/d_H$$
$$\overline{W}^2 = (W_1^2 + W_2^2)/2$$
$$\overline{W}^2 \geq (W_{th}^2 + W_2^2)/2 \tag{5-33}$$

where the hydraulic diameter, d_H, is computed as the average of the throat and tip values using Eq. (5-7). Note that the tip area used is normal to the flow direction, i.e., $A_2 \sin \beta_2$. The blade loading loss is given by

$$\bar{\omega}_{BL} = (\Delta W/W_1)^2/24 \tag{5-34}$$

where ΔW is given by Eq. (4-42). The hub-to-shroud loading loss is given by

$$\bar{\omega}_{HS} = (\bar{\kappa}_m \bar{b} \overline{W}/W_1)^2/6$$
$$\bar{\kappa}_m = (\alpha_{C2} - \alpha_{C1})/L$$
$$\bar{b} = (b_1 + b_2)/2$$
$$\overline{W} = (W_1 + W_2)/2 \tag{5-35}$$

It is seen that Eqs. (5-34) and (5-35) relate to the pressure gradients in the blade-to-blade and hub-to-shroud directions, respectively, which can be expected to contribute to the impeller loss. Indeed, when these gradients become large with respect to the velocity head, severe flow separation can occur. The impeller tip meridional velocity profile distortion factor defined by Eqs. (4-4) and (4-12) can be expected to contribute a loss (which occurs when the distorted flow mixes with the free stream flow) approximately equal to the well-known abrupt expansion loss (Benedict et al., 1966), i.e.

$$\overline{\omega}_\lambda = [(\lambda - 1)C_{m2}/W_1]^2 \tag{5-36}$$

Similarly, an abrupt expansion loss is a reasonable model to predict the loss due to blade wake mixing. To estimate the magnitude of the wake, the velocity at which separation takes place must be estimated. This is estimated from the equivalent diffusion factor, D_{eq}, defined in Eq. (4-41). The separation velocity is given by

$$W_{SEP} = W_2; \qquad D_{eq} \leq 2$$
$$W_{SEP} = W_2 D_{eq}/2; \qquad D_{eq} > 2 \tag{5-37}$$

Since wake mixing occurs outside the blades where no blade forces are active, the tangential velocity is controlled by conservation of angular momentum. Hence, the wake mixing loss involves only the meridional component of velocity. The meridional velocities before and after mixing are estimated from W_{SEP} and from conservation of mass, assuming gas density is approximately constant, i.e.,

$$C_{m,wake} = \sqrt{W_{SEP}^2 - W_U^2}$$
$$C_{m,mix} = C_{m2}A_2/(\pi d_2 b_2) \tag{5-38}$$

from which the wake mixing loss is given by

$$\overline{\omega}_{mix} = [(C_{m,wake} - C_{m,mix})/W_1]^2 \tag{5-39}$$

For open impellers, the blade-to-blade pressure difference and clearance gap leakage given by Eqs. (4-18) and (4-19) will yield a total pressure loss coefficient, given by

$$\overline{\omega}_{CL} = 2\dot{m}_{CL}\Delta p_{CL}/(\dot{m}\rho_1 W_1^2) \tag{5-40}$$

Using the relative total thermodynamic conditions at the midpassage, the local sonic velocity, W^*, is computed. Then the inlet critical Mach number corresponding to the onset of sonic velocity at the midpassage suction surface is estimated by

$$M'_{cr} = M_1 W^*/W_{max} \tag{5-41}$$

where W_{max} is given by Eq. (4-41). When the blade suction surface velocity is supersonic, shocks will form, producing a loss and likely inducing boundary layer separation. When this is the case, a supercritical Mach number loss is estimated from

$$\bar{\omega}_{cr} = 0.4[(M_1' - M_{CR}')W_{max}/W_1]^2 \qquad (5\text{-}42)$$

Now, using Eq. (5-24), the discharge relative total pressure can be computed with this set of loss coefficients. Since h_t' is known, all other relative total thermodynamic conditions can be computed with the chosen equation of state. The tip tangential velocity is known from the blade work input using Eqs. (4-2) and (3-1). The static enthalpy follows from Eq. (3-6). An isentropic process between h_t' and h yields all other static thermodynamic conditions. The convergence criterion is based on comparison of the calculated mass flow, $\rho_2\phi_2 U_2 A_2$, with the actual impeller mass flow. If adequate convergence is not obtained, ϕ_2 is updated and the process is repeated. After convergence, the absolute discharge total enthalpy and velocities are computed from

$$h_{t2}' = h_{t1}' + IU_2^2$$
$$C_{U2} = I_B U_2 + U_1 C_{U1}/U_2$$
$$C_2 = \sqrt{C_{m2}^2 + C_{U2}^2} \qquad (5\text{-}43)$$

All other absolute total thermodynamic conditions can be computed from the equation of state and the static conditions with isentropic calculations between h and h_t.

5.4 Vaneless Annular Passage Performance

Vaneless diffusers, crossover bends and inlet passages are the most common vaneless annular passages used in centrifugal compressor stages. Others may appear throughout the stage, basically connecting two other components. Consequently, a performance analysis for general annular passages is frequently required, often for several components in a stage. Aungier (1993a) provides a one-dimensional performance analysis for vaneless annular passages, which is described here. This analysis benefited substantially from an earlier three-dimensional vaneless passage performance analysis (Aungier, 1988b), which clarified the fundamental fluid dynamics governing these flow fields. But once this insight was gained, refinement of the method was accomplished through extensive comparison with experimental data. As is often the case, the end result is that the one-dimensional predictions now show better agreement with experimental data than do the three-dimensional method predictions. The governing equations for one dimensional flow in a vaneless passage, including wall friction forces, are

$$2\pi r \rho b C_m (1 - B) = \dot{m} \tag{5-44}$$

$$b C_m \frac{d(r C_U)}{dm} = -r C C_U c_f \tag{5-45}$$

$$\frac{1}{\rho} \frac{dp}{dm} = \frac{C_U^2 \sin \alpha_C}{r} - C_m \frac{dC_m}{dm} - \frac{C C_m c_f}{b} - \frac{d I_D}{dm} - I_C \tag{5-46}$$

$$h_t = h + \tfrac{1}{2} C^2 \tag{5-47}$$

Except for the last two terms in Eq. (5-46), this set of equations is conventional for a one-dimensional analysis (e.g., Johnston and Dean, 1966). The additional terms address loss contributions due to flow diffusion and passage curvature. Flow diffusion losses are modeled by a classical diffuser analogy. The data from Reneau et al. (1967) show the low loss regime can be identified by the divergence parameter

$$D = b_1 (A_R - 1)/L = 2 \tan \theta_C \tag{5-48}$$

where diffusion losses are low for values of D less than

$$D_m = 0.4 (b_1/L)^{0.35} \tag{5-49}$$

Equation (5-49) is an empirical fit of the data in figure (8b) of Reneau et al. (1967). The analogy used for the vaneless annular passage is

$$D = -\frac{b}{C} \frac{dC}{dm} \tag{5-50}$$

$$D_m = 0.4 (b_1/L)^{0.35} \sin \alpha \tag{5-51}$$

The flow angle term in Eq. (5-51) is an empirical factor derived from comparisons of predicted and measured loss data from more than 35 compressor stage tests. Based on this same comparison, an empirical diffusion efficiency model was formulated as follows:

$$\begin{aligned} E &= 1; & D &\le 0 \\ E &= 1 - 0.2(D/D_m)^2; & 0 &< D < D_m \\ E &= 0.8\sqrt{D_m/D}; & D &\ge D_m \end{aligned} \tag{5-52}$$

The diffusion term is given by

$$\frac{d I_D}{dm} = -2(p_t - p)(1 - E) \frac{1}{\rho C} \frac{dC}{dm} \tag{5-53}$$

In addition to this streamwise diffusion loss term, an excessive meridional gradient of the passage area can cause higher losses. Again, a diffuser analogy is used to check for this situation at each computing station. The maximum, stall-free, local area is estimated by

$$(rb)_m = (rb)_1[1 + 0.16m/b_1]$$ (5-54)

which is equivalent to a diffuser divergence angle, $2\theta_C$, of 9^0. If the local area exceeds this value, a second estimate of the diffusion term is generated

$$I_D = 0.65(p_t - p)[1 - (rb)_m/(rb)]/\rho$$ (5-55)

If this value exceeds the local value obtained by integrating Eq. (5-46), it replaces that lower value. The passage curvature loss term is given by

$$I_C = \kappa_m(p_t - p)C_m/(13\rho C)$$ (5-56)

Equation (5-56) was developed empirically from comparisons of predictions with test data for 35 different vaneless diffuser/return system combinations. It has a negligible effect on vaneless diffuser performance, but is always significant (and sometimes dominant) for crossover bends. Once the accuracy of the analysis was established for vaneless diffusers, Eq. (5-56) was developed to extend the analysis to crossover bends. Its validity is further supported by the successful use of this same analysis in overall stage performance prediction for other curved passages, such as the stage inlet passage and the exit turn from return channels.

The area blockage factor and the skin friction coefficient are computed using a simple boundary layer growth model, based on a 1/7th power law for the boundary layer velocity profiles

$$C_m = C_{me}(y/\delta)^{1/7}$$
$$C_U = C_{Ue}(y/\delta)^{1/7}$$ (5-57)

where subscript e designates a value at the boundary layer edge. If the thickness of the boundary layer, δ, is known and the two boundary layers are identical, integration across the passage for mass flow yields an expression for the area blockage, B, as a function of $2\delta/b$, i.e.

$$\int_0^b \rho C_m dy = \rho b C_{me}[1 - 2\delta/(8b)] = \rho b C_{me}(1 - B)$$

$$B = 2\delta/(8b)$$ (5-58)

Similarly, integrating for the angular momentum flux yields

$$\int_0^b r\rho C_m C_U dy = r\rho b C_{me} C_{Ue}[1 - 2\delta/(4.5b)] \qquad (5\text{-}59)$$

Now, note that Eq. (5-58) includes B, meaning the analysis will predict the boundary layer edge meridional velocity. But Eq. (5-59) will compute an average angular momentum flux, i.e., the predicted angular momentum is related to the boundary layer edge value by

$$r C_U = r C_{Ue}[1 - 2\delta/(4.5b)] \qquad (5\text{-}60)$$

Hence, if δ is known at the inlet, C_{Ue} can be computed from the inlet rC_U. Then as the analysis proceeds along the passage, the local predicted rC_U and the known rC_{Ue} (which must be conserved until the boundary layers fill the passage) can be used to compute the local boundary layer thickness. So if δ is specified at the inlet, Eqs. (5-58) and (5-60) provide a means for calculating the boundary layer growth to yield local values of δ and B; naturally, the limit $2\delta \leq b$ must be imposed. Fortunately, a high degree of accuracy is not required for specifying δ. When an impeller analysis precedes the first vaneless passage, a simple flat plate boundary layer thickness estimate is made in the impeller to start the present analysis. When a vaneless passage analysis is conducted by itself, an empirical equation offering reasonable agreement with the impeller prediction is

$$2\delta/b = 1 - (b/r)_{in}^{0.15} \qquad (5\text{-}61)$$

The influence of a finite boundary layer thickness lies primarily in its effect on c_f, which is computed using the pipe friction model described earlier in this chapter, using 2δ in place of the pipe diameter. This effect can be quite significant when $2\delta/b$ is quite small, as happens in very high flow coefficient compressors.

The vaneless annular passage analysis consists of integrating Eqs. (5-44) through (5-47) along the passage length, subject to the auxiliary relations presented above. The equations are cast into finite difference form and solved in a direct marching technique from inlet to discharge. This performance analysis is used for vaneless diffusers, crossover bends and other vaneless annular passages. Its validity has been established for a broad range of stage specific speeds and operating conditions, including cases where D is far in excess of D_m. Figures 5-7 through 5-9 compare predictions from this performance analysis with the vaneless diffuser test data used for this purpose in Aungier (1993a). Results, including predictions from this analysis for crossovers, will be shown later in this chapter after the return channel performance analysis has been described.

5.5 Vaned Diffuser Performance

Aungier (1990) presents a one-dimensional performance analysis for conventional thin-vaned or airfoil style vaned diffusers. A modified form of that analysis

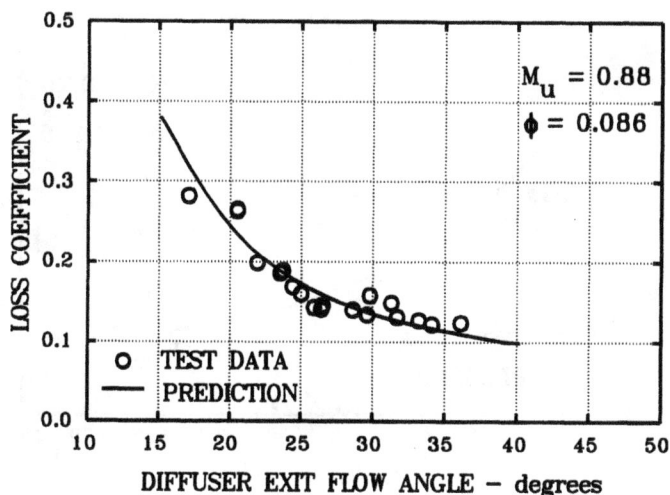

FIGURE 5-7. Loss in a Medium Flow Vaneless Diffuser

is described in this section, and has been generalized to treat vaned diffusers with nonparallel end walls and thick vanes. The generalization is relatively straight-forward, but it must be emphasized that the analysis has been qualified against experiments for the thin-vaned, parallel-walled diffuser type, only. On its occasional use for vane-island type diffusers, the analysis has appeared to give reasonable results; however, high quality test data were not available to really calibrate its accuracy in those cases.

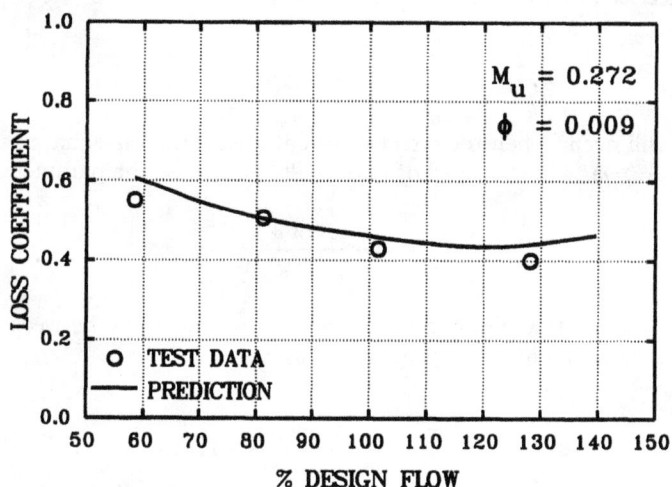

FIGURE 5-8. Loss in a Low Flow Vaneless Diffuser

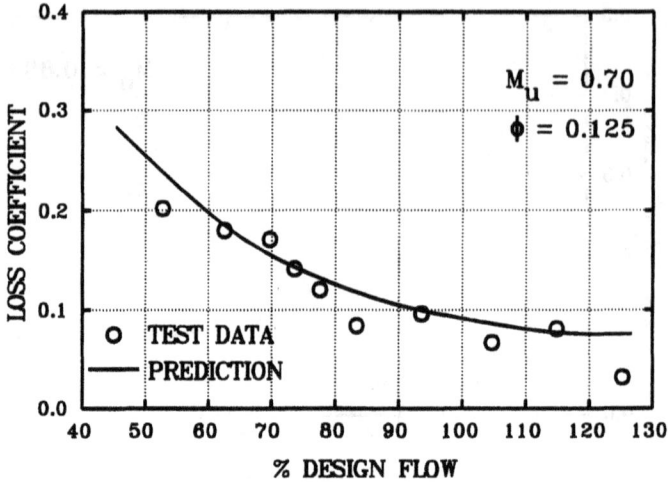

FIGURE 5-9. Loss in a High Flow Vaneless Diffuser

Figure 5-10 illustrates the basic vaned diffuser geometry to be analyzed. The vaned diffuser performance analysis is similar in concept to the impeller analysis. Specific flow calculations are accomplished at the vane leading edge, throat and trailing edge stations. The blade leading edge total thermodynamic conditions, mass flow and angular momentum (rC_U) are specified or supplied by a performance analysis of the upstream stage components. A simple mass balance then supplies all other vane leading edge fluid dynamics and thermodynamic conditions. The analysis starts by estimating the choke and stall flow limits for the vaned diffuser. Analogous to the impeller, viscous area blockage in the vane throat is estimated as a throat contraction ratio given by

$$C_r = \sqrt{A_3 \sin \beta_3 / A_{th}} \tag{5-62}$$

Choke will occur when the effective throat area ($C_r A_{th}$) is equal to the sonic flow area, $A^* = \dot{m}/(\rho^* C^*)$. Vaned diffuser stall is based on the parameter

$$K = -r \, \frac{\partial \cos \alpha}{\partial r} \tag{5-63}$$

evaluated between the inlet and the throat. An average value between the inlet and the throat is employed, approximated by

$$K = \frac{r_3}{h_{th}} \left[\frac{\cos \alpha_3}{\cos \alpha_{th}} - 1 \right]$$

$$\sin \alpha_{th} = A_{th}/A_3 \tag{5-64}$$

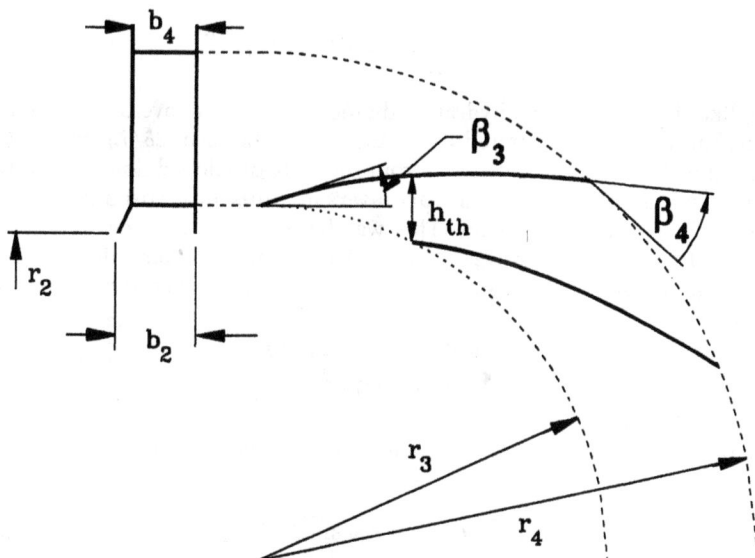

FIGURE 5-10. Vaned Diffuser Geometry

The specific approximations used are significant, since they were applied to develop the stall criterion from experimental data. Mach number effects are significant to the onset of vaned diffuser stall. This can be expected since as Mach number increases, K will assume nonzero values—even in a vaneless space. This "unguided" value of K can be expressed as

$$K_0 = \frac{M_3^2 \sin^2 \beta_3 \cos \beta_3}{1 - M_3^2 \sin^2 \beta_3} \tag{5-65}$$

which is derived from the basic fluid dynamics equation of motion in a radial, vaneless space of constant width with a flow angle equal to the vane inlet angle. After comparison with experimental stall limits, it has been found that stall in many vaned diffusers is well approximated by the following criterion:

$$K + K_0 = 0.39 \tag{5-66}$$

which can be solved to define the inlet flow angle, α_{3s}, or incidence angle corresponding to vaned diffuser stall.

The loss models used are also similar to the impeller loss system. A skin friction loss coefficient is given by

$$\overline{\omega}_{SF} = 4c_f(\overline{C}/C_3)^2 L_B/d_H/(2\delta/d_H)^{0.25} \tag{5-67}$$

where, like the impeller, the hydraulic diameter used is an average of the throat and discharge values, computed using Eq. (5-7). The term $2\delta/d_H$, in Eq. (5-67) corrects the skin friction coefficient from the fully-developed flow model, which was reviewed earlier because boundary layers in diffusers are usually not merged into fully-developed flow profiles. This was basically done in the same way for vaneless diffusers earlier in this chapter. The boundary layer thickness is estimated at midpassage from a simple flat-plate boundary layer approximation

$$2\delta/d_H = 5.142c_f L_B/d_H \tag{5-68}$$

$$2\delta/d_H \leq 1 \text{ required} \tag{5-69}$$

The optimum or minimum-loss incidence angle is defined as

$$\sin \alpha_3^* = C_{m3}/C_3^* = \sqrt{\sin \beta_3 \sin \alpha_{th}} \tag{5-70}$$

which represents a condition where the flow adjustments required to match the blade angle and the throat area are approximately balanced. For typical vanes, this corresponds to a modest negative incidence angle as the optimum condition. The minimum incidence loss for this optimum incidence is given by

$$\overline{\omega}_{i0} = 0.8[(C_3^* - C_{th})/C_3]^2 + [zt_{b3}/(2\pi r_3)]^2 \tag{5-71}$$

The first term can be recognized as 80% of an abrupt expansion loss (Benedict et al., 1966) between an entrance velocity defined by the minimum-loss inlet angle and the throat velocity, C_{th}. The second term accounts for the abrupt contraction in flow area at the leading edge due to the vane thickness. The off-design incidence loss is referenced to the velocities at the optimum incidence, C_3^*, and corresponding to the stall incidence, $C_{3S} = C_{m3}/\sin \alpha_{3S}$, as follows: If $C_3 \leq C_{3S}$, then

$$\overline{\omega}_i = 0.8[(C_3 - C_3^*)/C_3]^2 \tag{5-72}$$

whereas if $C_3 > C_{3S}$, it is assumed that 80% of the ideal pressure recovery for $C < C_{th} \ (C_3/C_{3S})$ is lost, i.e.,

$$\overline{\omega}_i = 0.8[((C_3/C_{3S})^2 - 1)C_{th}^2/C_3^2 + (C_{3S} - C_3^*)^2/C_{3S}^2] \tag{5-73}$$

The choking loss is identical to that used for the impeller, Eq. (5-32). Aungier (1988) presents a discharge area blockage correlation shown to be very effective in estimating the pressure recovery of a wide range of vaned diffusers (see Chapter 9). That model is the basis for a blockage loss coefficient, but with modifications to permit its application to thick vanes and variations in b; it employs two basic design parameters: the diffuser divergence angle and the blade loading parameter.

$$2\theta_C = 2 \tan^{-1}\{[(w_4 - t_{b4})b_4/b_3 - w_3 + t_{b3}]/(2L_B)\} \tag{5-74}$$
$$L = \Delta C/(C_3 - C_4) \tag{5-75}$$

where ΔC = average blade-to-blade velocity difference; and $w = (2\pi r \sin \beta)/z$. From simple potential flow

$$\Delta C = 2\pi(r_3 C_{U3} - r_4 C_{U4})/(zL_B) \tag{5-76}$$

It was observed that an abrupt deterioration in vaned diffuser performance occurs when $L > \frac{1}{3}$ or when $2\theta_C > 11°$. Hence, defining correction coefficients by

$$1 \leq C_\theta \geq 2\theta_C/11$$
$$1 \leq C_L \geq 3L \tag{5-78}$$

The discharge area blockage is defined as

$$B_4 = [K_1 + K_2(\overline{C}_R^2 - 1)]L_B/w_4 \tag{5-79}$$

where

$$\overline{C}_R = \frac{1}{2}\left[\frac{C_{m3} \sin \beta_4}{C_{m4} \sin \beta_3} + 1\right]$$

$$K_1 = 0.2[1 - 1/(C_L C_\theta)]$$

$$K_2 = \frac{2\theta_C}{125C_\theta}\left[1 - \frac{2\theta_C}{22C_\theta}\right] \tag{5-80}$$

The blockage correlation of Aungier (1988) contains an additional term in K_1 due to skin friction effects, which is omitted here since skin friction losses are handled separately in this analysis. Aungier (1990) imposed a correction term for this. Subsequent experience showed that simply neglecting the friction coefficient in K_1 works just as well. A wake mixing loss is included to account for excessive streamwise diffusion as well as vane discharge metal thickness. The flow is assumed to separate at a velocity defined by

$$C_{SEP} = C_3/(1 + 2C_\theta)$$
$$C_{SEP} \geq C_4 \text{ required} \tag{5-81}$$

As in the case of the impeller, only the meridional velocity is involved in the wake mixing process, since conservation of angular momentum governs C_U in the absence of blade forces. The meridional velocities before and after mixing are

$$C_{m,wake} = \sqrt{C_{SEP}^2 - C_{U4}^2}$$
$$C_{m,mix} = A_4 C_{m4}/(2\pi r_4 b_4) \tag{5-82}$$

and the wake mixing loss is given by

$$\bar{\omega}_{mix} = [(C_{m,wake} - C_{m,mix})/C_3]^2 \tag{5-83}$$

The vaned diffuser discharge total pressure is given by

$$p_{t4} = p_{t3} - (p_{t3} - p_3) \sum_i \bar{\omega}_i \tag{5-84}$$

Similar to the inlet guide vane analysis, the vane discharge flow angle is computed using axial-flow compressor correlations transformed to the radial plane. The minimum-loss deviation angle (Howell, 1947) is given by

$$\delta^* = \frac{\theta[0.92(a/c)^2 + 0.02(90^0 - \beta_4)]}{\sqrt{\sigma} - 0.02\theta} \tag{5-85}$$

where the location of the point of maximum camber, solidity and camber angle are given by

$$a/c = [2 - (\bar{\beta} - \beta_3)/(\beta_4 - \beta_3)]/3$$
$$\sigma = z(r_4 - r_3)/(2\pi r_3 \sin \bar{\beta})$$
$$\theta = \beta_4 - \beta_3 \tag{5-86}$$

The variation of the deviation angle with incidence is modeled by an empirical correlation of graphical data presented in Johnsen and Bullock (1965)

$$\frac{\partial \delta}{\partial i} = \exp[((1.5 - \beta_3/60)^2 - 3.3)\sigma] \tag{5-87}$$

and the vaned diffuser discharge flow angle is given by

$$\alpha_4 = \beta_4 - \delta^* - \frac{\partial \delta}{\partial i} (\beta_3 - \alpha_3) \tag{5-88}$$

The vaned diffuser performance analysis is an iterative process since many of the loss models depend on the discharge flow parameters. This involves repeating solutions for the discharge flow parameters, while updating the discharge meridional velocity to conserve mass. This process continues until convergence on C_{m4} is achieved.

5.6   Return Channel Performance

Aungier (1993a) provides a one-dimensional performance analysis for return channels, which will be reviewed here. Figure 5-11 illustrates the basic geometry to be analyzed. This analysis shares many common features with the vaned diffuser analysis reviewed in Section 5.5. A significant difference is the treatment of incidence losses, which are strongly influenced by the flow distortion imposed by the upstream crossover bend. Two estimates of aerodynamic area blockage at the vane entrance are made, and the larger value is used.

$$B_6 = 1 - (rb)_m/(r_6 b_6) \qquad (5\text{-}89)$$

$$B_6 = \frac{(\kappa_m b_6)^2}{12 + (\kappa_m b_6)^2} \qquad (5\text{-}90)$$

Equation (5-89) estimates blockage due to stall based on Eq. (5-54). Equation (5-90) is a simple inviscid flow estimate using the average curvature, κ_m, of the

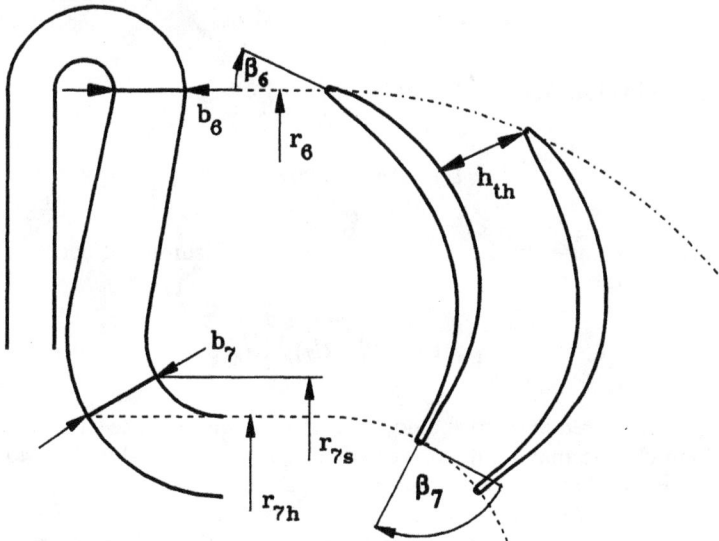

FIGURE 5-11. Return Channel Geometry

crossover bend. The minimum incidence loss is assumed to occur when the flow inlet angle is equal to the flow angle in the vane throat. Since this one-dimensional return channel analysis does not use area blockage directly when calculating the fluid dynamics data, this minimum loss incidence angle is adjusted to account for the effect of entrance blockage in a somewhat indirect manner. This yields

$$\tan \alpha^* = (1 - B_6) \tan[\sin^{-1}(A_{th}/A_6)] \tag{5-91}$$

where A_{th} = vane throat area. The incidence loss coefficient is given by

$$\bar{\omega}_{inc} = 0.8[1 - C_{m6}/(C_6 \sin \alpha^*)]^2 \tag{5-92}$$

Skin friction loss is computed by

$$\bar{\omega}_{SF} = 4c_f(\bar{C}/C_6)^2 L_B/d_H + C_{m6}C_{m7}|\alpha_{C6} - \alpha_{C7}|/(13C_6^2) \tag{5-93}$$

where $\bar{C}$ = average of the discharge velocity and either the inlet or throat velocity (whichever is larger); d_H = average of the throat and discharge hydraulic diameters; and c_f is computed from the pipe friction correlation discussed earlier. The last term in Eq. (5-93) is the curvature loss, derived from Eq. (5-56). The average blade-to-blade velocity difference is computed from the vane circulation

$$\Delta C = 2\pi(r_6 C_{U6} - r_7 C_{U7})/(zL_B) \tag{5-94}$$

and the blade loading loss coefficient is given by

$$\bar{\omega}_{BL} = [\Delta C/C_6]^2/6 \tag{5-95}$$

The maximum vane surface velocity is estimated assuming it occurs at midpassage for a midloaded vane

$$C_{max} = 0.5(C_6 + C_7) + \Delta C \tag{5-96}$$

and $C_{max} > C_6$ is required to include the more common case—i.e., when the return channel vane maximum surface velocity is the inlet value. When $C_{max} > 2C_7$, it is assumed that the flow will separate at a velocity of $C_{SEP} = C_{max}/2$. Otherwise C_{SEP} is set to C_7. The wake mixing loss is similar to the model used for the impeller and vaned diffuser. The meridional velocities before and after wake mixing is estimated from

$$C_{m,wake} = \sqrt{C_{SEP}^2 - C_{U7}^2}$$

$$C_{m,mix} = C_{m7}A_7/[\pi(r_{7s} + r_{7h})b_7] \qquad (5\text{-}97)$$

where A_7 includes the reduction due to the vane metal blockage. The wake mixing loss coefficient is given by

$$\overline{\omega}_{mix} = [(C_{m,wake} - C_{m,mix})/C_6]^2 \qquad (5\text{-}98)$$

The loss coefficient due to the exit turn into the eye of the next stage is given by an approximate solution for the friction loss and the curvature loss term of Eq. (5-56). Assuming that the radius of curvature of the mean streamline in the exit turn is equal to b_7, yields

$$\overline{\omega}_o = (4c_f + 1/13)|\alpha_{C7} - \alpha_{C8}|(C_{m7}/C_6)^2 \qquad (5\text{-}99)$$

For a more exact exit turn analysis, one can set α_{C7} to neglect this loss term and add a vaneless passage analysis after the return channel. The analysis also includes the same choking model as that used for the vaned diffuser analysis. Choke in a return channel is extremely rare, but it can happen, so a choking loss model is used for safety. The flow discharge angle is computed from the transformed axial flow compressor deviation angle model used for the vaned diffuser. The position of the point of maximum camber, the vane solidity and camber angle are estimated using the vane camberline vane angle at midchord, $\overline{\beta}$.

$$\frac{a}{c} = \left[\frac{2}{3} - \frac{(\overline{\beta} - \beta_6)}{3(\beta_7 - \beta_6)} \right] \qquad (5\text{-}100)$$

$$\sigma = z(r_6 - r_7)/(2\pi r_7 \sin \overline{\beta}) \qquad (5\text{-}101)$$

$$\theta = \beta_7 - \beta_6 \qquad (5\text{-}102)$$

and the reference deviation angle is (Howell, 1947)

$$\delta^* = \frac{\theta[0.92(a/c)^2 + 0.02(90 - \beta_6)}{\sqrt{\sigma} - 0.02\theta} \qquad (5\text{-}103)$$

Off-design incidence effects on flow deviation angle are given by the same empirical correlation as that used for the vaned diffuser (Johnsen and Bullock, 1965)

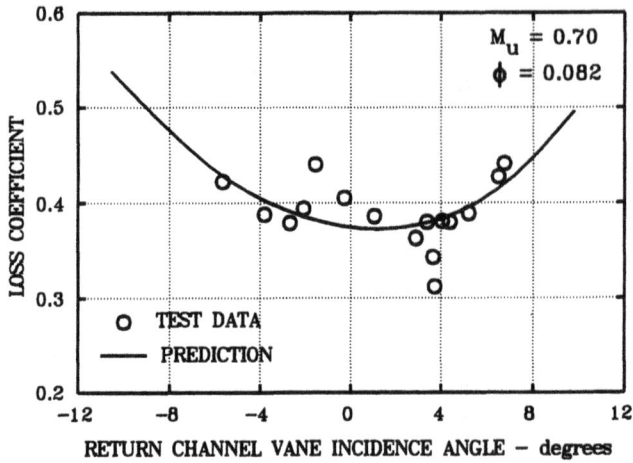

FIGURE 5-12. Loss in a Good Return System Design

$$\frac{d\delta}{di} = \exp[((1.5 - \beta_6/60)^2 - 3.3)\sigma]$$

$$\alpha_7 = \beta_7 - \delta^* - \frac{d\delta}{di}(\beta_6 - \alpha_6) \qquad (5\text{-}104)$$

The analysis is a simple iteration procedure, computing the losses and fluid turning while balancing mass at the discharge until convergence on the discharge meridional velocity is achieved. Figures 5-12 through 5-14 show typical loss

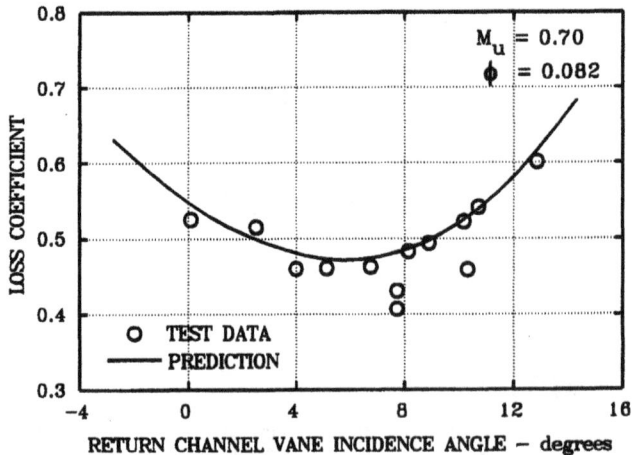

FIGURE 5-13. Loss in a Poor Return System Design

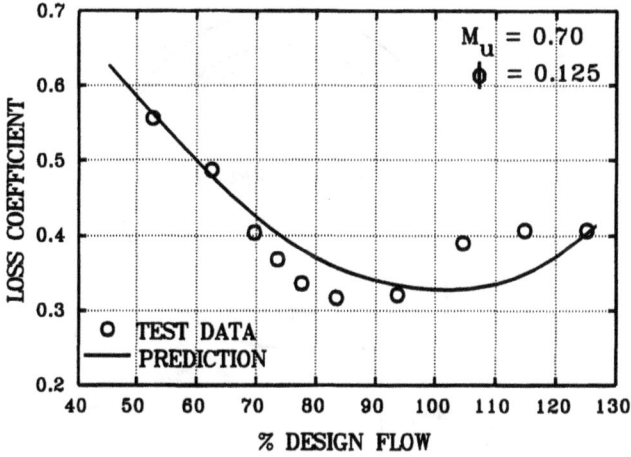

FIGURE 5-14. Loss in an Optimized Return System Design

coefficients for return systems (crossover and return channel together) obtained using the present vaneless passage and return channel performance analyses compared with test results from Aungier (1993a).

5.7 Volute and Collector Performance

Weber and Koronowski (1986) present a simple one-dimensional performance analysis for volutes and collectors. That is the basis for the method described here, although some improvements have been made in recent years. Figure 5-15

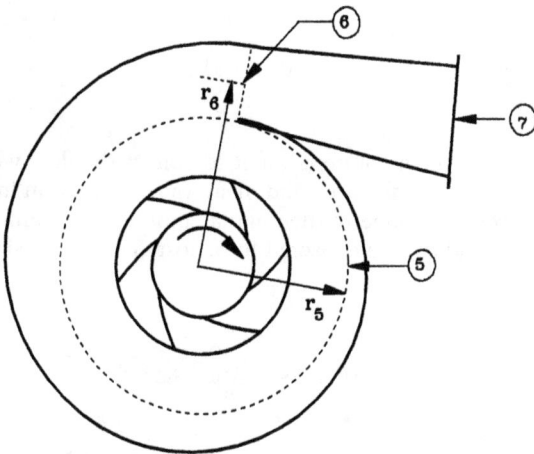

FIGURE 5-15. Volute Geometry

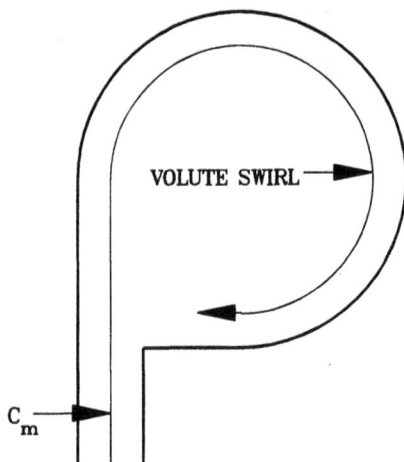

FIGURE 5-16. Volute Secondary Flow Pattern

shows the basic volute geometry and the station numbers employed. A collector is similar, except that the cross-sectional area of a collector is uniform at all circumferential locations. At Station 5, typically the diffuser exit, all fluid dynamics and thermodynamic parameters are assumed to be known. For either a volute or a collector, there is very little possibility of ever recovering the meridional velocity head. This velocity becomes a "swirl" component within the volute passage to be eventually dissipated in the downstream pipe, as illustrated in Fig. 5-16. Hence, the first loss considered corresponds to a complete loss of the meridional velocity head, i.e.,

$$\overline{\omega}_m = (c_{m5}/C_5)^2 \tag{5-105}$$

The ideal condition for a volute is for it to collect the flow while conserving angular momentum, rC_U. If the area and mean radius variation in the circumferential direction forces a change in the angular momentum, some portion of the tangential velocity head can be expected to be lost. The volute sizing parameter, SP, is defined as

$$SP = r_5 C_{U5}/(r_6 C_6) \tag{5-106}$$

where $SP = 1$ is the optimum value. Then a tangential velocity head loss is given by

$$\bar{\omega}_U = \frac{1}{2} \frac{r_5 C_{U5}^2}{r_6 C_5^2} \left[1 - \frac{1}{SP^2} \right]; \qquad SP \geq 1$$

$$\tag{5-107}$$

$$\bar{\omega}_U = \frac{r_5 C_{U5}^2}{r_6 C_5^2} \left[1 - \frac{1}{SP} \right]^2; \qquad SP < 1$$

For a collector, these relations apply only to the full-collection plane, i.e., Station 6. At the other extreme, when the flow collection process first starts, a collector attempts to diffuse the entering flow, essentially to zero velocity. This can be expected to cause a complete loss of the tangential velocity head, i.e., a local tangential head loss coefficient of unity. A corrected overall tangential velocity head loss for a collector is computed as the average of these two values, i.e.,

$$(\bar{\omega}_U)_{collector} = (1 + \bar{\omega}_U)/2 \tag{5-108}$$

A wall skin friction loss is computed from

$$\bar{\omega}_{SF} = 4c_f (C_6/C_5)^2 L/d_H$$
$$L = \pi(r_5 + r_6)/2$$
$$d_H = \sqrt{4A_6/\pi} \tag{5-109}$$

where L = average path length for the flow. Finally, an exit cone loss is given by

$$\bar{\omega}_{EC} = [(C_6 - C_7)/C_5]^2 \tag{5-110}$$

which is substantially greater than that used in Weber and Koronowski (1986); their model basically assumes that the exit cone's performance will be similar to that of an optimized ideal conical diffuser. Considering the distorted flow entering the exit cone, and the transition from a swirling annular flow (log-spiral flow path) to a linear flow path, that assumption is considered too optimistic.

The volute analysis follows the now familiar process of imposing the computed losses while iteratively converging on the full-collection plane velocity, C_6. Since gas density variations in a volute or collector are almost negligible, convergence is very rapid.

5.8   Overall Stage Predictions

The component performance analyses can be combined into an overall stage performance analysis. The most direct way is to develop a modular code such that any of the component analyses can be invoked for the particular order in which components appear in a given stage. By maintaining the total thermo-dynamic conditions, mass flow and angular momentum data after each compo-nent is treated, the starting data for the next component is always available. To illustrate the prediction accuracy achieved, a few representative examples will be given, comparing results with the test data reported in Aungier (1993b, 1995).

Figure 5-17 shows overall predictions for a very low flow coefficient stage with a vaneless diffuser and a return channel. This stage provides a severe test of the friction loss and parasitic work models, which really dominate its performance; clearly, excellent prediction accuracy is achieved. Figure 5-18 compares predic-tions and test data for a medium flow coefficient stage with a vaned diffuser and a return channel. The close agreement between the predicted vaned diffuser stall and the peak head with unstable operation is quite typical of this performance analysis. Figure 5-19 compares predictions and test data for a higher flow coeffi-cient and higher pressure ratio stage with a vaned diffuser and volute. Note the close agreement between predicted vaned diffuser stall and stage surge. Figure 5-20 shows results for a very high flow coefficient stage with a vaneless dif-fuser and return channel. There is rather good agreement between the predicted inducer stall and the rather obvious impeller stall indicated by the impeller total head curve. As noted in Aungier (1993b, 1995), this stage provides a particularly good test of the analysis, since it was designed following its development. Hence, it played no role in the "tuning" of the empirical models.

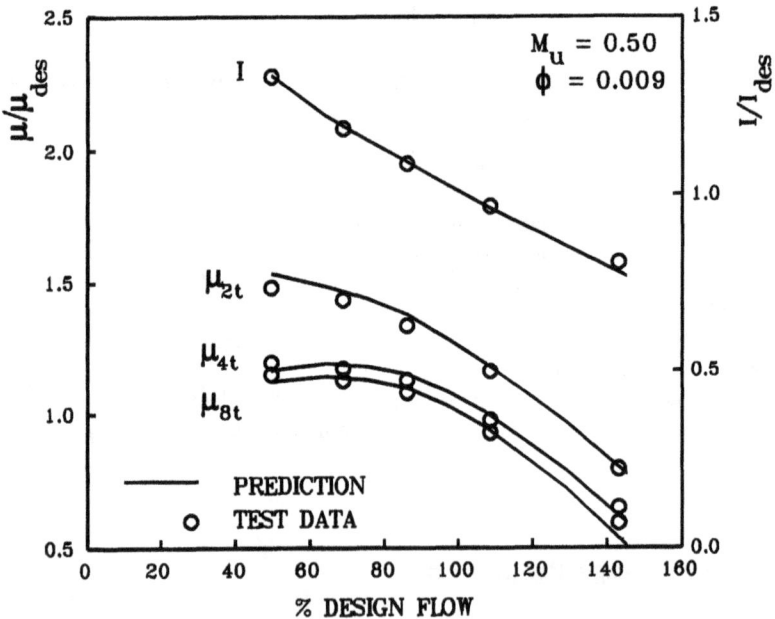

FIGURE 5-17. Performance of a Low Flow Stage

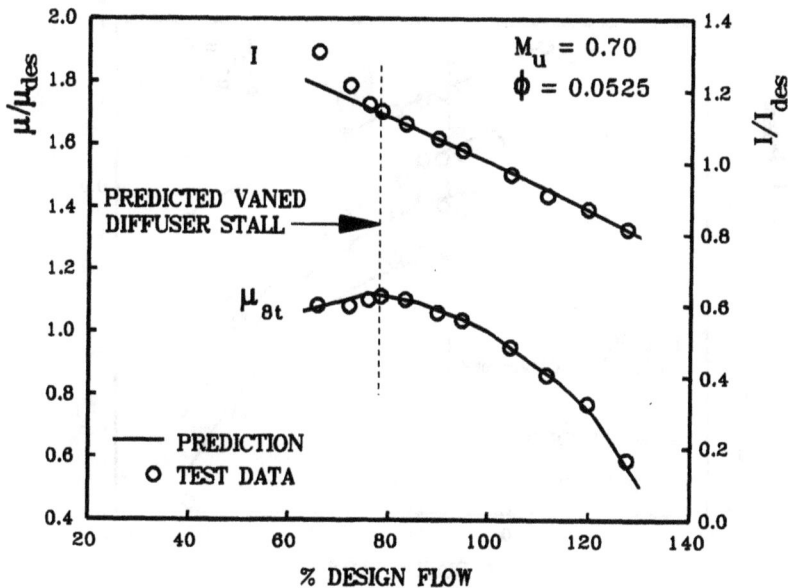

FIGURE 5-18. Performance of a Medium Flow Stage

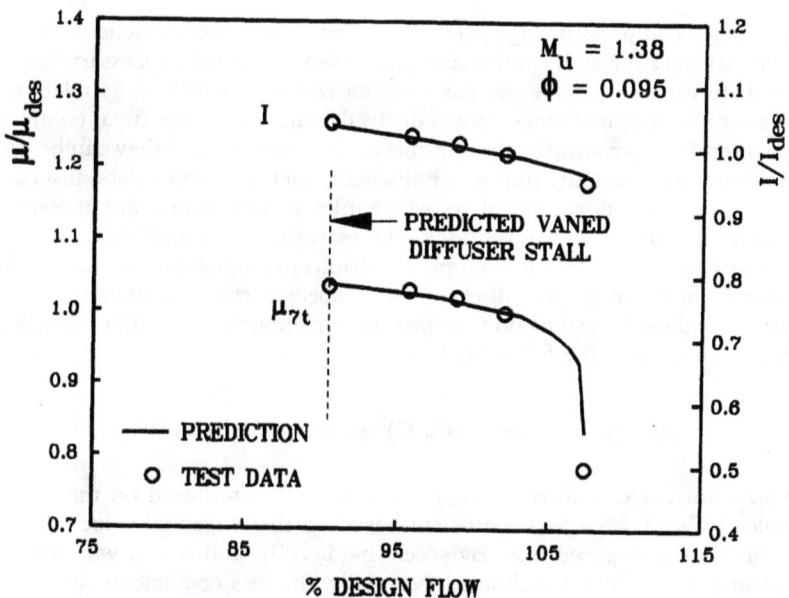

FIGURE 5-19. Performance of a Single Stage Air Compressor

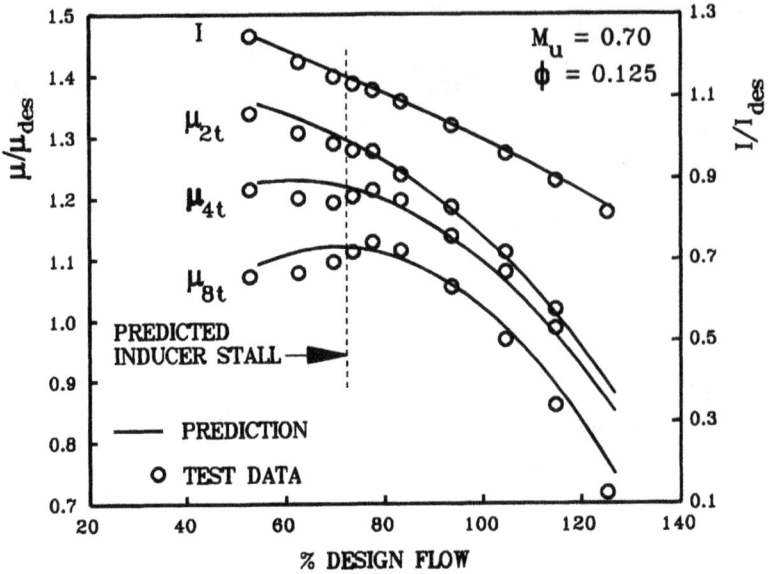

FIGURE 5-20. Performance of a High Flow Stage

5.9 Multistage Compressor Analysis

The centrifugal compressor stage performance analysis is easily extended to treat simple or "straight-through" multistage compressors. It is only necessary to supply the stage exit mass flow and gas thermodynamic conditions as inlet conditions to the subsequent stage. But industrial multistage centrifugal compressors are often far from simple, straight-through arrangements—they can be subject to numerous externally imposed influences, such as intercoolers, insertion or extraction of mass flow, special first-stage inlet arrangements, and unconventional interstage piping or connections. The versatility of a multistage analysis can be greatly increased by a few simple provisions to handle these features. This is easily accomplished by providing for some "special" stage components.

A simple and useful extension is to provide for imposing a total pressure loss anywhere in a stage in the following form:

$$p_{tc} = p_t - (p_t - p)[\overline{\omega}_m(C_m/C)^2 + \overline{\omega}_U(C_U/C)^2] - \Delta p_t \qquad (5\text{-}111)$$

where p_{tc} = corrected total pressure; $\overline{\omega}_m$ = loss coefficient based on the meridional velocity head; $\overline{\omega}_U$ = loss coefficient based on the tangential velocity head; and Δp_t is a user imposed loss. By specifying $\overline{\omega}_m$, $\overline{\omega}_U$ and Δp_t, a wide variety of loss sources can be approximated. Normally, the loss coefficients are specified with $\Delta p_t = 0$, but, Δp_t may be used when an experimental or specified pressure drop is the known quantity. For generality, loss coefficients are specified separately for the meridional and tangential velocity heads. For example, a step change in an annular passage's height will impose a meridional velocity head loss but the tangential velocity head will be conserved. The most common use of this total pressure loss "component" is to model a standard first-stage inlet (with a known loss coefficient) or to impose an exhaust loss, but it provides a means to include approximate treatments for unusual piping, interstage connections, etc. Note that by imposing $\overline{\omega}_m = \overline{\omega}_U = 1$ at the compressor exit, one obtains a total-to-static performance evaluation, which is often more meaningful than a total-to-total evaluation.

Another common interstage component encountered is an intercooler. The compressor aerodynamicist is not usually involved in heat-exchanger design; instead, the cooler's performance is usually a specified constraint. Most intercoolers can be modeled by specifying a total pressure loss coefficient, $\overline{\omega}$, and a flow area to be used in imposing it. A mass balance for the specified flow area yields the relevant density, ρ, and velocity, u. The corrected total pressure is then given by

$$p_{t,out} = p_{t,in} - \tfrac{1}{2}\overline{\omega}\rho u^2 - \Delta p_t \qquad (5\text{-}112)$$

where, again, Δp_t is included for cases where a pressure drop, rather than a loss coefficient, is known. The cooler exit temperature is normally the specified constraint, although a temperature drop may occasionally be the known quantity. Provision for either constraint is made by specifying the constants A and B in the following equation:

$$T_{t,out} = A T_{t,in} + B \qquad (5\text{-}113)$$

Usually, $A = 0$ and $B = T_{t,out}$ is used since most intercoolers are controlled to produce a specified discharge temperature. Alternatively, $A = 1$ and $B = \Delta T_t$ can be used if appropriate.

Provision for intercoolers necessitates another "pseudo-component," the liquid knockout or equilibrium flash calculation described in Chapter 2. A simple, common example of this need is a multistage air compressor, which may ingest humid air into its first stage. When the flow passes through the intercooler, some of the water vapor will condense and drop out of the flow. A liquid knockout calculation determines the corrected mass flow and the proper gas mixture composition to be used in the next stage. In the case of compressors applied to hydrocarbon mixtures, the liquid knockout calculation can become quite critical, dramatically altering the thermodynamic equation of state appropriate after an intercooler. Since a liquid knockout calculation can be useful under other conditions, it can be best provided for independent of an intercooler—for generality. For this calculation to be meaningful, it requires use of a nonideal gas equation of state and a multicomponent gas mixture.

Another common component encountered in industrial compressors is a side-load or side-stream flow added to the compressor's mass flow between stages. Figure 5-21 illustrates a typical side-load flow arrangement. Most compressor manufacturers use a fairly standard side-load flow arrangement for which basic

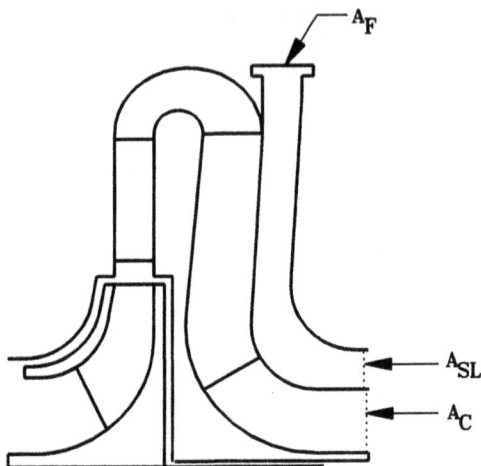

FIGURE 5-21. Side Load Flow Passage Geometry

performance parameters have been determined empirically. Typically, this takes the form of a loss coefficient defining the total pressure loss between the flange, p_{tF}, and the side-load exit total pressure, p_{tSL}. For generality, the performance analysis should specify the loss coefficient and the flow passage area to which it applies (since an empirical loss coefficient can be developed using various locations to define the relevant velocity head). Thus

$$p_{tF} = p_{tSL} + \tfrac{1}{2}\bar{\omega}\rho u^2 \qquad (5\text{-}114)$$

where ρ and u are computed from a mass balance at the specified flow area. The primary boundary condition imposed is that the side-load flow discharge static pressure must match the return channel exit static pressure. The passage area and two fluid dynamics parameters must be specified at the inlet flange. A convenient choice is the side-load total temperature and either the side-load mass flow or the flange static pressure. When the mass flow is specified, Eq. (5-114) provides a means to compute the flange pressure needed and therefore complete the thermodynamic definition of the side-load flow. If the flange static pressure is specified, the loss coefficient provides a means to calculate the mass flow. In either case, an iterative solution procedure is needed to match the boundary conditions at the flange and side-load discharge while satisfying Eq. (5-114). In general, the compressor discharge flow (at A_C in Fig. 5-21) and the side-load discharge flow (at A_{SL} in Fig. 5-21) will have different total temperatures, total pressures and velocities. Hence, a mixing loss calculation is required to determine the flow properties after mixing for use as the next stage inlet conditions. This can be accomplished by requiring the mixed-out flow to conserve the mass, momentum and total enthalpy supplied by the two unmixed flows, while maintaining a constant static pressure.

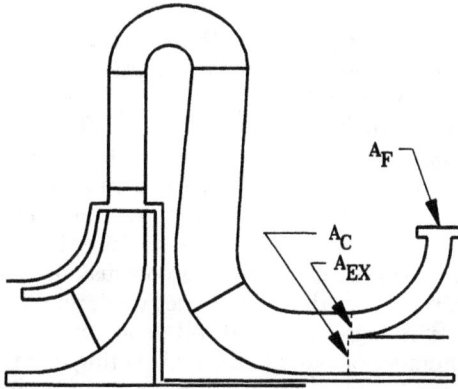

FIGURE 5-22. Extraction Flow Passage Geometry

Somewhat in contrast to the side-load flow, industrial compressors may also include an extraction flow, where a portion of the compressor flow is extracted for some other purpose. Figure 5-22 illustrates an extraction flow arrangement. Again, most compressor manufacturers will have a standard extraction design for which basic empirical performance has been determined. The process of modeling an extraction flow is quite similar to the side-load flow, but much simpler. Again, a loss coefficient and corresponding flow area are defined so Eq. (5-114) can be applied to the extraction flow. Now we need to know the relevant passage areas and either the flange pressure or the extraction mass flow. Since the extraction flow inlet conditions are just the stage exit conditions, the solution is relatively simple. If flange pressure is specified, an iterative solution is used to find the mass flow that matches the extraction inlet and flange pressures while satisfying Eq. (5-114). If the mass flow is specified, the solution is rather trivial, with Eq. (5-114) being used to predict the required flange pressure to achieve that extraction mass flow.

EXERCISES

5.1 Polishing surfaces in various stage components to reduce friction losses is common practice, but it can result in a significant increase in manufacturing cost. Explain how you could use results from a performance analysis to evaluate the need for improved surface finish of a compressor's various stage components for its specific operating conditions (without analyzing all combinations of different component surface finishes).

5.2 Stokes' theorem of vector field theory requires that for irrotational (potential) flow, the integral about any closed path of the velocity component tangent to the path must equal 0, i.e., $\oint \vec{C} \cdot \vec{dl} = 0$. Use this theorem to derive Eq. (5-76) or Eq. (5-94).

5.3 An industrial compressor product line consists of a series of stages, ranging in inlet flow coefficient from 0.005 to 0.15. All stages have identical impeller tip and vaneless diffuser discharge diameters. From Eqs. (5-46), which stages will be most influenced by diffuser friction losses? Which ones by diffuser diffusion losses?

5.4 When designing a vaneless diffuser, inlet mass flow and tangential velocity are boundary conditions set by the impeller. By adjusting the diffuser passage width, the designer has some freedom in setting the diffuser flow angles ($\tan\alpha = C_m/C_U$). When designing a low flow coefficient stage, should you favor high flow angles or low flow angles? How about for high flow coefficient stages? Do you expect similar conclusions to apply to other stage components? [Hint: if you didn't do Exercise 5.3, refer to it for background.]

5.5 A vaned diffuser is being designed. All geometrical parameters are fixed except for the number of vanes to be used. Noting that the throat area must remain constant, should you favor more vanes or less vanes if your objective is a wide stable operating range between the design flow and flow at the vaned diffuser stall?

Chapter 6

PRELIMINARY AERODYNAMIC DESIGN AND COMPONENT SIZING

The design of a centrifugal compressor stage starts with the selection of achievable design goals and sizing of the stage components to achieve those goals. Candidate stage geometry must be defined in sufficient detail to confirm that it can achieve the design objectives via the aerodynamic performance analysis presented in Chapter 5. An effective preliminary design system will generate the candidate stage design with the minimum input necessary. But it must possess sufficient flexibility to meet the wide range of objectives and constraints a designer may encounter. The preliminary design system should produce a candidate design for the complete stage by optimizing the stage components and matching them to one another. Attempts to accomplish the preliminary design directly to satisfy any arbitrary constraints and performance objectives have been rather ineffective. A performance analysis of the preliminary design usually indicates that the design objectives have not been met and the component geometry produced is usually totally impractical. The direct approach to preliminary design is not well suited to identifying a self-consistent combination of design constraints, performance objectives and key design parameters. Aungier (1995) presents a strategy for preliminary stage design that avoids most of these problems; an expanded description of that procedure is provided here.

It is essential that a preliminary design system be consistent with the detailed design and analysis methods used to evaluate and refine the design. The system described in this chapter is fully compatible with the various detailed design and analysis procedures covered in this book. The basic approach can be adapted to other design and analysis procedures; the reader must realize that this adaptation is necessary. The preliminary design system must provide a stage geometry that can achieve the performance goals and that has well-matched and optimized stage components as well as a practical component geometry. It provides the initial estimate of each component's geometry for evaluation by aerodynamic performance analysis and for refinement with detailed aerodynamic design methods. Substantial conflicts between the preliminary design system and the procedures used to evaluate and refine the preliminary design can easily make it ineffective.

NOMENCLATURE

A = passage area, and ellipse axial semiaxis

A_R = vaned diffuser area ratio

a = sound speed

B = ellipse radial semiaxis

B_A = fractional aerodynamic area blockage

B_M = fractional blade metal area blockage

b = passage width (hub-to-shroud)

C = absolute velocity

c_f = skin friction coefficient

d = diameter

h = enthalpy

I = work input coefficient, $(h_{t2} - h_{t0})/U_2^2$

I_B = impeller blade work input coefficient

I_{par} = parasitic work coefficient

i = incidence angle, $(\beta - \alpha)$

K = parameter defined in Eq. (6-17)

K_λ = parameter defined in Eq. (6-4)

L = impeller axial length; also vaned diffuser blade loading parameter

L_B = blade mean camberline length

m = meridional coordinate

$\dot{m}$ = mass flow

Q_0 = volume flow = $\dot{m}/\rho_{t0}$

R_c = mean stream surface radius of curvature

r = radius

SP = volute sizing parameter

U = impeller blade speed ωr

W = relative velocity

Z = number of vanes

z = axial coordinate

α = flow angle with tangential direction

β = blade angle with tangential direction

η = efficiency = stage head/work input head

θ = polar angle

$2\theta_C$ = vaned diffuser effective divergence angle

λ = impeller tip distortion factor = $1/(1 - B_{A2})$

μ = head coefficient, $\eta\,(h_{t2} - h_{t0})/U_2^2$

ρ = gas density

σ = slip factor

ϕ = stage flow coefficient = $Q_0/(\pi r_2^2 U_2)$

ω = rotation speed (rad/s)

ζ = normalized meridional distance

Subscripts

CO = crossover bend parameter

EX = return channel exit turn parameter

h = hub contour parameter
I = impeller parameter
m = meridional component
p = polytropic condition
s = shroud contour parameter
t = total thermodynamic condition
U = tangential component
VLD = vaneless diffuser parameter
VD = vaned diffuser parameter
0 = impeller eye condition
1 = impeller blade inlet condition
2 = impeller tip condition
3 = vaned diffuser inlet condition
4 = diffuser exit condition
5 = crossover or volute inlet condition
6 = crossover or volute exit condition
7 = return channel vane or exit cone exit condition
8 = return channel exit condition

Superscripts

$*$ = a sonic flow condition
$'$ = value relative to the rotating frame of reference

6.1 The Preliminary Design Strategy

Aungier's (1995) strategy works from reference design conditions that are consistent with good design practice and achievable performance objectives. This provides a practical starting point from which the reference designs can be adjusted to meet specific design constraints. To be effective, the preliminary design system must be directly interfaced with the aerodynamic performance analysis (Chapter 5) to permit immediate evaluation of the design and to provide direct guidance for adjusting the key design parameters. Simple iteration between the preliminary design system and the performance analysis quickly guides the designer to a consistent set of aerodynamic performance goals and key design parameters for the specific design constraints.

The reference designs are derived from geometric and aerodynamic performance correlations. They are somewhat subjective, reflecting what the author believes to be a state-of-the-art design practice. They were developed based on a series of rather successful industrial centrifugal compressor stage designs, including stage flow coefficients from 0.009 to 0.125 and stage pressure ratios up to 3.5. The performance prediction methods provided in Chapter 5 were used to extend the correlations to cover flow coefficients from 0.003 to 0.2. Polytropic (rather than adiabatic) performance data are used to minimize the dependence on pressure ratio as discussed in Section 2.2. No attempt was made to define

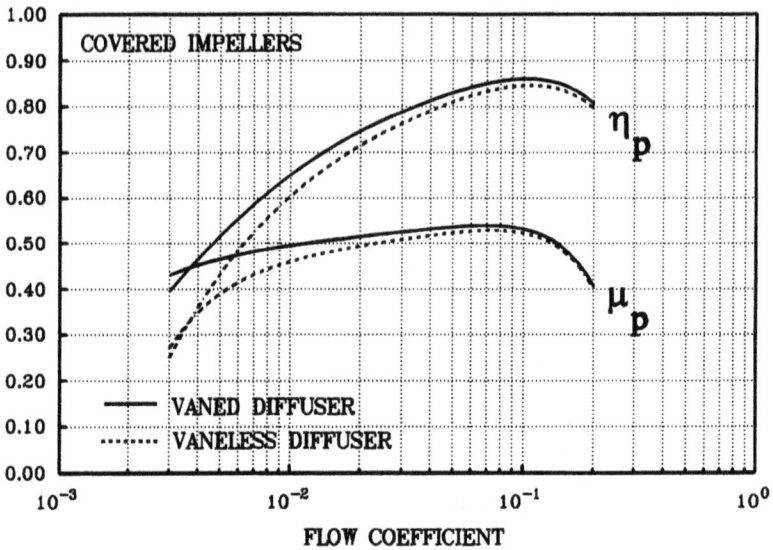

FIGURE 6-1. Covered Impeller Design Targets

the best achievable efficiency levels. Rather, these reference designs are intended to reflect a combination of good efficiency and stable operating range—that is considered to be readily achievable with the aerodynamic design procedures described in this book. The correlations for stage performance levels (head coefficient and efficiency) are shown in Figs. 6-1 and 6-2 for both covered and open impellers, respectively. The slight differences between the two impeller types are due to differences in the parasitic losses, particularly leakage flow (eye seal leakage versus clearance gap leakage). Different correlations are used for vaned and vaneless diffuser stages. Aungier (1995) suggests these correlations may be conservative for coefficients above 0.13, where mixed flow designs may exceed the levels shown. Limited experience since then has not provided evidence of that, and it remains an area of some uncertainty. Equation sets (6-1) and (6-2) express the correlations for covered and open impellers, respectively, in analytical form. Note that the work input coefficient, I, is assumed to be identical for both types of diffusers.

$$I = 0.62 - (\phi/0.4)^3 + 0.0014/\phi$$
$$(\mu_p)_{VD} = 0.51 + \phi - 7.6\phi^2 - 0.00025/\phi$$
$$(\eta_p)_{VLD} = (\eta_p)_{VD} - 0.017/[0.04 + 5\phi + (\eta_p)_{VD}^3] \qquad (6\text{-}1)$$
$$I = 0.68 - (\phi/0.37)^3 + 0.002/\phi$$
$$(\mu_p)_{VD} = 0.59 + 0.7\phi - 7.5\phi^2 - 0.00025/\phi$$
$$(\eta_p)_{VLD} = (\eta_p)_{VD} - 0.017/[0.04 + 5\phi + (\eta_p)_{VD}^3] \qquad (6\text{-}2)$$

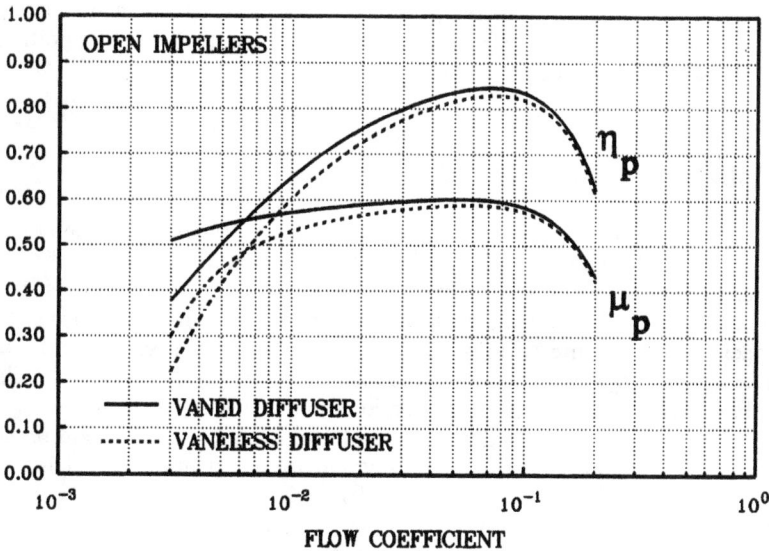

FIGURE 6-2. Open Impeller Design Targets

Based on past design experience, a correlation for impeller axial length, adequate to achieve the foregoing reference design performance levels, was developed. For consistency between various inducer styles, the axial length is measured from the impeller "eye" or axial entrance.

$$\Delta z_I/d_2 = 0.014 + 0.023 d_2/d_{0h} + 1.58\phi \tag{6-3}$$

Equation (6-3) differs from the form given in Aungier (1995). It was subsequently found that a dependence on the ratio d_{0h}/d_2 is needed for cases where this ratio is relatively small. Preliminary designs based on these reference design targets are consistently confirmed by the performance analysis presented in Chapter 5. Thus, they provide the desired starting point from which the designer can proceed to impose specific constraints for specific design objectives. Of course, if these reference design performance targets are not consistent with the actual design objectives, there is little point in generating a reference design. For example, if the actual design objective requires a head coefficient that is different from the reference design's value, it may as well be specified immediately. The reference design performance targets still play a role. They provide guidance in selecting a target efficiency. If the process starts with a realistic efficiency target, a preliminary design can be generated and analyzed—based on the methods provided in Chapter 5—to obtain a better estimate of the achievable efficiency. If the process starts with an unrealistic target efficiency, the preliminary design process may fail to arrive at a candidate design to be analyzed. The reference designs are also useful when the design performance objectives are not that well defined, e.g., when the designer is attempting to identify appropriate design targets for a specific application. In these cases, the reference designs provide a realistic starting point from which the designer can explore alternatives.

6.2 Simple Performance Correlations

The target values of μ_p and η_p define the required design work coefficient, I, which is critical to the preliminary design. Experience has shown that if realistic impeller tip flow conditions are generated, the preliminary design process is relatively insensitive to the specific performance targets. It is not necessary to know the specific losses in the various components to generate a viable preliminary stage design, but the impeller tip must be sized correctly. This requires knowledge of the blade work input coefficient, I_B, and a reasonable estimate of the impeller internal efficiency. As described in Chapter 4, estimating I_B from I requires some key impeller aerodynamic performance data, including the impeller tip distortion factor, the parasitic losses (windage, disk friction, leakage etc.) and the eye seal leakage mass flow for covered impellers. An efficient preliminary design system needs to be capable of estimating the data. Correlations to define these parameters were developed using the performance analysis of Chapter 5. For covered impellers, these parameters are given by

$$K_\lambda = 1 + [0.3 + (b_2/L_B)^2] \frac{b_2 A_2^2 \sin^2 \beta_2}{L_B A_1^2 \sin^2 \beta_1} \tag{6-4}$$

$$\lambda = K_\lambda + (0.00175/\phi)^2 + 0.0015/\phi - 0.022 \ln \phi \tag{6-5}$$

$$I_{par} = 0.0014/\phi \tag{6-6}$$

$$\dot{m}_{LEK}/\dot{m} = 0.005 + 0.475/(1 + 500\phi) \tag{6-7}$$

where Eq. (6-4) is a correlation using a form derived from the distortion factor model presented in Chapter 4. For open impellers, λ is increased by 2% over the value predicted by Eq. (6-5) due to clearance effects, and Eq. (6-6) is replaced by

$$I_{par} = 0.002/\phi \tag{6-8}$$

Figure 6-3 illustrates these correlations. The impeller internal or hydraulic efficiency has been correlated in the following form:

$$\eta_{2H} = 0.95 - 0.0005/\phi \tag{6-9}$$

Given I_B and η_{2H}, the impeller tip total thermodynamic conditions and the tip swirl velocity, C_{U2}, can be calculated to permit a valid mass balance at the impeller tip.

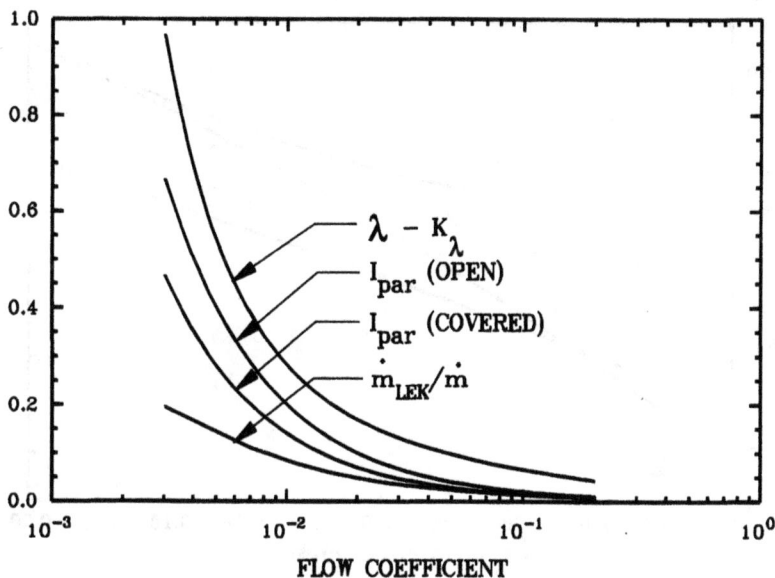

FIGURE 6-3. Parasitic Work Correlations

6.3 Component Matching

Proper matching of the various stage components with one another is necessary to obtain optimum performance from the stage under design operating conditions. The optimum matching conditions are a strong function of the design flow coefficient, and are quite dependent on the diffuser (vaned or vaneless) used. An exception is the impeller inlet relative flow angle, α_1'. A reasonable first estimate for this parameter is 30°, which (approximately) minimizes W_1. For vaneless diffuser stages, the choices of the impeller tip absolute flow angle, α_2, and the diffuser exit flow angle, α_4, are dependent on the design flow coefficient. For very low flow coefficients, stage performance is dominated by friction losses. To minimize these losses, the passages should be as wide as practical, leading to relatively low flow angles. For very high flow coefficients, diffusion and passage curvature losses are dominant factors, which require higher flow angles. Reasonable first estimates of these parameters are given by

$$\tan \alpha_2 = 0.26 + 3\phi \qquad (6\text{-}10)$$

$$\alpha_4 = 30° + (\phi/0.06)^2 \qquad (6\text{-}11)$$

For similar reasons, the return channel inlet angles are estimated by

$$\tan \alpha_6 = 0.32 + 1.7\phi \qquad (6\text{-}12)$$

Figure 6-4 illustrate the flow angles selected by these equations. Vaned diffuser

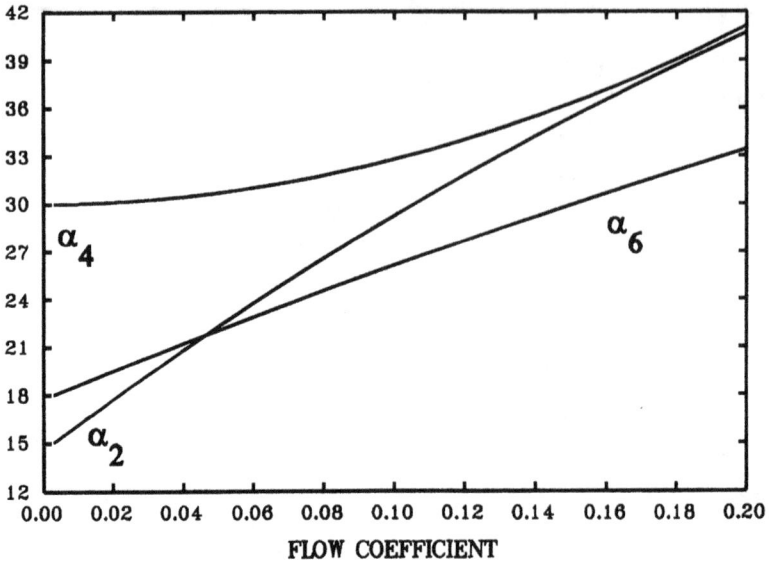

FIGURE 6-4. Vaneless Diffuser Stage Design Flow Angles

stages require different matching considerations. Both good performance and a wide stable operating range require low values of α_2 and α_3, but high flow coefficients require reasonably high values of α_2 for good impeller performance. For this reason, vaned diffusers cease to be effective at the higher flow coefficients. Precisely where this ineffectiveness occurs depends upon the designer's requirements for the flow range between the design flow and the vaned diffuser stall flow. As a first estimate for vaned diffuser stages, the design system of Aungier (1995) uses

$$\alpha_2 = 18° + 0.5 \ln \phi + 585\phi^2$$
$$\alpha_3 = 18° + (\alpha_2 - 18°)/4 \quad \text{if} \quad \alpha_2 \leq 18° \tag{6-13}$$
$$\alpha_3 = 18° \quad \text{if} \quad \alpha_2 > 18°$$
$$b_3 \leq b_2 \; required \tag{6-14}$$

and Eq. (6-12). The requirement that $b_3 \leq b_2$ will override the values of α_3 for higher flow coefficients, causing α_3 to be approximately equal to α_2. But as Mach numbers increase, compressibility effects will yield values somewhat lower than α_2, which is beneficial. When required, diffuser widths wider than the impeller tip width can be employed, but experience has shown the benefits from this are less than might be expected, and very difficult to predict. Hence, that situation is avoided during the preliminary design process. Figure 6-5 illustrates these flow angle selections, with a portion of the α_3 curve dashed to indicate values not likely to be selected due to the constraint on b_3.

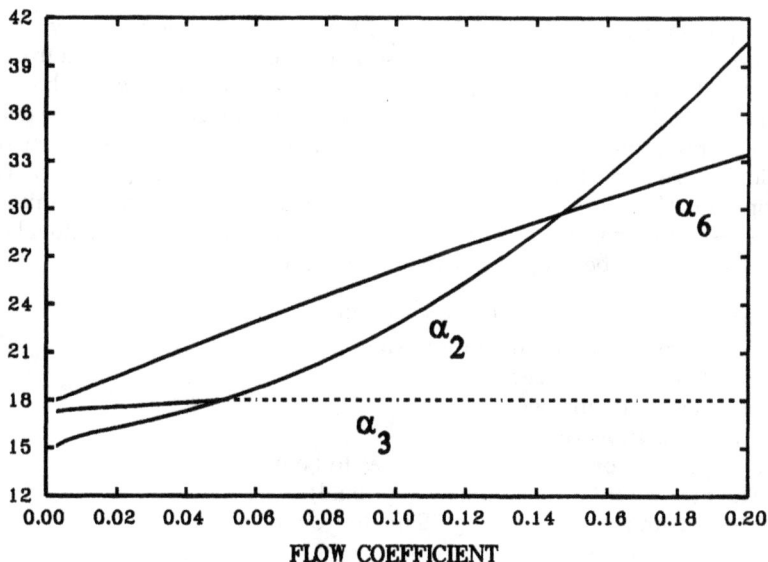

FIGURE 6-5. Vaned Diffuser Stage Design Flow Angles

For vaned components, the blade incidence angle, i, must be specified as a matching condition to define the inlet blade angle. Reasonable first estimates are $i_1 = 0$, $i_3 = -0.5°$, $i_6 = 4°$ for impellers, vaned diffusers and return channels, respectively. For computerized matching calculations for impellers and vaned diffusers, it is useful to impose a constraint on the incidence angles to obtain a reasonable range from design flow to choke flow. Otherwise, preliminary designs for higher rotational Mach number stages may yield a configuration that is choked at the design flow, precluding an evaluation by the performance analysis. Aungier (1995) recommends

$$\sin \beta_1 \geq 1.2 W_1 / a_1^{*\prime}$$
$$\sin \beta_3 \geq 1.2 C_3 / a_3^{*}$$

(6-15)

6.4 A Computerized Preliminary Design System

The preliminary design process is best accomplished in a computerized preliminary design system, with a direct interface to the aerodynamic performance analysis described in Chapters 4 and 5. Both are well suited to implementation on a personal computer of relatively modest capability. The reference design target performance and matching recommendations have all been presented in equation form to support that type of system. The preliminary design system sets these correlated parameters as default values. If none of them are modified by the designer, the reference design will be obtained. These default parameters always produce a self-consistent set of design specifications for a preliminary centrifugal compressor stage design consistent with good design practice. They are usually applicable (or at least good first estimates) to other design objectives and constraints. Several design parameters required for a performance analysis have little effect on the preliminary design process. These can conveniently be defined within the preliminary design system in accordance with the specific design organization's normal design practice and scaled with the impeller diameter where appropriate. These include blade thicknesses, seal geometry, surface finish and clearances. Other basic design specifications required to size a centrifugal compressor stage must be supplied by the designer, including

- Stage inlet total thermodynamic conditions
- Gas thermodynamic equation of state data
- Impeller rotation speed
- Impeller tip diameter
- Stage mass flow rate
- Specification of the component types to be used
- The minimum (shaft) diameter (optional)
- The maximum (outer casing) diameter (optional)

If the reference design is acceptable, it yields the candidate stage design directly, typically in a couple of minutes. Indeed, a useful role for the preliminary design system is to provide a quick evaluation of achievable performance levels and stage configurations appropriate to new applications. The designer may modify any of the default data to adjust the design to satisfy its specific objectives or constraints.

6.5   Impeller Sizing

The design system of Aungier (1995) treats two types of impeller blades. Three-dimensional, straight-line element blades are used for full-inducer or semi-inducer impellers; they are constructed with straight-line surface elements connecting any hub-and-shroud surface blade designs. Two-dimensional, axial-element blades are used for impellers having no inducer. For full-inducer impellers it is reasonable to assume an axial inlet with uniform flow, but for semi-inducer and no-inducer impellers, the leading edge meridional velocity distribution should be corrected for local passage curvature. If a series of linear "quasi-normals" are constructed from the inlet to the discharge, each approximately normal to the hub-and-shroud contours, simple potential flow calculations yield

$$C_{mh}\Delta m_h = \overline{C}_m \Delta \overline{m} = C_{ms}\Delta m_s \tag{6-16}$$

where Δm = local contour lengths between successive passage quasi-normals for the hub, mean and shroud contours. The impeller sizing requires an approximate blade design to ensure that blade rake angles (i.e., the angles between the blade leading and trailing edges and the meridional plane) are reasonable. An unacceptable blade rake angle is the most common inconsistency encountered during the detailed aerodynamic design of impellers from a preliminary design, particularly for semi-inducer stages. When an unacceptable rake angle is encountered, it can completely invalidate the preliminary stage design. For two-dimensional, axial-element blades the circular-arc mean camberline is a reasonable choice for a preliminary design. For three-dimensional blade styles, a more general approach is needed. The generalized hub-and-shroud blade angle distributions suggested in Aungier (1995) are a good choice. These blade angle distributions are defined in Eq. (6-17)

$$\beta_2 = \beta_{1S} + (\beta_2 - \beta_{1S})(3\varsigma^2 - 2\varsigma^3)$$
$$\beta_h = \beta_{1h} + A\varsigma + B\varsigma^2 + C\varsigma^3$$
$$\bar{\beta}_h = 90K + (1 - K)(\beta_2 + \beta_{1h})/2$$
$$A = -4(\beta_2 - 2\bar{\beta}_h + \beta_{1h})$$
$$B = 11\beta_2 - 16\bar{\beta}_h + 5\beta_{1h}$$
$$C = -6\beta_2 + 8\bar{\beta}_h - 2\beta_{1h} \tag{6-17}$$

These equations constrain the gradient of the blade angles, with respect to m, to be zero at the trailing edge, which reduces uncertainty with respect to work input prediction. This gradient is also set to zero at the inlet for the shroud blade angle distribution to reduce blade loading near the higher Mach number shroud leading edge. Blade design is clearly a matter for the detailed design process. But experience has shown that if this generalized blade style can be constructed, the detailed blade design process will almost certainly be successful. The parameter, K, adjusts the hub blade angle at midpassage to permit control over the blade rake angles. The leading and trailing edge rake angles are set equal and opposite in sign. If possible, these rake angles are held to less than $15°$, subject to limiting K to a maximum of 1. The leading edge blade angles are set from the local relative flow angles using the following constraints: A linear variation of blade angle from hub-to-shroud is assumed (which is a reasonable assumption for modest rake angles). The construction is constrained to match i_1 and to set the shroud incidence angle to 25% of the hub incidence angle. Figure 6-6 illustrates typical blade angle distributions generated in this fashion.

Hub contours are constructed to obtain the minimum contour curvature. This is accomplished by using the largest circular arc permitted (with linear extensions where needed) to match the impeller eye and tip slopes and coordinates. Shroud contours are constructed to match their end-point coordinates and slopes as well as the coordinates of one intermediate point. This can be accomplished using a three-point cubic spline fit curve. Both of these contour constructions are described in Chapter 7. When the hub contour requires a linear extension at the tip, a similar linear extension should be used on the shroud contour to obtain reasonable passage area distributions. By using this construction the passage area at three locations can be directly controlled. The selected areas set are at the blade leading and trailing edges and at the impeller eye. For full-inducer stages, where two of these locations are identical, the third location is set at midpassage.

The impeller preliminary design is accomplished by an iterative process where

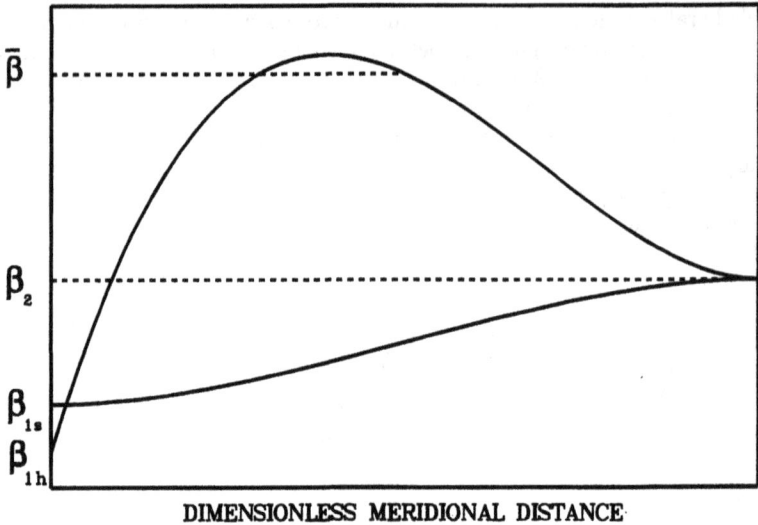

FIGURE 6-6. Impeller Blade Angle Distributions

the blades and hub-and-shroud contours are designed subject to the following constraints:

- Passage area varies linearly from inlet to discharge.
- The inlet blade angle is set to match leading edge incidence and flow angles.
- The impeller tip design must produce the tip velocity triangle.
- The leading and trailing edge rake angles must be controlled as discussed above.
- Mass must be conserved at the leading and trailing edges.
- The average passage width-to-radius of curvature ratio (b/R_c) can not exceed 0.5.

The last constraint results in a mixed-flow design for very high flow coefficient stages. When the impeller is a semi-inducer or no-inducer type, the following additional constraints are required to locate the blade leading edge:

- Limits on the magnitude of hub-and-shroud incidence angles.
- A 5% reduction in the passage area between the impeller eye and blade leading edge positions.

For semi-inducer and no-inducer impellers, the procedure is to locate the leading edge as far upstream as possible, consistent with the limits imposed on the incidence and rake angles.

The stage work input and parasitic losses are used to compute the impeller blade work input and tip tangential velocity, i.e.

$$I_B = I - I_{par} = C_{U2}/U_2 \qquad (6\text{-}18)$$

Following the impeller work input model of Chapter 4, the tip blade angle is computed by iterative solution of Eqs. (6-19) and (6-20), using the specific value of α_2.

$$\sigma = 1 - \sqrt{\sin \beta_2} \sin \alpha_C / Z^{0.7} \qquad (6\text{-}19)$$
$$I_B = \sigma(1 - \lambda C_{m2} \cot \beta_2 / U_2) \qquad (6\text{-}20)$$

The leading edge mean meridional velocity and blade angle are computed from the specified value of α'_1, and the local blade speed

$$C_{ml} = U_1 \tan \alpha'_1 \qquad (6\text{-}21)$$

The inlet passage width is set to conserve mass. The number of blades is estimated from the blade loading parameter

$$2\Delta W/(W_1 + W_2) \le 0.9 \qquad (6\text{-}22)$$

where ΔW is defined in Eq. (4-41) and Eq. (4-42) relates it to the number of blades. The iteration procedure continues until both the leading edge sizing and the number of blades converge.

6.6 Vaneless Diffuser Sizing

Vaneless diffuser sizing is accomplished following the design procedure of Aungier (1993a), described in Chapter 8. If the maximum (casing) radius has been specified, it is used to compute the discharge radius with allowance for the crossover or volute to follow. Otherwise, it is estimated from

$$r_4 = (1.55 + \phi)r_2 \qquad (6\text{-}23)$$

The discharge width is sized to yield the specified α_4. This includes a correction to the tangential velocity for wall friction effects computed from a simplified conservation of the angular momentum equation

$$\ln(r_4 C_{U4})/(r_2 C_{U2})] = \frac{-c_f(r_4 - r_2)}{\bar{b} \sin \bar{\alpha}} \qquad (6\text{-}24)$$

where the overbar designates average values in the passage. Equation (6-24) and conservation of mass yield b_4. The analysis imposes the requirement that $b_4 \le b_2$. Any adjustment in passage width between the impeller tip and diffuser exit is imposed on the shroud wall. For mixed flow stages, the slope of the hub contour at the impeller tip may be nonradial, requiring a gradual turn of this contour to radial. Circular arcs are used to construct the nonradial portions of the hub-and-shroud contours. They are imposed over the same portion of the passage length, not to exceed 50%. A minimum (shroud) radius of curvature equal to b_2 is used if the 50% length constraint permits. The larger hub radius of curvature is simply set to match any adjustment required.

6.7   Vaned Diffuser Sizing

Vaned diffuser sizing follows the design procedure presented in Aungier (1988a) and Chapter 9. The procedure used to set the vaneless diffuser exit radius also sets the maximum vaned diffuser exit radius. The vane leading edge radius is estimated by

$$r_3/r_2 = 1 + \alpha_3/360 + M_2^2/15 \tag{6-25}$$

This provides additional vaneless space to diffuse high Mach number flows to reduce the vane inlet Mach number. The passage width, b_3, is set to match the specified α_3 and conservation of mass, where C_{U3} is computed analogous to Eq. (6-24), and b_3 is constrained to $b_3 \le b_2$. Then, the specified i_3 yields β_3. Selection of the number of vanes is based on both aerodynamic and resonance considerations. The preferred choices are $Z_{VD} = Z_I \pm 1$. Since low values of Z_{VD} will improve the stall incidence range, $10 \le Z_{VD} \le 20$ is required. If the preferred choices of Z_{VD} are not in this range, $|Z_{VD} - Z_I| \ge 8$ is required. This basically defers the final selection of Z_{VD} and Z_I to the detailed design phase. Discharge sizing is based on the equivalent divergence angle, $2\theta_C$, the blade loading parameter, L, and the area ratio, A_R, with $b_4 = b_3$.

$$\tan \theta_C = \pi(r_4 \sin \beta_4 - r_3 \sin \beta_3)/(ZL_B) \tag{6-26}$$

$$L = \frac{2\pi(r_3 C_{U3} - r_4 C_{U4})}{ZL_B(C_3 - C_4)} \tag{6-27}$$

$$A_R = r_4 \sin \beta_4/(r_3 \sin \beta_3) \tag{6-28}$$

The design limits of Aungier (1988a) and Chapter 9 are used for these parameters, i.e., $L \le \frac{1}{3}$, $2\theta_C \le 11°$, with values close to these limits preferred. To select the best choices, values of $2\theta_C$ from 10.5° to 7° are checked. For each θ_C, A_R values from 2.4 to 1.4 are checked, ignoring values that require r_4 to be greater than the maximum value estimated above. The first combination of $2\theta_C$ and A_R to yield $L \le \frac{1}{3}$ is selected. If no suitable choice is found, the designer will need to modify the impeller tip velocity triangle. When the maximum (casing) radius is specified, it may exceed the value of r_4 computed here. In that case, a vaneless passage is inserted after the vaned diffuser. The diffuser vane is designed with the mean camberline presented in Aungier (1988a) and Chapter 9. The exit flow angle, α_4, is computed using the deviation angle model described in Section 5.5. Figure 6-7 shows a typical vaned diffuser design.

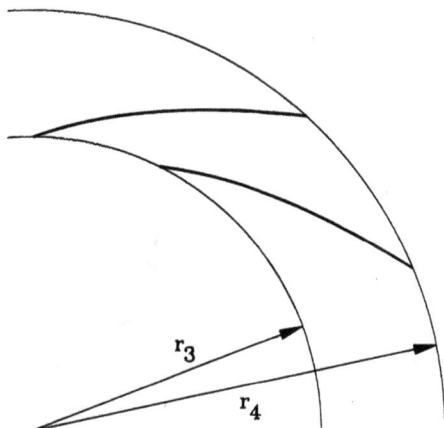

FIGURE 6-7. Typical Vaned Diffuser Design

6.8 Return System Sizing

The design procedure presented in Aungier (1993a) and Chapter 10 is used to design the crossover bend and the return channel. From the specified α_6, the crossover bend discharge passage width is computed

$$b_6 = b_4 \tan \alpha_4 / \tan \alpha_6 / (1 - B_{M6})$$
$$b_4 \leq b_6 \leq 2b_4 \ required \tag{6-29}$$

where the constraint may override the specified α_6. A circular-arc hub contour is used with a radius of curvature given by

$$R_{ch} = (b_6 + b_4)/2$$
$$R_{ch} \geq 0.8(b_8 - b_6) \ required \tag{6-30}$$

where b_8 = passage width for the eye of the next impeller. It is computed using the stage performance targets and the mass flow. This is a conservative choice, ensuring adequate axial length for the return system and reasonable values of b/R_c in the crossover bend passage. Usually, it is possible to reduce the axial length in the detailed design process. An elliptical contour is used for the crossover shroud wall, with axial and radial semiaxes given by

$$A_{CO} = R_{ch} + (b_4 + b_6)/2$$
$$B_{CO} = R_{ch} + b_4 \tag{6-31}$$

If the stage maximum (casing) radius is specified, the diffuser exit radius will have been set to accommodate this crossover. The exit turn is constructed with circular-arc hub-and-shroud contours, with the arc radii given by

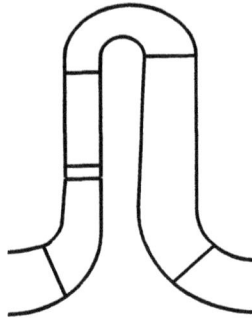

FIGURE 6-8. Typical Return System Cross-Section Design

$$R_{cEX_s} = b_8$$
$$R_{cEX_h} = 2b_8 \qquad\qquad (6\text{-}32)$$

The mean camberline and thickness distribution of Aungier (1993a) and Chapter 10 are used to construct the return channel vane. The assigned values of α_6 and i_6 determine β_6, and β_7 is set to 90°. The blade metal blockage, B_{M6}, in Eq. (6-29) depends on β_6, requiring a simple iteration process to converge on that parameter.

The return channel passage is constructed consistent with the detailed design procedure (Chapter 10), but simplified by requiring a vertical shroud wall. A typical cross-section of a return channel stage generated with the preliminary design system is shown in Fig. 6-8. Figure 6-9 shows the return channel vane design for this case.

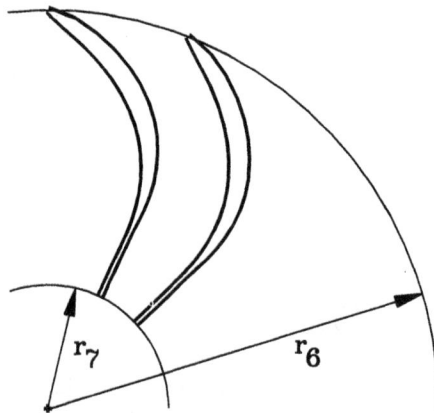

FIGURE 6-9. Typical Return Channel Vane Design

6.9 Volute Sizing

When the designer specifies the maximum (casing) radius for the stage, the volute cross-sectional shape is significant. Hence, both elliptical and rectangular cross-sections are considered. Volute sizing is based on the sizing parameter, SP, of Eq. (5-106). By definition, when $SP = 1.0$, the volute cross-sectional area is sized to conserve angular momentum between the volute centroid and the diffuser exit. For other values, the area calculated for $SP = 1.0$ is scaled by SP. The critical parameters are the area and mean radius at the circumferential position where all of the flow has been collected, A_6 and r_6. When no maximum radius is specified, a square or circular cross-section will be used, located above the diffuser discharge. When a maximum radius is specified, $r_6 = r_4$ is required, using a rectangular or elliptical cross-section while permitting an axial-to-radial aspect ratio up to 1.5. In this case, the diffuser exit radius will have been constrained to accommodate the volute. The full-collection area is given by

$$A_6 = SP(A_4 r_6 \tan \alpha_4 / r_4) \qquad (6\text{-}33)$$

Since r_6 is a function of r_4, A_6 and the aspect ratio, an iterative process is required. Figure 6-10 shows a typical volute stage designed by this procedure.

6.10 Implementation of the Design System

The author's preliminary design procedure is implemented as an iterative computer program for personal computers. The performance analysis program to which it is interfaced is also run on personal computers. By alternating between these programs, the designer can easily evaluate or customize the candidate centrifugal compressor stage preliminary designs. This preliminary design program also creates input data files for all of the detailed aerodynamic design systems

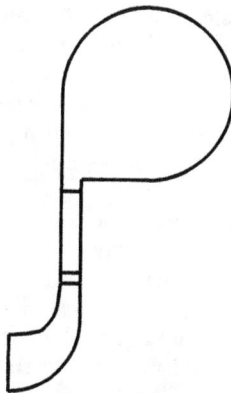

FIGURE 6-10. Typical Volute Stage Design

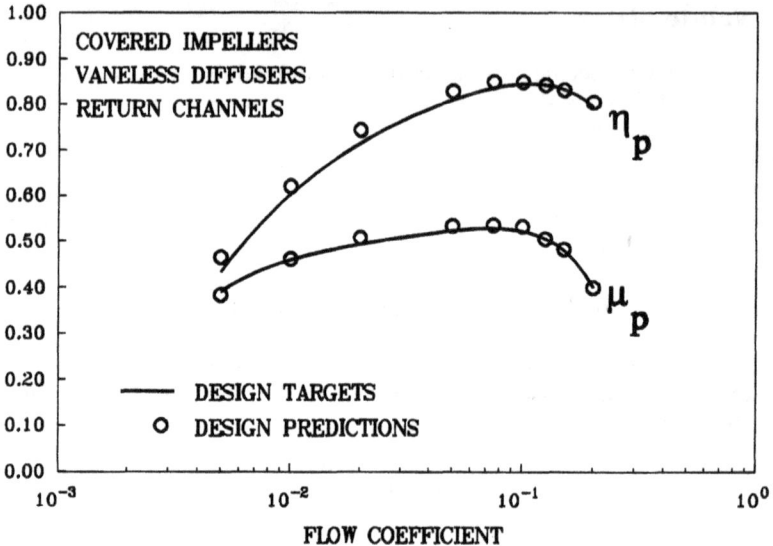

FIGURE 6-11. Qualification of Vaneless Diffuser Design Targets

for impellers, vaneless diffusers, vaned diffusers and return channels (Chapters 7 through 10)—all of which also run on personal computers. When a suitable preliminary design is established, the preliminary design geometry can be supplied directly for refinement by the more general, detailed aerodynamic design systems.

To demonstrate this design system, Aungier (1995) reports results for a series of 18 preliminary stage designs covering a range of stage flow coefficients and analyzed with the performance analysis. No-inducer impellers were used for $\phi \leq 0.05$, semi-inducer impellers for all others. Covered impellers and return channels were used. The default values of all design parameters were used (i.e., the reference designs). Figures 6-11 and 6-12 compare the performance prediction results for all stages with the design performance targets supplied in Figure 6-1. Note that the performance analysis indicates that nearly all preliminary designs should achieve the target performance levels. The performance analysis also confirmed that the components were well matched for each stage's design flow coefficient and predicted that a good stable operating flow range should be achieved.

One local weakness in the preliminary design system is apparent in Fig. 6-12. For ultra-low flow coefficient vaned diffuser stages, the design targets are not met. Indeed, this is a weakness in the vaned diffuser design procedure of Aungier (1988a). For ultra-low stage flow coefficients, wall shear forces in a vaneless diffuser yield nearly as much reduction in angular momentum as do the vaned diffusers, i.e., aspect ratio effects (not modeled in the design procedure) are quite significant. For vaned diffusers in ultra-low flow coefficient stages, modification to either the vaned diffuser design parameters (L and $2\theta_C$) or their design limits

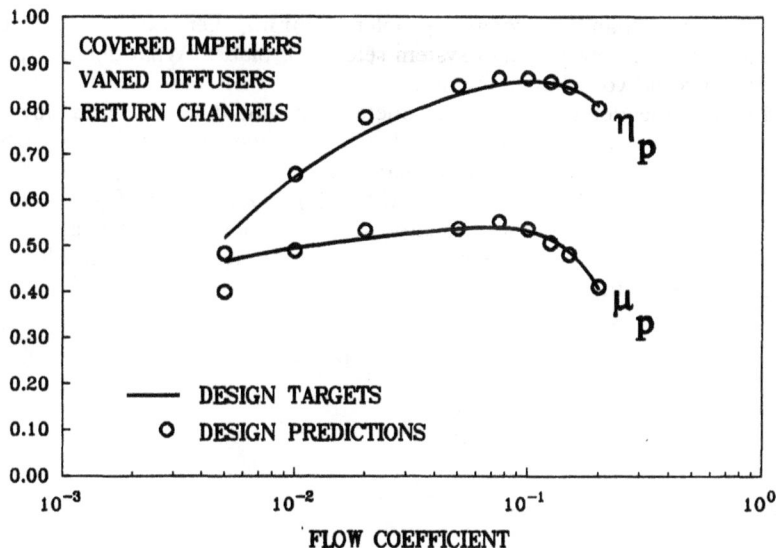

FIGURE 6-12. Qualification of Vaned Diffuser Design Targets

is needed. Some correction similar to Eq. (6-24) might be used to account for the additional diffusion supported by wall shear. For the present, this vaned diffuser procedure is not recommended for stage flow coefficient less than about 0.01.

These results also confirm a trend seen in Figs. 6-1 and 6-2, namely that there is little merit to using vaned diffusers for very high flow coefficient stages. Because relatively high impeller tip flow angles are needed for these stages, design of an effective vaned diffuser is really not possible. When lower impeller tip flow angles are used to favor improved vaned diffuser performance, the predicted impeller performance reduction is found to be greater than the gain provided by the better vaned diffuser design. Consequently, vaned diffusers can be expected to have little merit for very high flow coefficient stages.

EXERCISES

6.1 A performance analysis on your preliminary stage design shows a significant difference in both work input and efficiency relative to the design specifications. How would you correct this problem?

6.2 Why is it important to obtain reasonable agreement between the preliminary design performance specifications in Exercise 6.1 and the performance analysis results?

6.3 A performance analysis of your preliminary stage design shows that the stage cannot pass the design mass flow due to vaned diffuser choke. What default specification could you change to correct this?

6.4 Due to a mechanical resonance problem, your impeller *must* have 17 blades, but the preliminary design system selects 18 blades. What default specification could you change to correct this?

6.5 Identify the default design parameter you would change and how this should be done to acheieve the following changes in a preliminary design you have just generated: (a) Reduction in the impeller inlet passage width; (b) Increase in the impeller tip width; (c) Modification of a mixed-flow impeller style to a radial discharge style; (d) Increase in the range, from design flow to impeller choke flow; and (e) Reduction in friction losses in a vaneless diffuser.

Chapter 7

GENERAL GAS PATH
AND IMPELLER DESIGN

The term "gas path design" refers to the specification of the geometry of a component's surfaces that bound the passages through which fluid flows. This includes the hub-and-shroud contours and the blade surfaces for vaned components. This chapter describes some fairly general procedures for gas path design and their application to detailed impeller design. These gas path design procedures can be applied to the design of other stage components, for which very general design methods are needed. As will be seen in the following chapters, the detailed design of most centrifugal compressor components can be directly accomplished with methods specific to those components. Hence, the most common application of the methods of this chapter is impeller detailed design. Very little additional effort is required to implement the procedures in this chapter in a form applicable to any component consisting of an annular passage, with or without vanes, which covers most of the stage components of interest. On occasion, a more general design than that offered by the component-specific methods in the following chapters is needed. For example, there can be a need for three-dimensional vane geometry for vaned diffusers or even return channels, which is not offered in the procedures presented in Chapters 9 and 10. If implemented properly, the methods of this chapter will provide that more general, detailed design capability when required.

NOMENCLATURE

A = an area inside the blade passage
B = Bezier polynomial coefficient
$\vec{B}$ = vector along blade mean line defining element
b = hub-to-shroud passage width
h = distance between points on adjacent blade surfaces
i = point number
L = length of a line segment
m = meridional coordinate
N = number of points on a curve
n = distance along a quasi-normal
$\vec{P}$ = vector locating Bezier reference points
R_C = radius of curvature
r = radius

$\vec{S}$ = vector tangent to blade mean line in meridional surface
$\vec{T}$ = vector normal to blade mean surface
t_b = blade thickness
U = dimensionless parameter for curve generation
X = general Cartesian coordinate for curve construction or throat area calculation
x = Cartesian coordinate = $r \sin \theta$
Y = general Cartesian coordinate for curve construction or throat area calculation
y = Cartesian coordinate = $r \cos \theta$
z = axial coordinate
α_C = streamline slope angle with the axial coordinate
β = blade angle with respect to tangent
ϵ = deviation of a quasi-normal from a true normal
η = rotated Cartesian coordinate
θ = polar angle
ξ = the dimensionless meridional distance from the blade leading edge or along a line element
ϕ = angle of rotation of coordinates
χ = rotated Cartesian coordinate

Superscripts

$-$ = blade mean line parameter or an average value
$'$ = derivative of a function

Subscripts

B = a blade parameter
h = parameter on the hub contour
QN = parameter on a quasi-normal
s = parameter on the shroud contour
t = throat parameter
0 = impeller eye condition
1 = impeller blade inlet condition
2 = impeller tip condition

7.1 The General Gas Path Design Strategy

The gas path to be designed is always bounded by axisymmetric hub-and-shroud end-wall contours. When blades are included, three styles of blade geometry will be considered.

- Two-dimensional vanes whose blade mean camberline surfaces are constructed with line elements oriented in the axial direction, i.e., two-dimensional axial-element blades.

- Two-dimensional vanes whose blade mean camberline surfaces are constructed with line elements oriented in the radial direction, i.e., two-dimensional radial-element blades.
- Three-dimensional vanes whose blade surfaces are constructed with line elements connecting specified points on the hub-and-shroud contours, i.e., three-dimensional straight-line-element blades or "ruled-surface" blades.

These three types of blades are adequate for most centrifugal compressor design requirements. Two-dimensional axial-element blades are commonly used for vaned diffusers and return channels, as well as for low-flow coefficient impellers with no inducer or with a modest semi-inducer. They are relatively easy to manufacture—by simple-forming or three-axis milling methods. Two-dimensional radial-element blades offer the obvious advantage of minimal bending stresses for impellers since the centrifugal forces do not contribute. It was a very common blade style for many years, but with increased emphasis on efficiency and use of better materials, it is much less common today. The three-dimensional straight-line-element blade is very popular; because it has ruled surfaces, it is suitable for five-axis "flank milling," where the sides of the milling cutter can be used for metal removal. In comparison to the alternative "point-milling," this results in a substantial reduction in milling cost. It also simplifies pattern and tooling costs when manufacturing is accomplished by casting or hot forming.

For purposes of aerodynamic design, it is convenient to specify blade geometry in the form of distributions on specified surfaces, including the blade meanline geometry and a blade thickness distribution to be imposed on this mean line. The most convenient form is to specify the mean line blade angle, $\bar{\beta}$, and blade thickness, t_b, distributions as a function of a relevant length coordinate. Specifically, we will use

- Two-dimensional axial-element blades: $\bar{\beta} = \bar{\beta}(r)$, $t_b = t_b(r)$.
- Two-dimensional radial-element blades: $\bar{\beta} = \bar{\beta}(z)$, $t_b = t_b(z)$ for a specified value of r.
- Three-dimensional straight-line-element blades: $\bar{\beta} = \bar{\beta}(\xi)$, $t_b = t_b(\xi)$ for both the hub and shroud, where $\xi = (m - m_1)/(m_2 - m_1)$—the dimensionless meridional distance from the blade leading edge.

with the blade angles defined in the relevant surface, i.e.

$$axial\text{-}element: \quad \cot\bar{\beta} = \frac{r\partial\theta}{\partial r}$$

$$radial\text{-}element: \quad \cot\bar{\beta} = \frac{r\partial\theta}{\partial z} \qquad (7\text{-}1)$$

$$straight\text{-}line\text{-}element: \quad \cot\bar{\beta} = \frac{r\partial\theta}{\partial m}$$

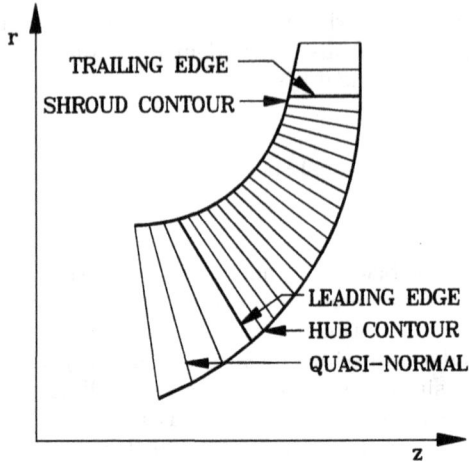

FIGURE 7-1. Meridional View of Gas Path

For the three-dimensional blade, the corresponding points on the hub-and-shroud blade profiles to be connected by straight-line elements must also be identified. This will be accomplished by defining a series of quasi-normals connecting the hub-and-shroud contours as illustrated in Fig. 7-1. Quasi-normals are straight lines that are approximately normal to these contours. The end points of these lines define corresponding locations on the hub-and-shroud blade profiles that are connected by the straight-line elements to form the three-dimensional ruled surfaces. Figure 7-2 illustrates the Cartesian and cylindrical coordinate systems employed in the gas path design process.

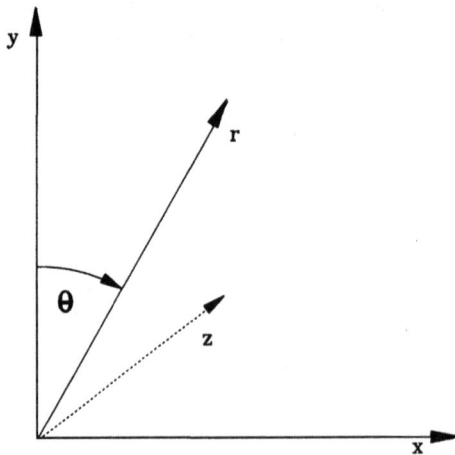

FIGURE 7-2. Coordinate Systems

This strategy reduces the gas path design problem to the specification of a series of curves used to construct the entire gas path. To be sure, some complex geometrical calculations are required, but those are easily handled by computers, so the designer is free to concentrate on the relevant aspects of the design process. Therefore, we need to specify the hub-and-shroud contours in the form $r = r(z)$, and the required blade profile curves in the forms discussed earlier. Before dealing with the mathematics of constructing the gas path from the curves, a number of curve forms found very useful for gas path design will be described.

7.2 Useful Curve Forms for Gas Path Design

The Bezier polynomial curve (Casey, 1983; Forrest, 1972) is a very general method used to generate smooth curves suitable for gas path design. It employs a set of reference points to define the curve in a parametric form. Let $(n + 1)$ be the number of reference points, with the points numbered 0 to n. The vector location of the kth reference point in a general Cartesian curve coordinate system (X, Y) is given by

$$\vec{P}_k = X_k \vec{i} + Y_k \vec{j} \tag{7-2}$$

The curve is defined as a function of the parameter, U, where U varies from 0 to 1 along the curve. The vector location of any point on the curve is defined as

$$\vec{R}(U) = \sum_{k=0}^{n} \vec{P}_k B_k^n(U) \tag{7-3}$$

and the Bezier polynomial is defined by

$$B_k^n = \binom{n}{k} U^k (1 - U)^{(n-k)} \tag{7-4}$$

$$\binom{n}{k} = \frac{n(n-1)\ldots(n-k+1)}{k!} \tag{7-5}$$

If the x and y components of the vector Eq. (7-3) are differentiated with respect to U, the curve slopes and curvatures can be computed using the derivatives $(dX/dU, dY/dU, d^2X/dU^2, d^2Y/dU^2)$

$$\frac{dY}{dX} = \frac{dY/dU}{dX/dU} \tag{7-6}$$

$$\frac{1}{R_c} = \frac{\dfrac{dX}{dU}\dfrac{d^2Y}{dU^2} - \dfrac{dY}{dU}\dfrac{d^2X}{dU^2}}{\left[\left(\dfrac{dX}{dU}\right)^2 + \left(\dfrac{dY}{dU}\right)^2\right]^{1.5}} \tag{7-7}$$

where R_c = radius of curvature. Special definitions are used for derivatives of the Bezier polynomials

$$\frac{dB_k^n}{dU} = n[B_{k-1}^{n-1}(U) - B_k^{n-1}(U)] \tag{7-8}$$

$$\frac{d^2B_k^n(U)}{dU^2} = n(n-1)[B_{k-2}^{n-2}(U) - 2B_{k-1}^{n-2}(U) + B_k^{n-2}(U)]$$

where for any k, the following special relations apply:

$$B_0^k(U) = 0; B_{-1}^k(U) = 0; B_{k+2}^k(U) = 0; B_{k+3}^k(U) = 0 \tag{7-9}$$

Casey (1983) provides a means of setting the radius of curvature at the end points by

$$R_c = \frac{n}{n-1}\frac{a^2}{b} \tag{7-10}$$

where a and b are illustrated in Fig. 7-3. The primary features of the Bezier polynomial that make it valuable for this application are

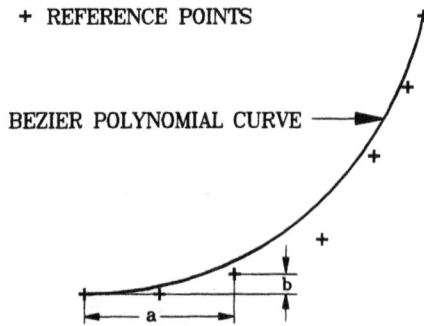

FIGURE 7-3. Bezier Polynomial Curves

- The curve passes through the two end reference points (but generally not through the other points).
- The slope of the curves at the end point is the same as that of a line joining the end reference point and its adjacent reference point.
- The end-point curvature is defined by the end reference point and its two adjacent reference points.
- The interior points "influence" the curve shape. In general, if an interior reference point is moved, the curve will be moved in the same direction but about $\frac{1}{3}$ as much.
- The curve is always continuous and smooth.
- Points can be distributed along the curve (e.g., quasi-normal end points) by selecting a distribution of discrete U values.

The disadvantage of the Bezier curve form lies in the fact that it must be constructed in an interactive graphics mode to be useful for gas path design. Typically, the reference points are displayed on the monitor screen with provision made to change the positions of the points and view the resulting curve. In this way, the curve can be fairly quickly shaped to the desired form. It is easily seen that use of seven or more reference points will permit control of the end points and their slopes and curvatures, while leaving at least one more point free to shape the curve without altering the curve at the end points. By implementing in a form where end-point slopes and curvatures can be automatically reset when the points are moved, all of the reference points can be used to shape the curve while still constraining the end-point slopes and curvatures. Figure 7-3 shows a typical Bezier polynomial curve, together with the reference points used to generate it.

Another very useful curve form is illustrated in Fig. 7-4. Here, the coordinates and slopes of the end points are specified together with optional linear segments located at each end point. Then, a circular arc with the largest possible radius of curvature is used to complete the curve. This construction generally requires changing the length of one of the linear segments. This curve construction is useful when it is desired to define a contour fitting within specified ranges of X and Y and having the lowest possible curvature. Indeed, this is the construction used in Chapter 6 for the impeller hub contour. From Fig. 7-4, the X, Y, α_C and L are specified at each end point. L is the minimum acceptable line segment length at the end point (which may be zero). Simple geometry yields

$$\Delta X = R_C(\sin \alpha_{C2} - \sin \alpha_{C1}) + L_2 \cos \alpha_{C2} + L_1 \cos \alpha_{C1} \qquad (7\text{-}11)$$

$$\Delta Y = R_C(\cos \alpha_{C1} - \cos \alpha_{C2}) + L_2 \sin \alpha_{C2} + L_1 \sin \alpha_{C1} \qquad (7\text{-}12)$$

By simply eliminating the arc radius, R_C, using Eq. (7-11) and (7-12), a necessary condition to construct a circular arc is obtained

$$L_2 - L_1 = \frac{\Delta Y(\sin \alpha_{C2} - \sin \alpha_{C1}) - \Delta X(\cos \alpha_{C1} - \cos \alpha_{C2})}{1 - \sin \alpha_{C1} \sin \alpha_{C2} - \cos \alpha_{C1} \cos \alpha_{C2}} \qquad (7\text{-}13)$$

Therefore, the circular-arc contour can be constructed if the appropriate line

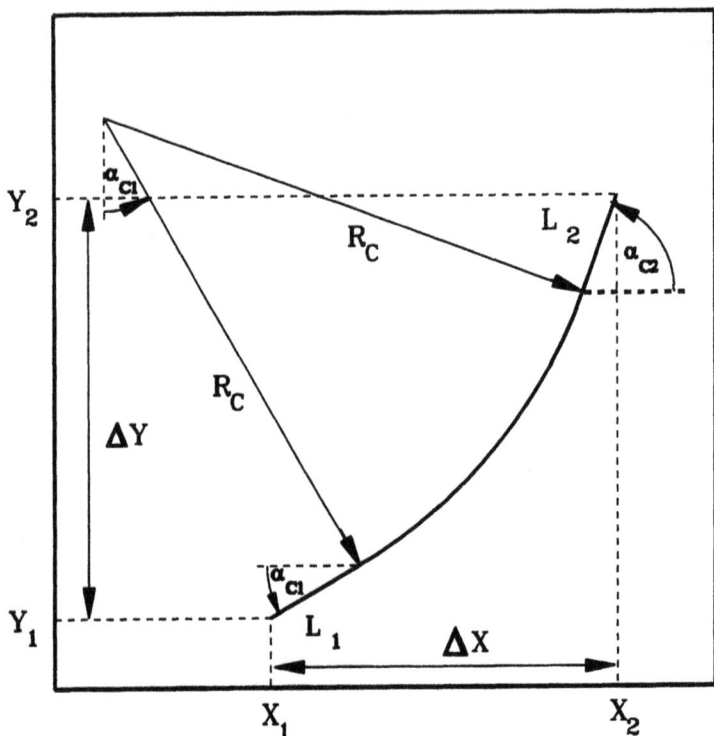

FIGURE 7-4. Circular-Arc Contour Geometry

segment length is increased to satisfy Eq. (7-13). R_C can be calculated from either Eq. (7-11) or Eq. (7-12) and the arc center from

$$X_C = X_1 + L_1 \cos \alpha_{C1} - R_C \sin \alpha_{C1}$$
$$Y_C = Y_1 + L_1 \sin \alpha_{C1} + R_C \cos \alpha_{C1} \tag{7-14}$$

from which the circular-arc and linear segments defining the curve are readily constructed.

Figure 7-5 illustrates a curve construction based on the well-known cubic spline fit (Walsh et al., 1962). For gas path design, it is most useful to employ a three-point spline fit with the slope of the curve specified for the two end points and optional line segments, L_1 and L_3. The spline fit employs a cubic polynomial between adjacent points while matching first and second derivatives of successive polynomials at each interior point. Three coefficients are required to define the two spline-connected cubics (Walsh et al., 1962). Denoting the end-point slopes by

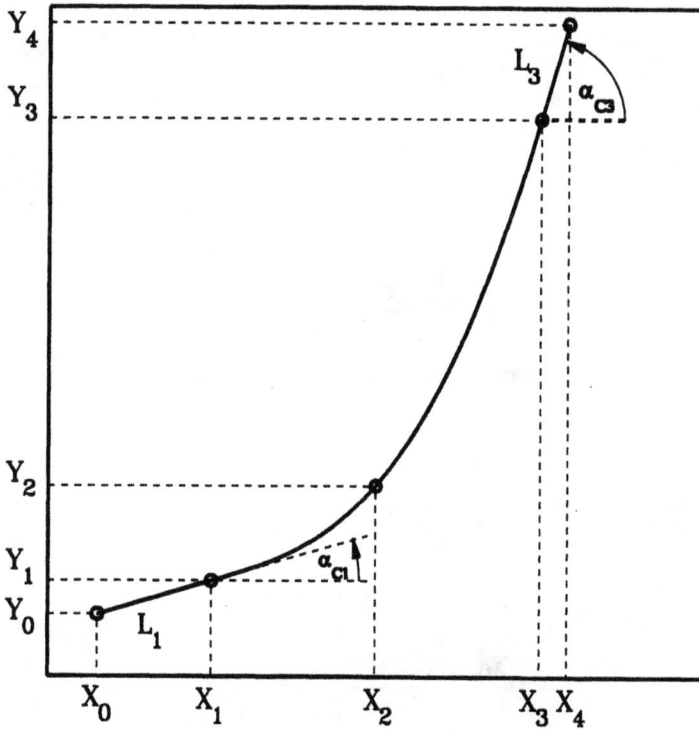

FIGURE 7-5. **Three-Point Cubic Spline Curves**

$$Y'_1 = \tan \alpha_{C1} \qquad (7\text{-}15)$$

$$Y'_3 = \tan \alpha_{C3} \qquad (7\text{-}16)$$

the three required coefficients are given by

$$M_2 = \frac{6(Y_3 - Y_2)}{(X_3 - X_2)(X_3 - X_1)} - \frac{6(Y_2 - Y_1)}{(X_2 - X_1)(X_3 - X_1)} + \frac{2(Y'_1 - Y'_3)}{X_3 - X_1} \qquad (7\text{-}17)$$

$$M_1 = \frac{3(Y_2 - Y_1)}{(X_2 - X_1)^2} - \frac{3Y'_1}{X_2 - X_1} - \frac{M_2}{2} \qquad (7\text{-}18)$$

$$M_3 = \frac{3Y'_3}{X_3 - X_2} - \frac{3(Y_3 - Y_2)}{(X_3 - X_2)^2} - \frac{M_2}{2} \qquad (7\text{-}19)$$

and the polynomial to be used is given by

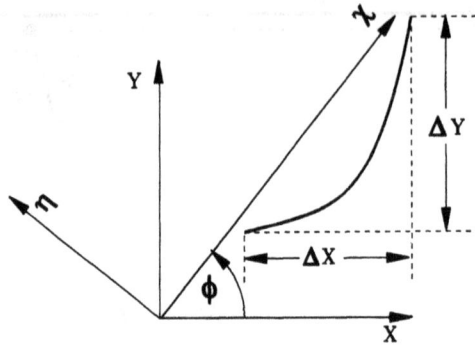

FIGURE 7-6. Rotation of Coordinates

$$
y = \frac{M_{k-1}(X_k - X)^3}{6(X_k - X_{k-1})} + \frac{M_k(X - X_{k-1})^3}{6(X_k - X_{k-1})}
$$
$$
+ \left[\frac{Y_k}{X_k - X_{k-1}} - \frac{M_k(X_k - X_{k-1})}{6} \right] (X - X_{k-1}) \tag{7-20}
$$
$$
+ \left[\frac{Y_{k-1}}{X_k - X_{k-1}} - \frac{M_{k-1}(X_k - X_{k-1})}{6} \right] (X_k - X); \ \ where \ X_{k-1} \le X \le X_k
$$

Consequently, the entire curve is easily generated based on the coordinates of three points and the end-point slopes. This curve form can be convenient when attempting to set passage areas at inlet and discharge while controlling the area at some intermediate location. This is precisely the application described in Chapter 6 relative to the impeller shroud contour definition.

While the circular-arc and cubic spline curve generation outlined above is quite direct, centrifugal compressor gas path design often involves curves with $\alpha_C = 90°$, which will result in an infinite slope. To avoid this problem, the simple rotation of coordinates illustrated in Fig. 7-6 should always be included as the first step in generating the curves. Defining

$$
\sin\phi = \frac{\Delta Y}{\sqrt{(\Delta X)^2 + (\Delta Y)^2}} \tag{7-21}
$$

$$
\cos\phi = \frac{\Delta X}{\sqrt{(\Delta X)^2 + (\Delta Y)^2}} \tag{7-22}
$$

a new set of coordinates are defined by

$$
\chi = X \cos\phi + Y \sin\phi
$$
$$
\eta = Y \cos\phi - X \sin\phi \tag{7-23}
$$

Except for 180° bends such as the crossover, this rotation of the axis will remove the problem of infinite slopes. By simply correcting all assigned values of α_C by subtracting ϕ, the curve generation process is carried out in (χ, η) coordinates. Then, the curve is rotated back to (X, Y) coordinates by

$$X = \chi \cos\phi - \eta \sin\phi$$
$$Y = \eta \cos\phi + \chi \sin\phi$$

(7-24)

The next curve form to be discussed is most easily generated in this rotated coordinate system. Figure 7-7 illustrates a curve constructed with a third-degree polynomial curve matched to the coordinates and slopes at the two end points, including optional line segments L_1 and L_2. In terms of (χ, η), the end-point slopes are

$$\eta_1' = \tan(\alpha_{C1} - \phi)$$
$$\eta_2' = \tan(\alpha_{C2} - \phi)$$

(7-25)

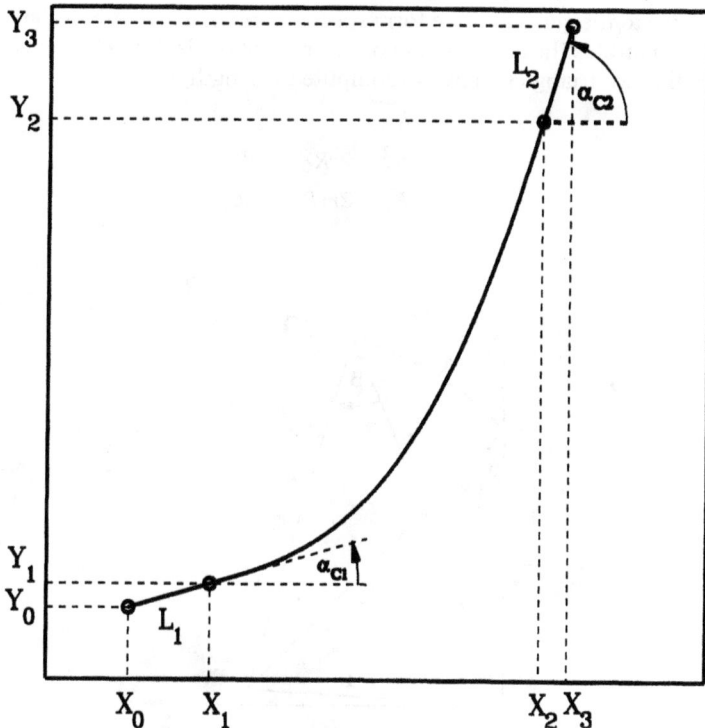

FIGURE 7-7. Third-Order Polynomial Curves

and we generate the curve as a function of the dimensionless parameter, U

$$U = (x - x_1)/(x_2 - x_1)$$
$$\eta = A + BU + CU^2 + DU^3 \tag{7-26}$$

where the coefficients are given by

$$
\begin{aligned}
A &= \eta_1 \\
B &= (x_2 - x_1)\eta_1' \\
C &= -(x_2 - x_1)(2\eta_1' + \eta_2') \\
D &= (x_2 - x_1)(\eta_2' + \eta_1')
\end{aligned}
\tag{7-27}
$$

and the (X, Y) coordinates are given by Eq. (7-24).

One popular two-dimensional axial-element blade style is the circular-arc mean line, primarily because it is easily manufactured. In the context of modern design, this is less significant, but the circular-arc blade style is often a good starting point when axial element blades are used, and adequate from an aerodynamic performance point of view, also. It is convenient to provide for use of this blade style as an alternative to the general curve form specification in a gas path design system. Figure 7-8 illustrates the construction of this type of blade. If the law of cosines is applied to the two triangles shown to compute the length, L

$$
\begin{aligned}
L^2 &= r^2 + R_C^2 - 2rR_C \cos\beta \\
L^2 &= r_1^2 + R_C^2 - 2r_1 R_C \cos\beta_1
\end{aligned}
\tag{7-28}
$$

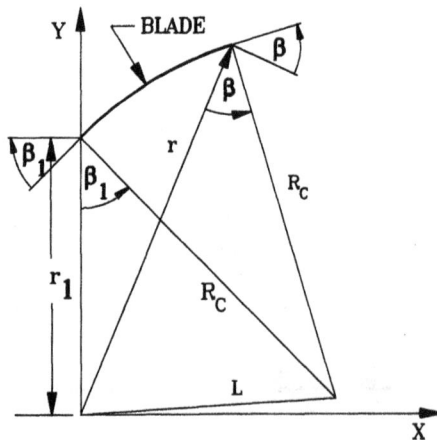

FIGURE 7-8. Circular-Arc Blade Geometry

Eliminating L from these two equations yields

$$\frac{1}{R_C} = \frac{2(r \cos \beta - r_1 \cos \beta_1)}{r^2 - r_1^2} \tag{7-29}$$

By specifying the inlet and discharge radii and blade angles (in a plane of constant z) Eq. (7-29) can be used to calculate the arc radius. Then the equation can be used to predict the mean blade angles as a function of radius. A simple generalization of this concept is to provide one or more intermediate radii and blade angles to construct a multiple circular-arc blade. Note that when $1/R_C = 0$, this process yields a straight-line blade mean line.

The above selection of curve types is sufficient for most gas path design applications, if properly implemented. The most important additional feature that consistently proves useful is to provide for forming a composite curve, joining two or more separate curves. Since all the curve forms reviewed permit setting coordinates and slopes at the curve's end points, two curves can be joined with matched slopes quite easily. For generality, provision to specify a series of points to define a curve should be included, but with the alternatives available from among the forms reviewed, it is almost never necessary to resort to that approach.

7.3 End-Wall and Quasi-Normal Construction

When suitable curves are defined for the end-wall contours, they are used to construct the annulus geometry. In addition to the end-wall contours themselves, it will be necessary to construct a grid system so that the blade can be constructed in the passage, blade passage throat area computed, etc. This permits exporting geometry data from the gas path design system for use in aerodynamic analysis, drafting, numerical controlled machining, stress analysis, etc. The grid structure normally consists of a series of stream surfaces, typically containing equal annulus areas between them, and straight-line quasi-normals extending from the hub to the shroud and approximately normal to them. Figure 7-9 illustrates a grid structure of this type.

To construct this grid, an equal number, N, of discrete points are distributed along the predefined hub-and-shroud contour curves. The quasi-normals are constructed as straight lines connecting corresponding points. The meridional distance at any point along either contour is given by direct integration along the curve, i.e.

$$m_i = \int_{z_1}^{z_i} \sqrt{1 + \left(\frac{\partial r}{\partial z}\right)^2} \, dz \tag{7-30}$$

The integration will be accomplished numerically, raising the possibility of an infinite partial derivative in Eq. (7-30). An axis rotation similar to that shown in Fig. 7-6 can avoid this problem, but one of the more useful numerical concepts

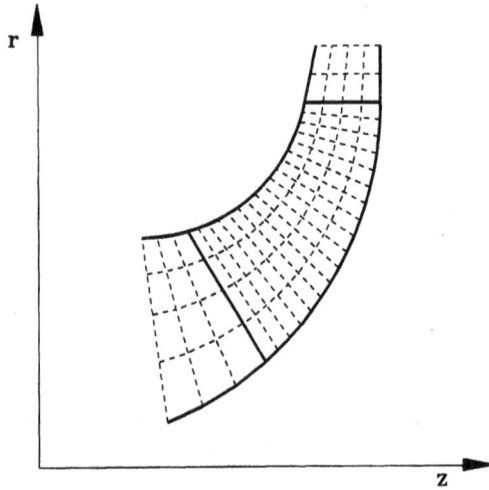

FIGURE 7-9. Meridional Grid Structure

for gas path design is to perform such calculations parametrically. And the most convenient parameter is the point number, i. In this manner, the meridional distance is given by

$$m_i = \int_1^i \sqrt{\left(\frac{\partial r}{\partial i}\right)^2 + \left(\frac{\partial z}{\partial i}\right)^2} \, di \tag{7-31}$$

and very accurate numerical approximations to the partial derivatives can be used with no possibility of singularities. This simple numerical "trick" is useful for many such calculations required for gas path design. Now the contour "cone" angles and the quasi-normal angles illustrated in Fig. 7-10 are calculated

$$\sin \alpha_C = \frac{\partial r}{\partial m} \tag{7-32}$$

$$\tan \alpha_{QN} = \frac{\Delta z}{\Delta r} \tag{7-33}$$

where a consistent convention must be established to handle cases where these angles move through different "quadrants." The angle, ϵ, between a normal to the quasi-normal and a tangent to the contour is a measure of how close the quasi-normal is to a true normal

$$\epsilon = \alpha_{QN} - \alpha_C \tag{7-34}$$

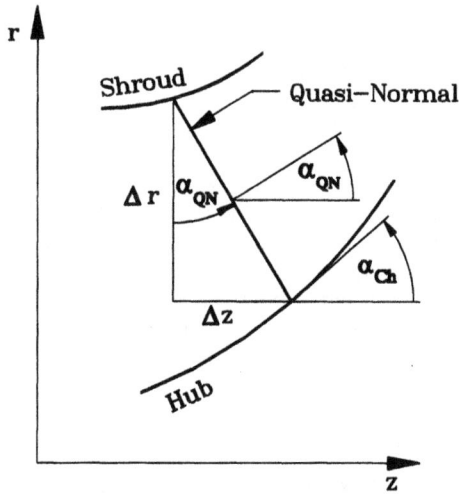

FIGURE 7-10. Quasi-Normal Geometry

All gas path calculations and other analyses using these quasi-normals should recognize and correct for the fact that usually ϵ will not be zero, but it is always beneficial to maintain reasonably small values of ϵ when possible. In positioning points along the hub and shroud to define the quasi-normals, the desired result is that (approximately) $\epsilon_h = -\epsilon_s$, which can be the basis for an automated quasi-normal construction, but usually that is neither necessary nor particularly useful. A simple interactive-graphics-type process for selecting the points along the curve is more effective. By locating a few quasi-normal end points interactively and spacing others linearly between the selected points, a good quasi-normal structure is usually quite easily established. Once this grid is established, all required parameters can be easily computed at all grid points using the logic already described as well as standard interpolation procedures, e.g., m, z, r, α_C, etc.

7.4 Blade Mean Line Construction

The blade mean line consists of the blade angle distribution defined as in Eq. (7-1)—as a function of the relevant coordinate for the particular blade style, using one of the curve forms described. For construction of the blade mean line in the actual gas path, these equations are integrated to provide the polar angle, θ, as a function of the relevant coordinate (r, z, or ξ). For any grid point on the hub or shroud contour, all local values of θ can be recovered from the appropriate, computed θ distribution by interpolation. Then, all other blade mean line data can be computed from

$$x = r \sin \theta$$

$$y = r \cos \theta \qquad (7\text{-}35)$$

$$\cot \bar{\beta} = \frac{r \partial \theta}{\partial m}$$

For two-dimensional blade styles, the correlations, $\theta(r)$ or $\theta(z)$, can be used to compute the mean line coordinates on any of the intermediate stream surfaces in the grid structure. For three-dimensional blades, treating these surfaces is more complicated. The interior blade geometry is constructed by connecting the hub-and-shroud mean line points with a straight line. If ξ is the dimensionless distance along this line element, measured from the hub, the intermediate mean line coordinates are given by

$$x = x_h + (x_s - x_h)\xi$$
$$y = y_h + (y_s - y_h)\xi \qquad (7\text{-}36)$$
$$z = z_h + (z_s - z_h)\xi$$

If the stream surface coordinates are denoted as (z_C, r_C), it is necessary to locate the intercept of this curve with the line element. To avoid problems with infinite slopes, a rotation of coordinates completely analogous to Eqs. (7-21) to (7-23) is defined from the end points of the stream surface. Only the coordinate χ is needed for this calculation. The stream surface is now represented as a function of χ, i.e., $r_C = r_C(\chi)$. The line element coordinates can also be defined as a function of χ. After considerable algebra, it can be shown that

$$r = \sqrt{r_h^2 + A\xi + B\xi^2}$$
$$\chi = E + F\xi + G\xi^2 \qquad (7\text{-}37)$$

where the constants are defined by

$$A = 2[x_h(x_s - x_h) + y_h(y_s - y_h)]$$
$$B = (x_s - x_h)^2 + (y_s - y_h)^2$$
$$C = z_s - z_h$$
$$E = z_h \cos \phi + r_h \sin \phi \qquad (7\text{-}38)$$
$$F = C \cos \phi + A \sin \phi$$
$$G = B \sin \phi$$

The blade mean line coordinates can then be computed numerically by an iterative process seeking to reduce the difference, $r_C(\chi) - r(\chi)$ to zero, using ξ as the variable. If the line element and curve cross, this difference must be monotonic—increasing or decreasing—so the iterative process converges quite rapidly and reliably. Once ξ and r are determined at the intercept, x, y and z are known from Eq. (7-36) and

$$\theta = \tan^{-1}(x/y) \tag{7-39}$$

7.5 Blade Surface Construction

The blade thickness distribution is obtained from the predefined curve form. But for two-dimensional blades, this may be too restrictive in some cases. Normally, the designer wants the blade thickness at the leading edge to be relatively small, yet the leading edge may not lie on a constant value of the independent coordinate for which t_b is assigned. Usually, it is best to provide the capability to impose a leading edge taper on the base t_b distribution supplied by one of the curve forms over a specified distance. The blade thickness must be imposed on the mean line coordinates in a direciton normal to the mean blade surface to obtain the blade surface coordinates. Define unit vectors, $\vec{S}$, tangent to the blade mean line in the stream surface, $\vec{B}$, along the defining line element of the blade and $\vec{T}$, normal to the blade mean surface. Clearly, the third vector can be computed by taking the cross product of the first two vectors. Hence

$$\vec{S} = S_x\vec{i} + S_y\vec{j} + S_z\vec{k}$$
$$\vec{B} = B_x\vec{i} + B_y\vec{j} + B_z\vec{k} \tag{7-40}$$
$$\vec{T} = \vec{S} \times \vec{B} = T_x\vec{i} + T_y\vec{j} + T_z\vec{k}$$

where $\vec{i}, \vec{j}$ and $\vec{k}$ = usual Cartesian unit vectors. After considerable algebra, it can be shown that

$$S_x = \sin\theta \sin\alpha_C \sin\bar{\beta} + \cos\theta \cos\bar{\beta}$$
$$S_y = \cos\theta \sin\alpha_C \sin\bar{\beta} - \sin\theta \cos\bar{\beta} \tag{7-41}$$
$$S_z = \sin\alpha_C \sin\bar{\beta}$$

The vector along the defining line element depends upon the blade style. For the three-dimensional straight-line-element blade

$$B_x = (x_s - x_h)/L$$
$$B_y = (y_s - y_h)/L \tag{7-42}$$
$$B_z = (z_s - z_h)/L$$

where L = length of the line element. For two-dimensional axial-element blades

$$B_x = 0$$
$$B_y = 0 \tag{7-43}$$
$$B_z = -1$$

and for two-dimensional radial-element blades

$$B_x = \sin \theta$$
$$B_y = \cos \theta \qquad \qquad (7\text{-}44)$$
$$B_z = 0$$

As noted in Eq. (7-40), the vector normal to the mean surface follows directly from the definition of the cross product. This vector normal, directed toward positive θ, is

$$T_x = S_z B_y - S_y B_z$$
$$T_y = S_x B_z - S_z B_x \qquad \qquad (7\text{-}45)$$
$$T_z = S_y B_x - S_x B_y$$

Then, the blade surface coordinates are given by

$$x = \bar{x} \pm \tfrac{1}{2} t_b T_x$$
$$y = \bar{y} \pm \tfrac{1}{2} t_b T_y \qquad \qquad (7\text{-}46)$$
$$z = \bar{z} \pm \tfrac{1}{2} t_b T_z$$

For the two-dimensional blade styles, this completely defines the blade surfaces over the entire grid structure. Surfaces of the three-dimensional blades are defined only on the hub-and-shroud contours. To define the blade surfaces on intermediate stream surfaces, corresponding points on the hub-and-shroud blade surfaces are connected by straight lines using the same procedure as that for the blade mean surface in Section 7.4.

At the discharge of an impeller, usual practice is to maintain a constant radius blade trailing edge on any meridional surface. This can be accomplished by noting that a unit vector in the θ direction is given by

$$\vec{e}_\theta = \cos \theta \vec{i} - \sin \theta \vec{j} \qquad \qquad (7\text{-}47)$$

and the modified blade thickness in the θ direction is given by

$$t_\theta = t_b / |T_x \cos \theta - T_y \sin \theta| \qquad \qquad (7\text{-}48)$$

and θ on the two blade surfaces at radius r is given by

$$\theta = \bar{\theta} \pm \tan^{-1}[t_\theta/(2r)] \qquad \qquad (7\text{-}49)$$

from which all other coordinates can be computed.

7.6 Blade Passage Throat Area

An accurate throat area calculation is critical to the aerodynamic design process. In addition to its obvious role in establishing the choke flow limit, Chapter 5 has noted the important role of the throat area in several of the loss models used and for the impeller stall limit. The most complicated step in this process is computing the blade-to-blade throat width, h_t, in a stream surface. This width lies on a three-dimensional surface with both r and θ varying across the passage. This is best handled by using a conformal transformation to map the (m, θ) plane into a Cartesian plane (X, Y). Such a path-independent conformal transformation is defined by

$$X = \int_{m_1}^{m} \frac{dm}{r}$$

(7-50)

$$Y = \theta$$

Figure 7-11 illustrates the throat width as viewed in the transformed coordinate plane. Since angles are preserved in a conformal transformation, Fig. 7-11 shows that the throat width has a constant angle over its length, i.e.

$$\frac{\partial Y}{\partial X} = \frac{r\partial\theta}{\partial m} = constant$$

(7-51)

Since this angle is constant along h_t, the throat width can be calculated by

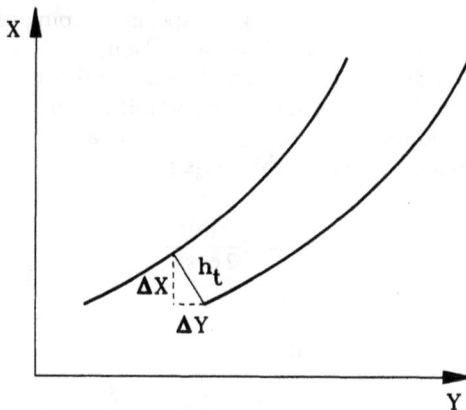

FIGURE 7-11. Throat in Transformed Plane

$$h_t = \int_{m_1}^{m} \sqrt{1 + \left(\frac{r\partial\theta}{\partial m}\right)^2}\, dm = \sqrt{1 + \left(\frac{\Delta\theta}{X}\right)^2}\, \Delta m \qquad (7\text{-}52)$$

where $\Delta\theta$ and Δm are differences in values between points on the opposing blade surfaces and X is evaluated on the left blade surface in Fig. 7-11. Indeed, Eq. (7-52) can be applied to compute the distance between any two points on the opposing blade surfaces. Hence, it is used to compute distances for all grid points on the left surface in Fig. 7-11 from the fixed point on the right-hand surface and the minimum represents the throat width on the particular stream surface. The distribution of computed values can be differentiated numerically (say, as a function of point number or of X) and the result interpolated for a derivative of zero, such that the throat can be correctly located when it lies between points on the chosen grid structure. Now, if n is the distance along quasi-normals, the blade passage throat area between two stream surfaces can be estimated by

$$\Delta A_t = \overline{h_t}\, \overline{\cos \epsilon \Delta n} \qquad (7\text{-}53)$$

where an average of h_t on the two stream surfaces and an average of $\Delta n \cos \epsilon$ on the opposing blade surfaces are used. Note that $\cos \epsilon$ accounts for the fact that the quasi-normal may not be normal to the stream surfaces. By summing for all passages between all stream surfaces, the total blade passage throat area is obtained. The accuracy of the calculation improves as more stream surfaces are employed.

7.7   The Blade Leading Edge

In many cases it is necessary to refine the blade leading edge region relative to the base thickness distribution developed by a curve form. More precise definition may be needed for five-axis milling, or special shaping of the leading edge region may be desired for aerodynamic reasons. Hence, it is useful to provide for imposing a specific leading edge radius and blending this to the base blade profile supplied by the thickness distribution curves. If the distance along the blade mean line is designated by s, then $t_b = t_b(s)$, and the largest leading edge radius, R, that can be imposed on the base thickness is given by

$$R_{\max} = \frac{t_b(0)}{2 \cos \delta}$$

$$(7\text{-}54)$$

$$\tan \delta = \tfrac{1}{2}\, t_b'(0)$$

i.e., the base thickness and its gradient at the start of the base profile limits this parameter. To impose this or any smaller value of R and blend it to the base

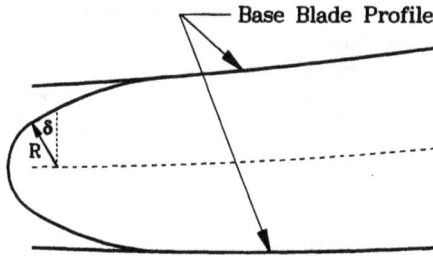

FIGURE 7-12. Blended Leading Edge Radius

thickness distribution, the following procedure can be used; Define the distance over which the blending process will occur by

$$s_* = t_b^2(0)/(4R) \tag{7-55}$$

and modify the base thickness in this region by

$$t_b(s) \rightarrow f(s)t_b(s)$$
$$f(s) = k + (1 - k)(2 - s/s_*)s/s_* \tag{7-56}$$

This function will smoothly blend the new thickness with the base thickness at $s = s_*$ as can be seen from the fact that $f(s_*) = 1, f'(s_*) = 0$. With some simple algebra, it can be shown that the new thickness distribution will be matched to the leading edge radius if the following relations are satisfied:

$$k = 2R \cos \delta / t_b(0)$$
$$\tan\delta = t_b'(0)k/2 + t_b(0)(1 - k)/s_* \tag{7-57}$$

These two equations can be solved for k by a predictor-corrector numerical scheme. Then, the actual leading edge radius is added with its center located at $s = R \sin \delta$. Figure 7-12 shows a typical result of this blending process.

7.8   A Computerized Gas Path Design System

The procedures described in this chapter form the basis for a very efficient computerized gas path design system. In concept, the process is quite simple. The design system must permit the user to generate the basic curves defining the hub-and-shroud contours and the blade angle and thickness distributions. It must then perform the tedious calculations required to construct the gas path, create the grid structure, compute the throat area, etc. The design system should include extensive monitor-screen graphics for the designer to view plots of the gas path and key geometrical distributions, e.g., blade angle, cone angle, curvature, polar angle, etc. To fully benefit from such a design system, it should have the provision to export geometry data to the various aerodynamic analyses for immediate feedback on the quality of the design. Geometry data exporting for stress analysis, computer-aided drafting and numerical-controlled machining can also yield substantial productivity benefits. The following subtle features, learned from experience, can be valuable:

- The Bezier polynomial curve generation should include an interactive graphics option, which lets the designer move the reference points and view the results on the monitor screen. All curve forms discussed in this chapter benefit from screen graphics support, particularly in the edit mode.
- For all curve forms, both a generation mode and an edit mode are needed. The two primary functions of this process are defining the curve shape, and establishing the distribution of data points along the curve to be used as quasi-normal end points on contours or for interpolation for distribution curves.
- Multiple options for positioning the quasi-normal end points along the curve are needed. Direct editing of values, equal spacing of points between specified points, positioning specific points at specific coordinates (e.g., blade leading and trailing edge points), etc. can greatly simplify the design process.
- The capability to edit one curve while displaying another in the background is quite useful, e.g., for hub-and-shroud contour generation. The second curve can also be used to supply data useful for the process, e.g., passage area distributions.
- The curve generation process needs support for filing and labeling curves so subsequent trials do not erase the previous attempts unless requested. The design system needs the capability to select curves from this file as the designer chooses. It should permit storing several alternate curves for the same purpose, to give the designer freedom to experiment with alternatives, yet return to an earlier curve if appropriate.
- Provision to change the relative difference in the leading or trailing edge hub-and-shroud polar angles for three-dimensional blades should be included. This lets the designer change the blade "rake" angles between the blade line elements and the meridional plane.
- If properly formulated, the design system will permit changing the number of stream surfaces used in the grid structure with no compromise to the basic design.
- The design system must recognize the impact of changes made and maintain only valid geometrical data. Hence, if a new curve is selected for a contour or distribution, all geometrical constructions based on the old one must be eliminated automatically. The design system should not be capable of supplying geometrical output that is not valid for the set of curves selected. It should advise the designer of the status of the design's construction process on request.
- Simple features such as automatic scaling of the geometry or reversing

the direction of rotation can be quite useful on occasion and are easily incorporated.

- When exporting data to other aerodynamic design and analysis programs, use a geometry update approach, i.e., change only geometry data available from this gas path design. Usually, other input data for those programs remain valid, which allows immediate execution of those other analyses after the update.
- It is fairly simple to make the provision to generate a Bezier polynomial curve as a curve fit to another existing curve. By distributing a number of points along the curve equal to $(n + 1)$, Eq. (7-3) can be used to generate $(n + 1)$ equations with $(n + 1)$ unknowns for both X_k and Y_k, which can be solved by a simple matrix inversion. Matching a Bezier polynomial curve to an existing curve is useful when greater generality is needed. A simple variant is to also match the end-point slopes using the properties of this curve discussed earlier.

When creating a general gas path design system of this type, it may be advisable to create a parallel geometry analysis system for use on existing gas path designs. Many of the gas path construction and data exporting procedures can be common to both systems if planned in advance. There are many functions that a geometry analysis system can accomplish that are not part of the gas path design process. Imposing new shroud contours to change the flow capacity, trimming or extending impeller tips to change the head capability, scaling geometry, etc., are examples. A geometry analysis system also makes available the many capabilities discussed in this chapter for existing designs, including designs not accomplished with the gas path design system. The principle difference is that the geometry analysis starts with contours and distributions specifically defined by a series of points rather than by general curve forms. The gas path design system should supply its first input data file automatically.

7.9 Impeller Detailed Design

Since the most common use of a general gas path design system is for impeller detailed design, that application will be described in this chapter. A basic flow chart of this process is provided in Fig. 7-13. Note that the preliminary design procedure presented in Chapter 6, with its interaction with the performance analysis of Chapter 5, starts the impeller design process. It should create the initial input data files (curve and basic design data) for the gas path design system. The impeller detailed design process will refine the relatively crude impeller geometry supplied by the preliminary design system to achieve desirable internal flow profile characteristics as well as the desired performance. The designer can evaluate some aspects of the design from the gas path design system itself, e.g., end-wall curvatures, passage area distribution, throat area, etc. But most of the designer's guidance for the impeller detailed design process will come from various internal flow analyses, with periodic checks with the performance analysis given in Chapter 5. Methods for conducting internal flow analysis are covered in Chapters 12 to

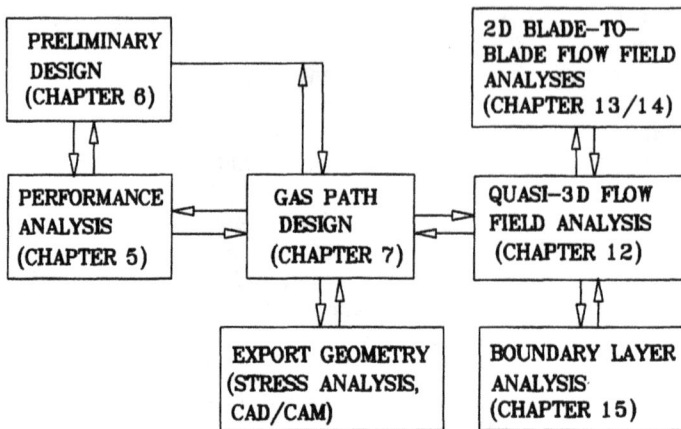

FIGURE 7-13. Impeller Detailed Design Flow Chart

15; the details of how these analyses function are not required for the present discussion.

A basic evaluation of the passage curvature can be accomplished before any aerodynamic analysis is accomplished. Consider the momentum equation in the normal direction, Eq. (3-22). If the impeller inlet flow is isentropic and has no swirl component, the right-hand side of that equation will vanish. For simplicity, neglect the effects of swirl and assume W_m varies linearly from hub to shroud. Then, Eq. (3-22) can be integrated to yield

$$\frac{2(W_{ms} - W_{mh})}{W_{ms} + W_{mh}} = -\kappa_m b \qquad (7\text{-}58)$$

which shows that if $-\kappa_m b = 2$, the hub velocity would be zero, i.e., the inviscid flow would separate. Viscous effects will result in flow separation under much less severe curvature-induced pressure gradients. As a general rule, designers should limit the passage mean curvature by

$$|\bar{\kappa}_m b| \leq 1 \qquad (7\text{-}59)$$

and the gas path design system should allow evaluation of this parameter during the design process. On completion of the preliminary design process, the designer should be able to establish the minimum throat area required. As the design is refined, this minimum may change somewhat, but the designer can monitor this parameter in the gas path design process to avoid submitting a clearly unacceptable design for aerodynamic analysis.

The basic performance analysis provides a number of key design parameters that can be monitored. Arguments very similar to those for the passage curvature limit lead to a need for controlling the blade-to-blade velocity difference at the design flow conditions. A preferred design limit is

$$\frac{\Delta W}{\overline{W}} \le 0.8$$

(7-60)

$$\overline{W} = (W_1 + W_2)/2$$

where ΔW is given by Eq. (4-42). It is not always practical to achieve the limit of 0.8, but values exceeding 1.0 should definitely be avoided. Similarly, the overall impeller diffusion needs to be maintained within reasonable limits. The ratio of the relative velocity across the impeller, W_2/W_1, is a convenient parameter to monitor. At design flow, values around 0.75 are desirable; values less then about 0.65 should never be used. Since this velocity ratio decreases as inlet mass flow decreases, a reasonable stable operating range with acceptable efficiency levels requires limiting its value at the design flow. The equivalent diffusion factor of Eq. (4-41) is an indicator of the blade surface diffusion level. Performance can be expected to deteriorate significantly when this parameter exceeds values of about 2, so significantly lower values should be used at design flow conditions, say around 1.6 to 1.7 to provide for a reasonable efficiency at lower mass flows. The inducer stall limit of Eq. (5-29) also must be monitored. It is unwise to violate this criterion for any anticipated operating condition since an inducer stall will usually lead to unstable operation, if not surge. These design parameters can be evaluated from the basic mean streamline performance analysis as the design progresses. To a very large degree, they will have to be established in the preliminary design process. Once the inlet and tip geometry is established, they will not change too much, but the detailed design process always requires adjustments that introduce some differences. Thus, periodic checks are necessary as the design progresses. If the gas path design system has the capability to update the geometry for the performance analysis, these checks are easily accomplished.

The primary guidance for the impeller detailed design process is obtained from an internal flow field analysis, such as that described in Chapter 12, preferably augmented by boundary layer analyses, such as that described in Chapter 15. A key benefit is an accurate assessment of the hub-to-shroud blade incidence angle distribution, including an accurate prediction of the passage curvature effects. The inviscid flow analysis also supplies velocity distribution data along the physical boundaries. The primary goal of the detailed design process is to control the quality of these profiles to avoid excessive diffusion, large velocity gradients, separation, etc. The process is necessarily qualitative in nature, depending very much on the designer's judgment as to the acceptability of flow profiles. Use of two-dimensional and three-dimensional, axisymmetric boundary layer analyses as discussed in Section 3.4, does not significantly complicate the internal flow analysis and does add some interpretation of viscous effects. But judgment continues to be required, since the important secondary flow influence is not included. One of the most critical considerations is the velocity profiles on blade surfaces. This is commonly viewed as a blade loading diagram, consisting of the velocity or pressure distributions on the suction (low pressure) and pressure (high pressure) surfaces in the various meridional surfaces through the blade row. Figure 7-14 illustrates a typical blade loading diagram from the quasi-three-dimensional flow analysis presented in Chapter 12. Here, the ratio of blade

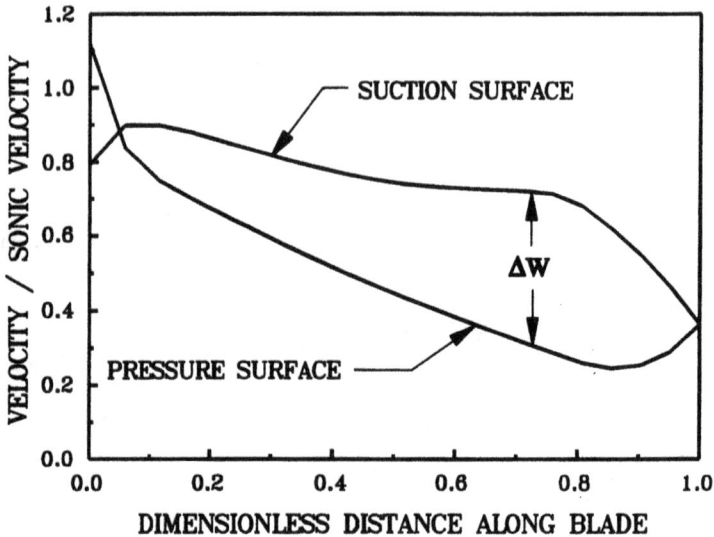

FIGURE 7-14. Shroud Blade Loading Diagram

surface relative velocity to local sonic velocity is used to evaluate blade load-ing. The "crossing" of the loading diagram at the leading edge shows that this impeller was designed with a fairly significant negative incidence angle, proba-bly to achieve the desired stable operating range objective. The designer seeks to maintain smooth velocity distributions, while avoiding large velocity gradients or excessive diffusion. Again, the ratio of the velocity difference to the local mean velocity should be maintained at reasonable levels, similar to the limit used for the average loading in Eq. (7-60), although locally higher values may be toler-ated. This velocity difference (or pressure difference) is a driving force behind the secondary flows and substantially impacts the velocity gradients imposed. It can be controlled by employing a sufficient number of blades or an adequate blade length.

Figure 7-15 shows a similar velocity diagram—a dimensionless meridional velocity loading diagram—used to evaluate the hub-to-shroud loading. Again, the designer seeks smooth profiles, with reasonable velocity gradients. Overall dif-fusion levels are also a key concern here. Ratios of the minimum-to-maximum meridional velocities should always exceed 0.5, with 0.6 to 0.7 being preferred lower limits. The difference between the hub-and-shroud velocities on this plot is a direct consequence of the passage curvature, as discussed earlier in this sec-tion. This loading diagram evaluates the passage curvature effects in terms of fluid dynamics parameters.

Figure 7-16 shows results from boundary layer analyses for the same impeller near the shroud. The boundary layer shape factor, H, defined as the ratio of dis-placement thickness to momentum thickness (δ^*/θ in the nomenclature of Chap-ter 3) is used as a key indicator of the proximity to flow separation. For two-

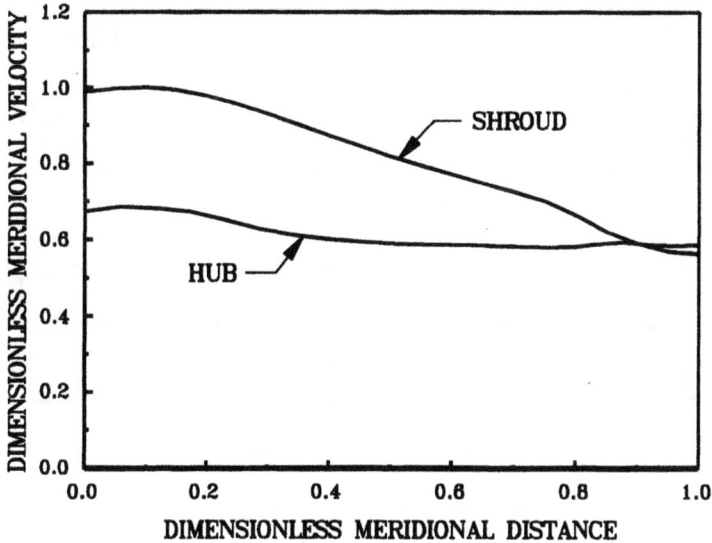

FIGURE 7-15. Hub-Shroud Loading Diagram

dimensional boundary layers, separation can be expected when H is in the range of about 2.2 to 2.4. Figure 7-16 shows that the shroud end-wall boundary layer is of good quality. The blade suction surface boundary layer is also well behaved except close to the trailing edge, where the fairly large velocity gradient seen in Fig. 7-14 can be expected to induce separation. The pressure surface boundary layer appears to be at risk of separation over most of the blade length. But the experienced designer will recognize that secondary flows improve the quality of the pressure surface boundary layer while causing a corresponding deterioration in the suction surface boundary layer. This is due to the migration of low momentum fluid from the high pressure zones to the low pressure zones. Thus, the designer will be far more concerned with the suction surface and shroud end-wall boundary layer quality than with the pressure surface. This type of boundary layer analysis cannot be interpreted too literally, but it does add a recognition of viscous effects in a gross fashion. When a designer has experience gained from calibrating this type of analysis by applying it to good and poor performing impellers, it is an effective guide to optimizing the gas path boundaries.

The internal flow analysis is employed similar to the performance analysis. The gas path design system should be capable of updating the geometry for the flow analysis such that it can be easily accomplished for a candidate design. Through an iterative design process, the gas path contours can be refined until acceptable internal flow characteristics are achieved. Returning to our flow chart of the design process, one additional evaluation tool is noted—two-dimensional blade-to-blade flow analysis. For reasons of computational speed, the quasi-three-dimensional flow analysis presented in Chapter 12 employs a linearized blade-to-blade flow model. While almost always adequate for impeller analysis,

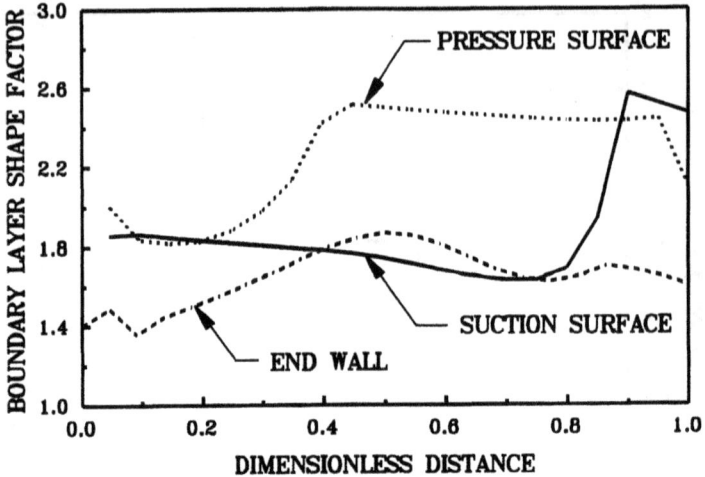

FIGURE 7-16. Boundary Layer Analysis

it is wise to conduct a final check on the blade loading with a more exact method, such as those described in Chapters 13 and 14. If the designer is fortunate enough to have access to a suitable computation fluid dynamics (CFD) viscous solver, a final evaluation will be made with that analysis to obtain a more accurate evaluation of the internal flow quality. This may indicate areas where improvement is desirable, leading to further iterations with the design system. Use of CFD as a final evaluation tool for impeller design is relatively common today, and is certainly a step to be highly recommended.

EXERCISES

7.1 As the least general of the curves presented in this chapter, the third-order polynomial is often overlooked by designers, who often fail to exploit it to simplify their design problems. This curve is defined by coordinates, slope and line segment lengths at the two end points, making it easy to use and very convenient for certain functions. As an example, develop the curve specifications required to model a blade thickness distribution as a function of dimensionless meridional distance ($0 \leq x \leq 1$). The thickness is to be constant for $x \geq 0.5$, equal to 0.2. The leading edge thickness ($x = 0$) is 0.1. The distribution should vary smoothly over the range $x \leq 0.5$ and not exceed an upper limit of 0.2. What curve specification is available for modification to change the shape of the curve?

7.2 For a valid internal flow analysis of an impeller, it is necessary to include vaneless extensions upstream and downstream of the blade passage. These could be included as part of the gas path design, but it is often more convenient not to, so that the blade passage inlet and discharge contour coordinates can be precisely controlled. Instead, special vaneless extension curves can be developed for combining with the blade passage contours (i.e., the "composite curve" discussed in this chapter) to define the complete end-wall contours. To do this, the extension curves *must* match the blade passage end-point coordinates as well as the slope. Using the third-order polynomial curve type, develop specifications for two vaneless extension curves to match impeller blade passage contours at the tip, where

Hub: $z = 0$, $r = 15$, cone angle = $90°$.
Shroud: $z = -1$, $r = 15$, cone angle = $80°$.

The curves should smoothly blend to a radial parallel-wall vaneless diffuser with the hub wall defined by $z = 0$ and shroud wall defined by $z = -0.5$. Which parameter is undefined by these boundary conditions? Is this an advantage or a disadvantage when constructing these extensions in an interactive gas path design system?

7.3 The preliminary design system of Chapter 6 exports its end-wall contours, blade geometry distributions, etc., for an impeller to a gas path design system. Normally, it is found that the three-point spline curve used to model the shroud contour serves to define the basic contour needed, but it is not sufficiently smooth or produces unacceptable passage area distributions for a final, detailed impeller design. Similarly, the blade angle distributions supplied from the preliminary design system are usually a very good starting point, but not sufficient for detailed design. Assuming the gas path design systems has all of the features recommended in Section 7.8, what is the simplest way to obtain a smooth curve that closely approximates the basic preliminary design system's curves, but which can be readily modified to improve the design?

7.4 The geometry of a radial in-flow turbine impeller is similar to that of a centrifugal compressor. Similarly a quasi-three-dimensional internal flow analysis capable of treating one type of impeller should be capable of treating the other. What special features, could you add to an existing centrifugal compressor gas path design system to permit it to also treat radial in-flow turbine impellers? including supplying the geometry to the internal flow analysis?

Chapter 8

VANELESS DIFFUSER DESIGN

In this chapter, a detailed design procedure for vaneless diffusers based on the design and analysis procedures of Aungier (1993a) is presented. The major component of this design system has already been reviewed in Chapter 5, i.e., the one-dimensional performance analysis for vaneless components. The present design procedure differs from the original reference in form, but not in substance. Aungier (1993a) uses a centrifugal compressor stage performance analysis and a manual procedure to develop the vaneless diffuser design. Subsequently, a more formal design procedure was formulated to apply this concept in a more systematic fashion.

The early attempts to apply fluid dynamics principles to vaneless diffusers employed one-dimensional methods with wall friction effects included (e.g., Stanitz, 1952; Johnston and Dean, 1966). These methods worked reasonably well if the wall skin friction coefficient was a specified parameter, and if the designer had the skill or experience to specify a suitable value. It should be no surprise that the designer rarely had a good basis for making this specification, resulting in rather unsatisfactory performance prediction accuracy. As fluid dynamics analysis procedures matured, many investigators sought to improve this situation by using three-dimensional boundary analysis techniques (e.g., Aungier, 1988b; Davis, 1976; Jansen, 1964; Schumann, 1985; Senoo et al., 1977). Overall, this was quite successful, particularly with regard to the more recent methods. For example, Aungier's (1988b) method provided a dramatic improvement in performance prediction accuracy over the one-dimensional method in use by this author at that time. As mentioned in Chapter 5, this three-dimensional boundary layer analysis technique clarified the fundamental fluid dynamics governing this problem to a point where the one-dimensional performance analysis models of Chapter 5 could be formulated and refined through comparison with experimental data. The result was that the one-dimensional performance analysis became superior to the more fundamental three-dimensional boundary layer technique in terms of prediction accuracy. Consequently, the one-dimensional method is, again, the basis for detailed design and performance analysis in this author's design and analysis system.

NOMENCLATURE

b = hub-to-shroud passage width
C = absolute velocity

C_m = absolute meridional velocity
C_U = absolute tangential velocity
c_p = static pressure recovery coefficient
p = pressure
r = radius
z = axial coordinate
α = flow angle with respect to tangent
α_C = streamline slope angle with axis
$\bar{\omega}$ = total pressure loss coefficient

Subscripts

h = parameter on the hub contour
s = parameter on the shroud contour
t = total thermodynamic condition
2 = impeller tip condition
3 = point where diffuser walls become linear
4 = diffuser exit condition

Superscripts

$'$ = partial derivative with respect to r

8.1 Geometric Construction

Following Aungier (1993a), the geometrical construction used for the vaneless diffuser is illustrated in Fig. 8-1. The hub contour matches the impeller tip location and slope (z_{2h}, r_{2h}, α_{C2h}) and adjusts to a radial contour extending from r_3 to r_4, which are both specified by the designer. The passage widths at these two locations, b_3 and b_4, are also specified. The hub contour from the impeller tip radius to the blend radius, r_3, is defined by a second-order polynomial, which matches the boundary conditions (two slopes and one point location).

$$z_h = A + Br + Cr^2 \tag{8-1}$$

$$C = \frac{\cot\alpha_{C2h}}{2(r_{2h} - r_3)} \tag{8-2}$$

$$B = -2r_3C \tag{8-3}$$

$$A = z_{2h} - Br_{2h} - Cr_{2h}^2 \tag{8-4}$$

For most cases, $\alpha_{C2h} = 90°$ and the entire hub contour will be a radial surface. For mixed flow impellers, Eq. (8-1) will predict z_{3h} to match the boundary conditions.

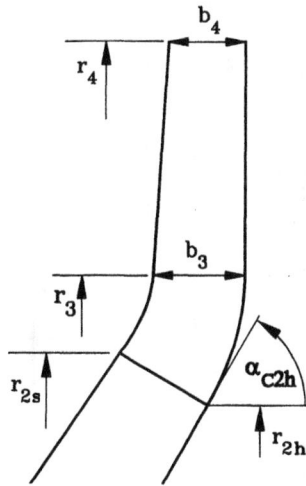

FIGURE 8-1. Vaneless Diffuser Construction

For the shroud contour, z_{3s} and z_{4s} are known from the hub coordinates and the specified passage widths. A linear shroud contour is used between r_3 and r_4, with a slope given by $z'_{3s} = (b_4 - b_3)/(r_4 - r_3)$. Again, a second-order polynomial construction is used to blend with the linear section. But here, the coordinates of the two end points and the slope at the blend radius determine the constants (i.e., the diffuser shroud contour is not required to match the impeller tip shroud contour slope).

$$z_s = D + Er + Fr^2 \tag{8-5}$$

$$F = \frac{z_{2s} - z_{3s} + z'_{3s}(r_3 - r_{2s})}{(r_3 - r_{2s})^2} \tag{8-6}$$

$$E = z'_{3s} - 2r_3F \tag{8-7}$$

$$D = z_{2s} - r_{2s}E - r_{2s}^2F \tag{8-8}$$

This geometrical construction is adequate for almost all vaneless diffuser designs. In the event a more general construction is needed, the general gas path design procedure of Chapter 7 can be used.

8.2 The Design Procedure

An effective detailed design procedure is easily implemented in a simple computerized design system that incorporates the aerodynamic performance analysis procedure presented in Section 5.4. The basic design conditions to be specified are

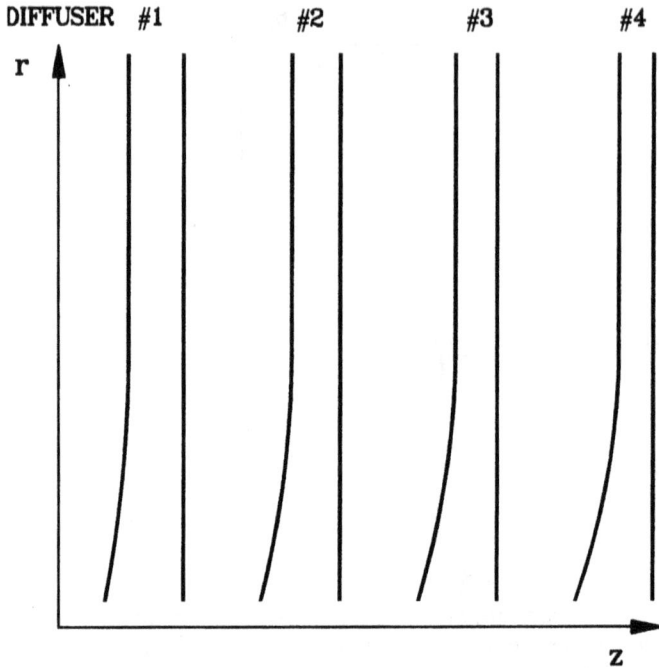

FIGURE 8-2. Candidate Vaneless Diffuser Designs

- Impeller tip geometry, z_{2h}, r_{2h}, α_{C2h}, z_{2s} and r_{2s}.
- Impeller tip flow data p_{t2}, T_{t2}, C_2, and α_2.
- Upper and lower limits of desired impeller tip flow angle operating range.

Design data to be optimized are

- Diffuser exit geometry, r_4 and b_4.
- Diffuser geometry at the blend point, r_3 and b_3.

The process is most easily carried out by comparing the candidate designs using the performance analysis. Figures 8-2 through 8-5 illustrate how this process is accomplished. Figure 8-2 shows four different candidate designs, where b_4 is varied, with $b_3 = b_4$. Figure 8-3 shows the predicted total pressure loss coefficient, $\overline{\omega}$, as a function of inlet flow angle for these four diffusers.

$$\overline{\omega} = \frac{p_{t2} - p_{t4}}{p_{t2} - p_2} \tag{8-9}$$

It is seen that diffusers #2 and #3 achieve the lowest loss at the design flow angle. Design #1 does not appear to be a good choice since the higher losses at lower flow angles may limit the stage's stable operating range, yet not benefit

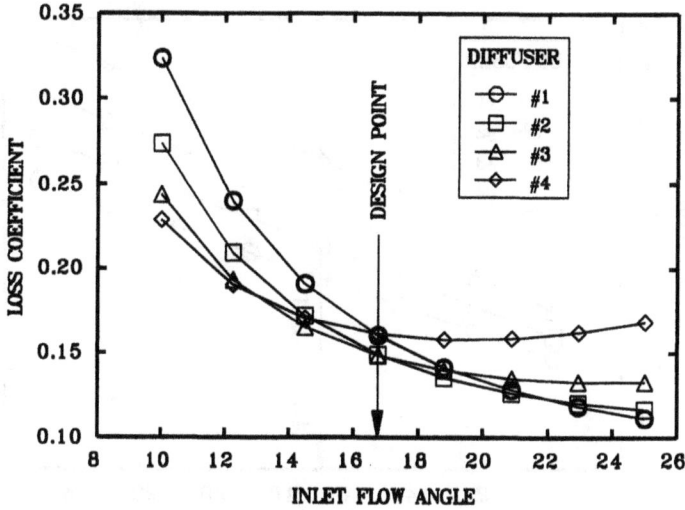

FIGURE 8-3. Loss Coefficient Comparison

performance at the design condition. Diffuser #4 achieves its minimum loss at about the design point, which is attractive from a stable operating range point of view, but this narrower diffuser shows a higher loss than designs #2 and #3 at the design flow angle. This example illustrates an important feature of vaneless diffuser design. Designing a vaneless diffuser to operate at a minimum on its loss curve at the design flow does not necessarily yield the lowest loss.

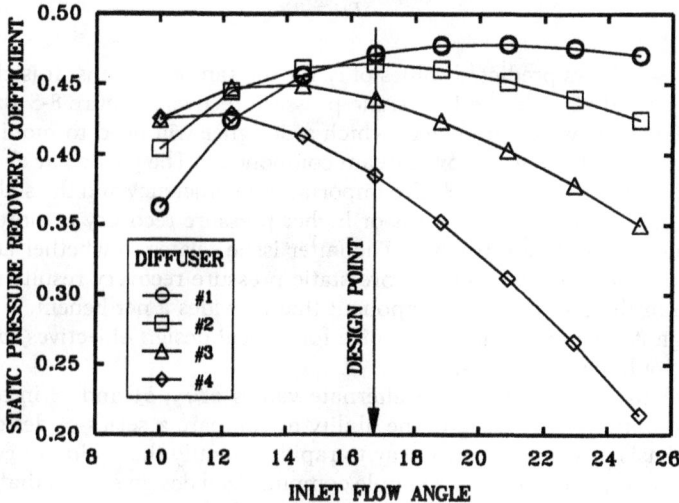

FIGURE 8-4. Pressure Recovery Comparison

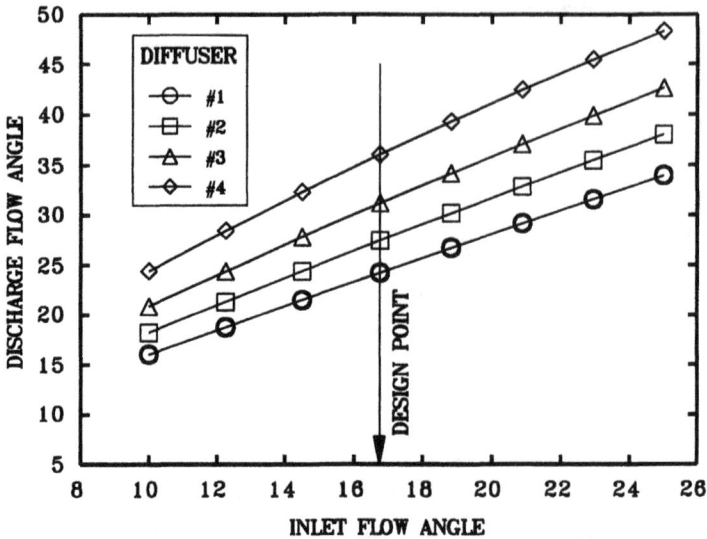

FIGURE 8-5. Discharge Angle Comparison

While diffuser loss is an important consideration, the primary purpose of a diffuser is to convert kinetic energy to static pressure to reduce losses in the downstream stage component. This is conveniently evaluated by the diffuser's static pressure recovery coefficient, c_p.

$$c_p = \frac{p_4 - p_2}{p_{t2} - p_2} \tag{8-10}$$

Figure 8-4 shows predicted values of c_p for the sample designs. It is clear that diffusers #1 and #2 yield the best static pressure recovery. Figure 8-5 shows the diffuser exit flow angles predicted, which a designer will need to monitor relative to matching the flow to downstream components. The particular choice of a design will depend upon the relative importance of efficiency and the stable operating range and whether lower loss or higher pressure recovery is more critical to the stage's overall performance. The latter issue relates to whether accepting a higher diffuser loss to achieve more static pressure recovery results in a loss reduction in the downstream component that provides a net benefit. It is likely that design #2 would be the best choice for typical design objectives, but there certainly could be exceptions.

It is relatively easy to explore alternate values of r_3, b_3 and r_4 in a similar fashion to optimize the design. The ability to compare a series of designs on a common basis is a very effective way to rapidly identify the optimum geometry. Implementation in the form of a simple computerized design system that permits the user to modify the design parameters interactively, while viewing the results illustrated in Figs. 8-2 through 8-5 on the monitor screen, provides a very effective design method. This vaneless diffuser design system should also be capable of updating the geometry for the overall stage performance analysis to allow a rapid evaluation of the diffuser design with respect to the overall stage performance.

8.3 Rotating Stall Considerations

The designer must be aware of the potential for rotating stall in vaneless diffusers. This is a periodic unsteady flow pattern consisting of high and low pressure zones rotating in the direction of the impeller rotation at subsynchronous speeds, typically, 5%–20% of the impeller rotation speed. Van den Braembussche (1987) provides an excellent overview of this phenomenon and the technology available to recognize and avoid it. In addition to its adverse effect on performance, rotating stall can be a source of mechanical excitation for the blades and shaft. At sufficiently high inlet pressures, severe vibration problems may limit the range of operation. As a general rule, a properly designed vaneless diffuser should encounter this problem only at very low flows, close to the compressor surge line. However, a designer who is not conscious of this potential can easily design a diffuser with a rotating stall well within the expected compressor operating range. Figure 8-6 shows the critical diffuser flow angle at which the onset of rotating stall can be expected, as predicted in Senoo and Kinoshita (1978). This curve applies for relatively long diffusers (making it conservative) and for moderate Mach number levels. The key factor here is that rotating stall is to be expected at rather low flow angles, well below any reasonable design flow angles.

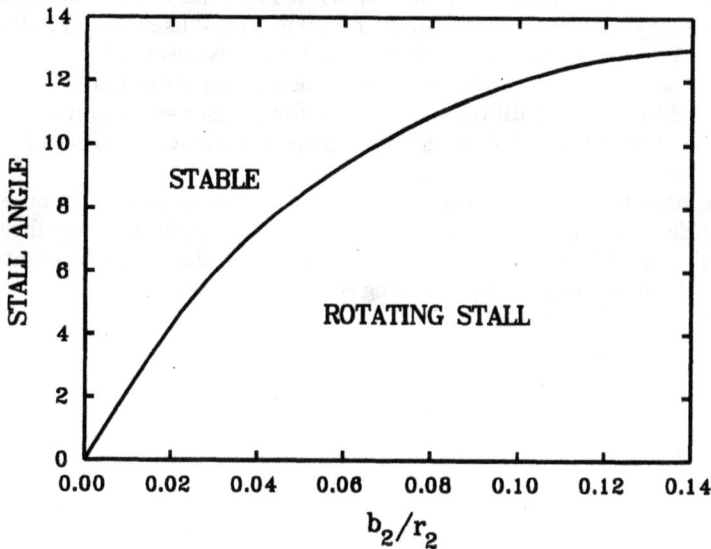

FIGURE 8-6. Vaneless Diffuser Stability

Thus, vaneless diffuser rotating stall should not be encountered in the expected operating range if the design procedures presented in this chapter are followed. For flow angles at the levels shown in Fig. 8-6, rather high diffuser losses will be predicted as can be seen from Figs. 8-3 and 8-5. Hence, if the diffuser is properly sized for the intended application, a good basic performance analysis will indicate that the stage is near its stable operating limit as its head characteristic approaches a maximum. This problem is of greater concern when the diffuser is not properly sized, such that it approaches stall while other stage components are sufficiently stable to maintain a rising head characteristic as flow is reduced. In those cases, vaneless diffuser rotating stall may occur well before surge, causing a region of rough, noisy operation and possible vibration problems. A common example is when the stage is fitted with variable inlet guide vanes, which can lead to operation at part load that is far from the diffuser's design conditions.

Rotating stall in vaneless diffusers has received a lot of attention due to some unfortunate and expensive problems encountered before the problem was well understood. It appears to continue to be an area of significant concern to many investigators, but the designer can easily minimize the risk by sound basic design. It is also true that this specific rotating stall is by far the easiest type to correct. A simple width reduction should directly shift the onset of rotating stall to lower flows, should the designer have the misfortune to encounter it.

EXERCISES

8.1   Some designers, seeking to optimize stage efficiency, follow the practice of designing vaneless diffusers such that the loss coefficient passes through a minimum at the design flow. Why is this practice questionable? In what stage flow coefficient range (high or low) is it most likely to fail in its objective? What is the major benefit derived from this practice?

8.2   Refer to Figs. 8-2 through 8-5. State the design objectives that would be best served by choosing diffuser #1? Repeat for diffuser #4. Is diffuser #4 likely to be chosen if the following component is a volute or collector? Why or why not?

8.3   You have to design a vaneless diffuser for a stage consisting of an existing impeller and an existing volute, which have not been used together previously. Should you give priority to the vaneless diffuser's performance or to matching the volute [i.e., obtaining SP of Eq. (5-106) close to unity] to minimize the volute losses?

Chapter 9

VANED DIFFUSER DESIGN

This chapter describes a systematic procedure for the aerodynamic design of conventional vaned diffusers based on Aungier (1988a, 1990). The term conventional vaned diffusers refers to the various thin-vaned or airfoil style vaned diffusers (e.g., see Aungier, 1988a, 1990; Yoshinaga et al., 1980; Mischina and Gyobu, 1978; Pampreen, 1972) as illustrated in Fig. 9-1. The conventional vaned diffuser relies on standard cascade technology where the swirl velocity component is reduced by turning vanes. Two other basic types of vaned diffuser are in common use today. The thick-vaned styles includes the popular vane island or channel diffuser (Runstadler and Dean, 1969; Rogers, 1972; Conrad et al., 1979; Runstadler et al., 1975) and the pipe diffuser (Kenny, 1979). These styles are patterned more after classical exhaust diffusers, where the rate of increase in passage area is controlled by increasing the vane thickness with radius. At the exit of the vanes, there is an abrupt and substantial area increase due to the large vane thickness. In recent years, the low-solidity vaned diffuser has received considerable attention (Senoo, 1981; Senoo et al., 1983; Osborne and Sorokes, 1988; Hohlweg et al., 1993). Here, very short vanes, or few in number, are used such that the vanes are too far apart to form an aerodynamic throat.

Few areas of centrifugal compressor technology involve as much difference of opinion as the vaned diffuser. There are many strong advocates of each of the above vaned diffuser styles being the "best" style to use. Often, it is difficult to separate fact from opinion where this subject is debated. A few definite differences can be identified: The thick-vaned style requires a substantially larger discharge-to-inlet radius ratio to achieve the same pressure recovery as the conventional style. Indeed, the primary advantage of the conventional vaned diffuser is its very compact size. For many years, the vane-island diffuser was favored because it was the only style for which a systematic design procedure was available. Aungier (1988a, 1990) largely eliminated that advantage by offering a very systematic and reliable design method for the conventional vaned diffuser. Some investigators claim the vane-island diffuser achieves better static pressure recovery, but this author has seen no definitive evidence to support that claim. The low-solidity vaned diffuser can offer a wider operating range by virtue of its lack of an aerodynamic throat. Like the conventional style, the low-solidity vaned diffuser is a direct fluid-turning device, although these vanes probably function more like isolated air foils than a cascade of vanes. Some care in assessing claims reported in the literature is necessary. Except for Hohlweg et al. (1993), no systematic comparison of low-solidity vaned diffusers with conventional or vane-island styles has been reported. In Hohlweg et al. (1993), it was found that the

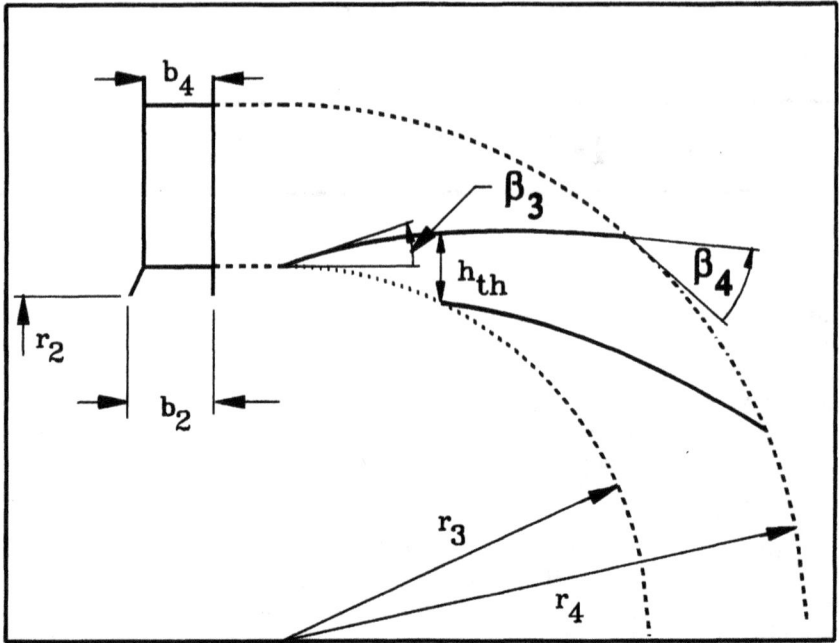

FIGURE 9-1. Vaned Diffuser Geometry

conventional vaned diffuser provided better efficiency gains over a vaneless diffuser than did the low-solidity vaned diffuser. The low-solidity vaned diffuser was capable of a wider flow operating range in most (but not all) cases. Hohlweg et al. (1993) notes that this comparison is somewhat suspect since the conventional vaned diffusers benefited from the systematic design procedure presented in this chapter, while the procedure for designing the low-solidity vaned diffuser was far less sophisticated. But it remains the only comparative evaluation reported to date.

The present design procedure applies to radial, parallel-walled vaned diffusers. It is not difficult to define logical extensions to address nonradial designs, radial variations in passage width and thick vanes, much as was done for the performance analysis in Chapter 5. But without validation against experiment, it is not considered appropriate to present such extensions as a "design procedure." Centrifugal compressor diffusers are basically always radial except for very high flow coefficient stages. As pointed out in Chapter 6, vaned diffusers provide no significant benefit in those cases due to the relatively high impeller tip flow angles required. Hence, the restriction to a radial passage is not a significant limitation. After using this method for more than 10 years, the author has yet to find the need to use a variable passage width within the vaned portion of the passage. Thus, there is no obvious motivation to deviate from the radial parallel-walled construction, which has been extensively validated (Aungier, 1988a, 1990).

NOMENCLATURE

A = an area inside the blade passage
A_R = area ratio, W_4/W_3
B = fractional area blockage
b = hub-to-shroud passage width
C = velocity
c = chord length
C_L = empirical blockage correction factor
C_p = vane pressure surface velocity
C_r = radial velocity component
C_s = vane suction surface velocity
C_U = tangential velocity component
C_θ = empirical blockage correction factor
c_p = static pressure recovery coefficient
D = diffusion factor
L = blade loading parameter
L_B = length of blade mean camberline
m = meridional coordinate
P_R = ideal (no loss) ratio of inlet-to-discharge velocity pressures
p = pressure
R = radius ratio, r_4/r_3
r = radius
t_b = vane thickness
x = distance along the chord
W = effective blade-to-blade passage width
z = number of vanes
α = flow angle with respect to tangent
β = blade angle with respect to tangent
η = r/r_3
ξ = distance along blade camberline
$2\theta_C$ = equivalent divergence angle
$\overline{\omega}$ = total pressure loss coefficient

Subscripts

ave = average value
B = a blade parameter
h = parameter on the hub contour
i = ideal (no loss) condition
p = blade pressure surface parameter
s = shroud contour or blade suction surface parameter
t = total thermodynamic condition
2 = impeller tip condition
3 = vane inlet condition
4 = diffuser exit condition

Superscripts

$*$ = sonic flow condition

9.1 Vaned Diffuser Performance Parameters

The conventional vaned diffuser relies on two basic diffusion mechanisms to achieve its static pressure recovery. The effective passage area increases from inlet to discharge, providing direct streamwise diffusion similar to that of a simple exhaust diffuser. This is augmented by the fluid turning or blade loading effect, which provides the additional diffusion achieved by the vaned diffuser relative to the vaneless diffuser. An effective vaned diffuser must make full use of both of these diffusion mechanisms to achieve the maximum pressure recovery. Much like simple exhaust diffusers, the vaned diffuser achieves its best performance when diffusion levels from both basic mechanisms are close to the maximum achievable. But exceeding those limits will result in a substantial performance penalty. Aungier (1988a) presents key design parameters and criteria to address both diffusion mechanisms. Defining the blade passage length, L_B, and the effective passage width, W, by

$$L_B = \int_{r3}^{r4} \frac{dr}{\sin \beta} \tag{9-1}$$

$$W = (2\pi r \sin \beta)/z \tag{9-2}$$

the first key design parameter is a direct analogy with the well-known divergence angle of classical diffuser technology (Reneau et al., 1967)

$$2\theta_C = 2\tan^{-1}[(W_4 - W_3)/(2L_B)] \tag{9-3}$$

To evaluate blade loading, the average blade-to-blade velocity difference is computed from simple potential flow theory, which relates it directly to the change in angular momentum.

$$\overline{\Delta C}L_B = \int_0^{L_B} (C_s - C_p)d\xi = 2\pi(r_3 C_{U3} - r_4 C_{U4})/z \tag{9-4}$$

At design flow, the inlet and discharge flow angles are approximately equal to the blade angles, which yields

$$\overline{\Delta C} = C_3 W_3(\cot\beta_3 - R C_{r4}\cot\beta_4/C_{r3})/L_B \tag{9-5}$$

where $R = r_4/r_3$ = diffuser radius ratio. Then, the blade loading parameter, L, is defined as

$$L = \frac{\overline{\Delta C}}{C_3 - C_4} \qquad (9\text{-}6)$$

The blade loading parameter can be interpreted as an approximation to the ratio of the average vane-to-vane pressure difference to the inlet-to-discharge pressure difference. Indeed, this is exactly the case for incompressible flow through a vaned diffuser with uniform blade loading. When considering experimental data of substantially different Mach number levels, a useful parameter was found to be the ratio of the ideal (no loss) inlet-to-discharge velocity pressures, P_R.

$$P_R^2 = \frac{p_{t3} - p_3}{p_{t3} - p_{4i}} \qquad (9\text{-}7)$$

For incompressible flow, $P_R = A_R = W_4/W_3$. As was done for the vaneless diffuser in Chapter 8, the static pressure recovery coefficient will be used to evaluate the diffuser's capability to convert kinetic energy to static pressure

$$c_p = \frac{p_4 - p_3}{p_{t3} - p_3} \qquad (9\text{-}8)$$

9.2 Design Criteria

Aungier (1988a) investigated experimental results for 18 different vaned diffusers to develop design criteria for vaned diffusers. Sixteen of these are reported in detail in Yoshinaga et al. (1980). The other two (VD-17 and VD-18) are special research vaned diffusers tested in essentially incompressible flow and summarized in Aungier (1988a). Table 9-1 summarizes the primary geometrical and performance parameters for these vaned diffusers. The approach taken in Aungier (1988a) was to compute the discharge area blockage, B_4, required to match the experimental pressure recovery, assuming isentropic flow and perfect guidance of the flow by the vanes. Since this deals with static pressure recovery, discharge blockage is expected to be a reasonable parameter to use for correlation purposes. The growth of boundary layers and wakes in the blade passage would be expected to limit the static pressure rise. Subsequent mixing of the low-momentum viscous flow with the inviscid core flow will result in losses, but should not substantially alter results inferred from the blockage alone. Simple boundary layer arguments suggest that the blockage should correlate with skin friction and diffusion terms in the form

$$B_4 = [K_1 + K_2(D - 1)]L_B/W_4 \qquad (9\text{-}9)$$

where K_1 should be related to the skin friction coefficient; K_2 is an empirical constant to account for flow diffusion; and D is a diffusion factor defined by

Table 9-1. Design Data for Tested Varied Diffusers

Diffuser	R	A_R	B_4	L_B/W_3	$2\theta_C$	L	c_p
VD-1	1.350	1.54	0.0722	6.05	5.11	0.1455	0.56
VD-2	1.350	1.77	0.1752	5.43	8.11	0.2483	0.58
VD-3	1.350	1.92	0.1832	6.15	8.56	0.2745	0.64
VD-4	1.350	1.89	0.2665	3.88	13.09	0.4070	0.53
VD-5	1.350	1.89	0.1592	4.79	10.62	0.3297	0.65
VD-6	1.350	1.89	0.1909	5.70	8.93	0.2770	0.62
VD-7	1.350	2.07	0.1993	5.48	11.15	0.3271	0.68
VD-8	1.350	2.17	0.3483	4.31	15.46	0.4394	0.55
VD-9	1.350	2.17	0.3119	5.14	13.00	0.3684	0.60
VD-10	1.420	2.05	0.2630	6.14	9.77	0.2562	0.61
VD-11	1.420	2.18	0.2897	5.16	13.05	0.3326	0.63
VD-12	1.350	1.79	0.1658	5.12	8.82	0.2710	0.60
VD-13	1.450	1.89	0.2099	7.03	7.24	0.1777	0.60
VD-14	1.450	2.17	0.2953	6.43	10.40	0.2503	0.62
VD-15	1.600	2.19	0.3101	8.88	7.67	0.1427	0.61
VD-16	1.600	2.40	0.3628	8.43	9.49	0.1774	0.62
VD-17	1.442	2.70	0.3918	7.17	13.50	0.3258	0.63
VD-18	1.395	2.50	0.3518	6.41	13.35	0.3565	0.62

$$D = (P_R + 1)^2/4 \tag{9-10}$$

The validity of the proposed model is demonstrated in Fig. 9-2, where the line designated as "empirical equation" is obtained from Eq. (9-9), using $K_1 = 0.005$ and $K_2 = 0.044$. Most of the vaned diffusers investigated correlate with this empirical model. A few diffusers showed substantially higher discharge blockage levels than are explained by the simple boundary layer model of Eq. (9-9). In all cases, those diffusers violate one or both of the following design criteria (and are shown as the solid symbols in Fig. 9-2).

$$2\theta_C \leq 11^o \tag{9-11}$$

$$L \leq \tfrac{1}{3} \tag{9-12}$$

From this comparison, it can be concluded that Eqs. (9-11) and (9-12) express important limits that should be observed when designing a vaned diffuser. As discussed previously in this chapter, it can be expected that a good vaned diffuser will be designed very close to these design limits. Table 9-1 tends to confirm this. Diffusers VD-5 and VD-7 are clearly the most effective of the series and are designed almost exactly at both of the design limits. By contrast, look at VD-16, which has a significantly higher area ratio. This should yield higher pressure recovery levels than for VD-5 and VD-7, yet it achieves a rather unimpressive pressure recovery. This diffuser's blade loading parameter is rather low,

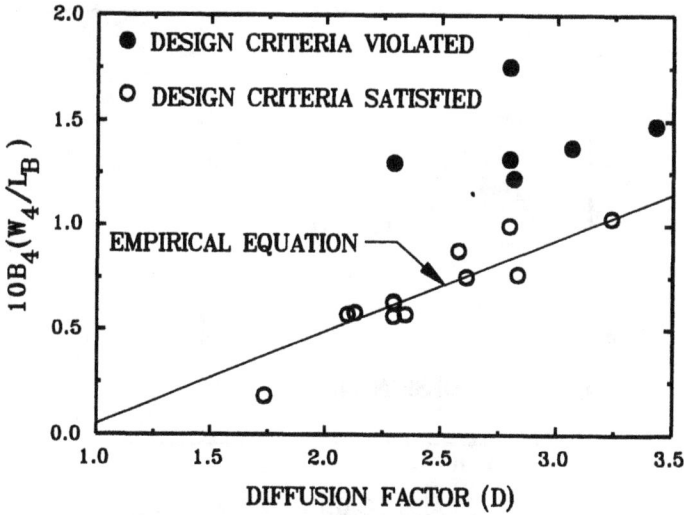

FIGURE 9-2. Basic Blockage Correlation

showing that this design features too little support from the blade loading diffusion mechanism to be a truly effective design. Figures 9-3 and 9-4 also illustrate this point—note the clear trend toward peak pressure recovery very close to the design limits.

Aungier (1988a) carries this model somewhat further, to generalize it to vaned

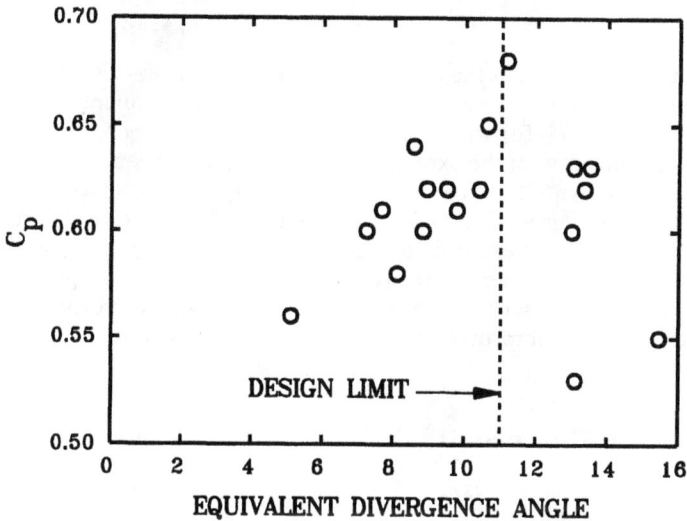

FIGURE 9-3. The Effect of Divergence Angle

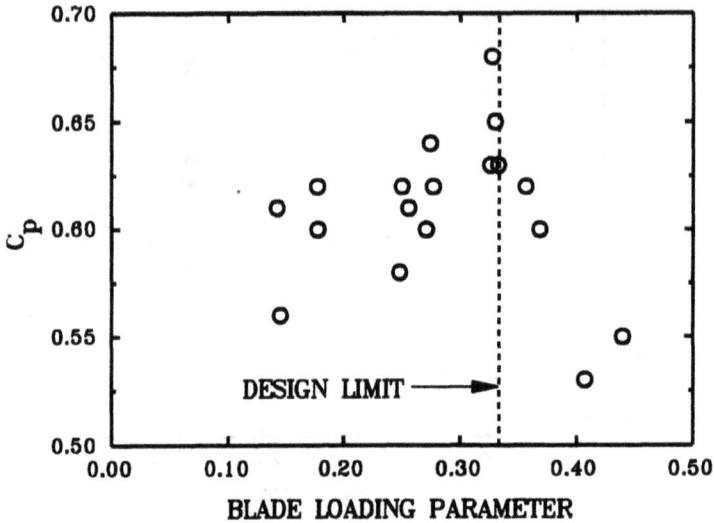

FIGURE 9-4. The Effect of Blade Loading Parameter

diffusers that violate the above design limits. This was accomplished by defining

$$K_1 = 0.005 + [1 - 1/(C_L C_\theta)]/5 \qquad (9\text{-}13)$$

$$K_2 = 2\theta_C[1 - 2\theta_C/(22C_\theta)]/(125C_\theta) \qquad (9\text{-}14)$$

$$1 \le C_L \ge 3L \qquad (9\text{-}15)$$

$$1 \le C_\theta \ge 2\theta_C/11 \qquad (9\text{-}16)$$

which can be recognized as the basis for the vaned diffuser blockage loss of Chapter 5. Figure 9-5 shows improved agreement with the experiment achieved when Eqs. (9-13) through (9-16) are combined with Eq. (9-9). To more clearly illustrate the good agreement with the experiment, Fig. 9-6 compares predicted and measured values of c_p with the experimental uncertainty band reported by Yoshinaga et al. (1980) as background. It is seen that all but three of the 18 vaned diffusers are predicted within the experimental uncertainty. This simple performance correlation was used as a means for evaluating vaned diffuser designs in Aungier (1988a). This is no longer necessary, since the more exact performance analysis of Aungier (1990), as improved in Chapter 5, is now available.

9.3 Vaned Diffuser Stall

Subsequent to Aungier (1988a), the performance analysis of Aungier (1990, Chapter 5) was finalized to include a prediction of the vaned diffuser stall incidence angle. This is an important addition to the vaned diffuser design method,

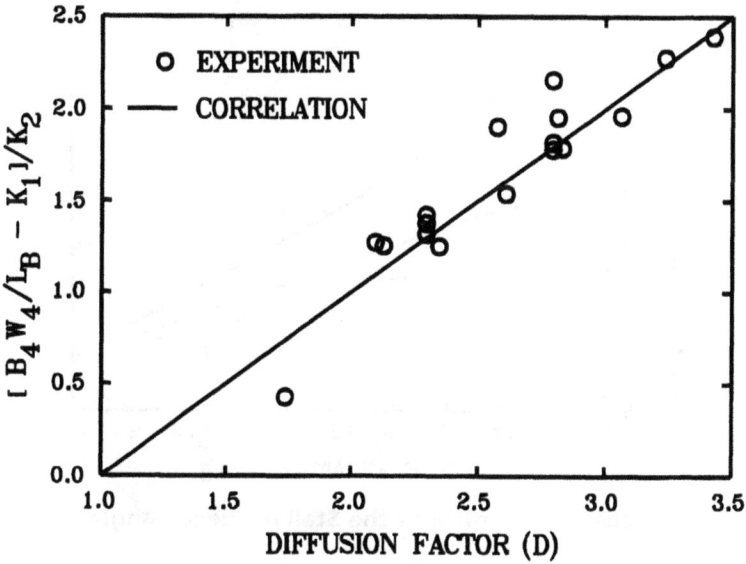

FIGURE 9-5. Modified Blockage Correlation

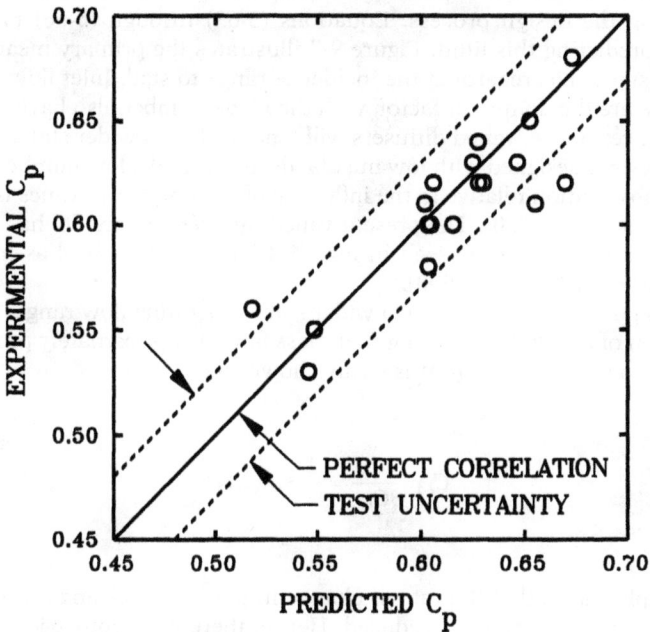

FIGURE 9-6. Validation of the Blockage Correlation

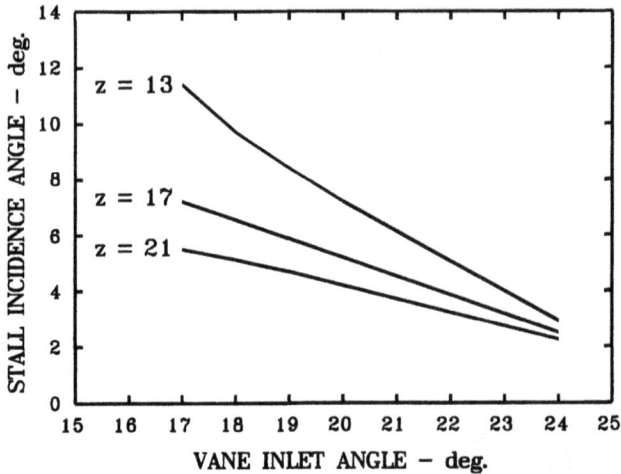

FIGURE 9-7. Control of the Stall Incidence Angle

since vaned diffuser stall is relatively severe. It almost always leads to flow instability. At or above moderately high Mach numbers, it will normally cause the
compressor to surge. Consequently, the designer needs to be aware of the stall
limit during the design process. Equations (5-63) through (5-66) provide the
means of predicting this limit. Figure 9-7 illustrates the primary means of control the designer has regarding the incidence range to stall. Inlet flow angle and
blade angle are the dominant factor, with the blade number also having a significant influence. Hence, vaned diffusers will tend to show a wider stable operating
range if they are designed with low inlet blade angles and a low number of vanes.
A similar observation relative to the influence of the number of vanes is reported
in Yoshinaga et al. (1980). The present vaned diffuser stall model has proven to
be quite accurate as can be seen in Figs. 5-17 and 5-18, as well as from other
cases reported in Aungier (1990).

The designer is always seeking a wider stable operating flow range, expressed
as a fraction of design flow. Noting that mass flow is approximately proportional
to C_{r3} and that $\tan\alpha = C_r/C_U$, it is easily shown that

$$C_{r3} \frac{\partial \alpha_3}{\partial C_{r3}} = \frac{1}{2} \sin(2\alpha_3) \qquad (9\text{-}17)$$

which simply shows that the rate of change in the incidence angle with percent
mass flow is reduced as α_3 is reduced. Hence, there is a geometrical benefit to
improving stability by using low design inlet flow angles as well as a benefit to
the stall incidence angle.

9.4   Vaned Diffuser Inlet Design

The first step in the detailed design of a vaned diffuser is to establish the vaned leading edge location, r_3, blade angle, β_3, and the vaneless space between the impeller tip and the vane leading edge. From the preceding discussion, it is apparent that low inlet blade angles are highly desirable. The more successful vaned diffuser designs will generally feature $16° \leq \beta_3 \leq 22°$. The lower limit follows from the fact that the upstream vaneless space losses increase as flow angle is reduced, as seen in Fig. 8-3. Also, the possibility of a vaneless diffuser rotating stall, discussed in Chapter 8, cannot be ignored. While this is very unlikely in the short vaneless space preceding the vaned diffuser, if flow angles get too low it certainly can happen.

But the designer's primary constraint is the need to match the vaned diffuser to the impeller tip flow. Specifically, at the design flow, C_{U2} is fixed by the impeller detailed design. Here, the designer's choices are limited to selecting the inlet passage width, b_3, and the design incidence angle, $i_3 = \beta_3 - \alpha_3$. Conservation of mass and angular momentum, using the vaneless annular passage performance analysis of Chapter 5, will basically determine all other fluid dynamics and geometrical data at the leading edge. There are secondary effects involved, since the vaned diffuser incidence angle is best defined with the flow inside the vane passage, i.e., with vane metal area blockage included. Hence the number of vanes and the vane leading edge thickness have an influence. This means refinement of the leading edge design may be required as the design of the vanes progresses. The vaned diffuser design incidence angle should usually be around $-1°$, except as compromises are required to obtain the desired flow range to stall and, for higher Mach number levels, the desired flow range to choke. Thus, for practical purposes, the designer's primary control over setting β_3 is to select b_3. This will not pose a major problem for the design of a complete stage if proper preliminary design procedures have been followed. In that case, the impeller design process will have recognized the need to match with a vaned diffuser, e.g., through Eqs. (6-13) and (6-14). The more difficult problem arises when retrofitting an existing vaneless diffuser stage with a vaned diffuser. Designers should recognize that it may not be possible to obtain a good vaned diffuser design in those cases.

As a general principle of good design practice, $b_3 \leq b_2$ should be required. Clearly, by relaxing that constraint, the designer may gain flexibility when attempting to match a vaned diffuser to an existing impeller. If done properly, this author's experience has shown that this causes no serious performance problems. The designer should resist the temptation to use a gradual increase in passage width. This raises the risk of a flow separation initiating within the desired compressor operating range with the distinct possibility of a flow instability. But no evidence of any instability has been observed when a "step change" in passage width is imposed at the impeller tip to achieve $b_3 > b_2$. It appears that a separation present at all operating conditions is quite stable. This will impose a total pressure loss, which can be estimated from classical abrupt expansion loss models (Benedict et al., 1966) applied to the radial velocity component, i.e.

$$\overline{\omega} = \sin^2 \alpha_2 (1 - b_2/b_3)^2 \qquad (9\text{-}18)$$

This is a very mild loss penalty, since α_2 is relatively small. From classical experiments with two-dimensional flow over a step, the separation zone would be expected to extend a distance downstream, approximately equal to the step height. But this author's experience indicates the zone extends much further downstream in these radial, diffusing passages. Hence, designers should recognize that the benefits of this type of step change are likely to be much less than expected from the basic fluid dynamics analysis procedures presented in this book.

The other key decision required to establish the vaned diffuser inlet is the location of the leading edge, i.e., r_3. As a general guideline $1.06 \leq r_3/r_2 \leq 1.12$ is recommended. The lower limit provides space for the distorted impeller flow profiles to smooth out and the blade wakes to decay before the flow enters the vanes. For higher impeller tip Mach number levels, a longer vaneless space may be needed to reduce the Mach number level before entering the vanes. Due to the low-flow angles involved, vaneless space loss levels can be relatively high, even in the short distances involved, as evidenced by Fig. 8-3. Thus, a vaneless space that is longer than needed will impose unnecessary losses. As a first estimate, Eq. (6-25) from the preliminary design procedure can be used.

9.5   Vaned Diffuser Sizing

The next step in the design process is to establish the vane discharge geometry, which basically requires sizing the diffuser. Here, the basic design criteria of Eqs. (9-11) and (9-12) provide the primary guidance. The process is usually constrained by factors other than aerodynamic performance. The available radial space may be limited by overall casing size design limits. Usually the choice of the number of vanes to be used is not totally arbitrary. The mechanical designer will normally work with the aerodynamic designer to select a vane number compatible with the number of impeller blades from vibration and resonance considerations. Within these external constraints, the designer should seek to design close to, but not exceeding, the design limits of Eqs. (9-11) and (9-12). Where possible, vaned diffusers should be designed within the following ranges:

$$10^{o} \leq \theta_C \leq 11^{o}$$

$$0.3 \leq L \leq 0.33$$

(9-19)

Aungier (1988a) also introduces a design parameter, E, defined by

$$E = \frac{R^2(A_R^2 - 1)}{A_R^2(R^2 - 1)}$$

(9-20)

which evaluates the effectiveness of the vaned diffuser design relative to a vaneless diffuser. For incompressible flow, it is easily shown that E is the ratio of

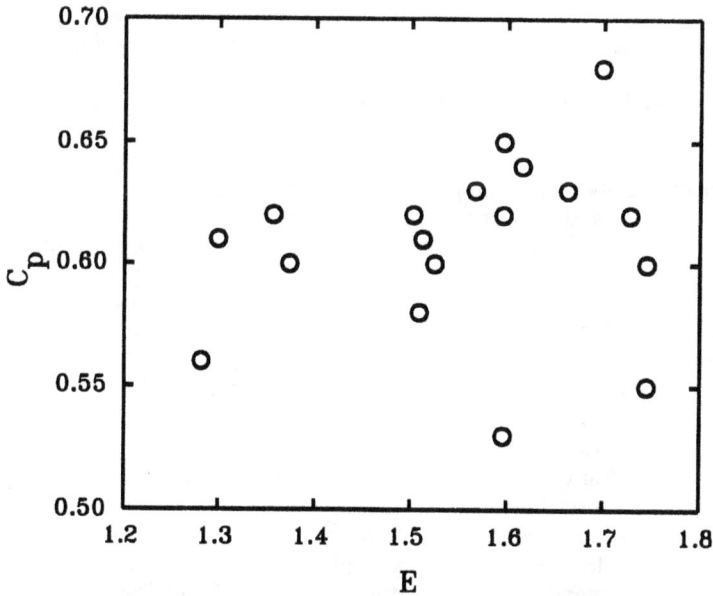

FIGURE 9-8. Vaned Diffuser Effectiveness Parameter

the ideal pressure recovery coefficients for a vaned diffuser to a vaneless diffuser designed for the same radius ratio. Figure 9-8 shows the experimental pressure recovery for the 18 vaned diffusers presented in Table 9-1 as a function of this parameter. Clearly, good performance is achievable to values of E up to at least 1.7. Aungier (1988a) recommends designing to values of E in the range of 1.5–1.7. This parameter is not really a design criterion so much as just a measure of whether the design has practical merit. Unless the vaned diffuser substantially improves the pressure recovery level over a vaneless diffuser, it is unlikely that the additional cost and probable reduction in stable operating range can be justified. The vaned diffuser area ratio is also a design parameter of some significance. Classical two-dimensional diffuser performance maps (Reneau et al., 1967) show that area ratios in the range of 2.2–2.4 are about optimum for the inlet blockage and profile distortion levels to be expected in a vaned diffuser.

It should be evident at this point that not all of these design guidelines can normally be satisfied. The vaned diffuser sizing process basically establishes r_4 (or R), β_4 and z. There are simply not enough degrees of freedom to satisfy all of the preferred design constraints. As usual, the designer's task is to select the best compromise. At this phase of the design, the detailed vane design is not available, so an approximate relation is used for vane length.

$$L_B \approx 2r_3(R - 1)/(\sin \beta_3 + \sin \beta_4) \tag{9-21}$$

Typical practice is to try alternate values of R, A_R and z until the best combination of $2\theta_C$, L and A_R are obtained. In a computerized design system, the process is aided by providing alternate options for the design specifications, including

- Select R, A_R and z—compute $2\theta_C$, L and E.
- Select R, A_R and $2\theta_C$—compute z (rounded to integer), L, E and revise A_R or $2\theta_C$.
- Select R, z and $2\theta_C$—compute A_R, L and E.
- Select R, β_4 and z—compute A_R, $2\theta_C$, L and E.
- Select R, β_4 and $2\theta_C$—compute z (rounded to integer), L, A_R, E and revise $2\theta_C$.

which let the designer establish the parameter most critical to his specific design constraints while exploring alternatives.

It was noted in Chapter 6 that this design procedure loses significance at ultralow stage flow coefficients (say, less than 0.01). In these cases, the wall friction forces may reduce the tangential velocity more than a vaned diffuser designed according to the above procedure. Here, the designer must recognize that the purpose served by a vaned diffuser is not really an improved static pressure recovery. In these stages, the dominant vaneless diffuser loss mechanism is wall friction due to the very small passage widths. A vaned diffuser can permit the use of wider passage widths without the risk of vaneless diffuser rotating stall, even when it is designed beyond the preferred design limits, such that its pressure recovery is compromised. It is highly likely that the reduction in friction losses due to a vaned diffuser designed beyond the present design limits will be very beneficial to stage efficiency. In those cases the designer must disregard these design limits and concentrate on increased passage width as the design goal.

9.6 Vane Design

Once the leading and trailing edge geometry is established, the actual vane design can be accomplished by imposing a vane thickness distribution on a mean camberline. The general mean camberline of Aungier (1988a) is convenient for this purpose. Defining $\eta = r/r_3$, the camberline is given by

$$\theta(\eta) = A \ln(\eta) + B(\eta - 1) + C(\eta^2 - 1) + D(\eta^3 - 1) \tag{9-22}$$

Four boundary conditions are needed to compute the constants in Eq. (9-22). Two are supplied by the leading and trailing edge blade angles since

$$\cot\beta = \eta \, \frac{d\theta}{d\eta} = A + B\eta + 2C\eta^2 + 3D\eta^3 \tag{9-23}$$

The other two boundary conditions are selected to control the form of the blade loading distribution. This is accomplished by specifying the parameter, K, at the leading and trailing edges, where

$$K = \frac{R - 1}{\cot\beta_4 - \cot\beta_3} \frac{d\cot\beta}{d\eta} \tag{9-24}$$

The differential form of Eq. (9-4) illustrates the control over the blade loading distribution provided by the parameter K, i.e.,

$$C_s - C_p = -\frac{W}{r} \frac{\partial(rC_r\cot\beta)}{\partial r} \tag{9-25}$$

It is easily shown that these boundary conditions yield

$$D = \frac{(\cot\beta_4 - \cot\beta_3)(K_3 + K_4 - 2)}{3(R - 1)^3} \tag{9-26}$$

$$C = \frac{(\cot\beta_4 - \cot\beta_3)(K_4 - K_3)}{4(R - 1)^2} - \frac{9D(R + 1)}{4} \tag{9-27}$$

$$B = \frac{K_3(\cot\beta_4 - \cot\beta_3)}{R - 1} - 4C - 9D \tag{9-28}$$

$$A = \cot\beta_3 - B - 2C - 3D \tag{9-29}$$

To complete the vane design, it is only necessary to impose the desired vane thickness distribution on the mean camberline. This author's vaned diffuser design system offers two choices for vane thickness distribution:

- A constant thickness vane with an optional linear taper to a smaller leading edge thickness over a specified portion of the camberline length.
- An airfoil shape closely approximating the NACA 66-006 airfoil.

The NACA 66-006 airfoil has been used for vaned diffusers by a number of investigators (e.g., Yoshinaga et al., 1980). This author has seen unpublished experimental results comparing the constant thickness/tapered vane and NACA 66-006 airfoil shape using the same camberline for two different vaned diffuser designs. One was a research diffuser tested in a low-speed test rig, and the other was in a compressor with a design pressure ratio of 2.5 : 1. In neither case was there any measurable difference in performance. Consequently, this author bases this choice on manufacturing cost, stress levels and vibration considerations. When the airfoil shape is used, an analytical approximation is used to support a computerized design system; the approximation used is given by

$$
\begin{aligned}
t_b/t_{b\max} &= t_0 + (1 - t_0)(2x/c)^n; & \text{for } x/c \leq 0.5 \\
t_b/t_{b\max} &= t_0 + (1 - t_0)(2 - 2x/c)^n; & \text{for } x/c > 0.5 \\
t_0 &= [t_{b3} + (t_{b4} - t_{b3})x/c]/t_{b\max}
\end{aligned} \tag{9-30}
$$

where the maximum thickness is specified by the designer; c = chord length; x = distance along the chord; t_0 = influence of leading and trailing edge radii, and

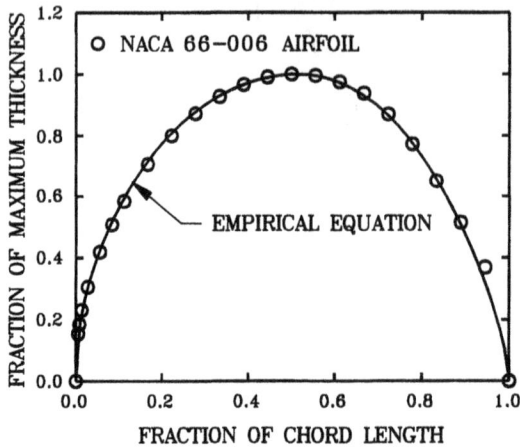

FIGURE 9-9. NACA 66-Series Blade Thickness Equation

$$n = 0.755(0.57 - x/c); \qquad \text{for } x/c \leq 0.539$$

$$\text{(9-31)}$$

$$n = 1.225(x/c - 0.52); \qquad \text{for } x/c > 0.539$$

Figure 9-9 compares the base thickness distribution (i.e., with $t_0 = 0$) from Eqs. (9-30) and (9-31) with the actual NACA 66-006 airfoil thickness distribution.

Once the camberline and thickness distribution are defined, the vane can be constructed and the throat area calculated following the procedures described in Chapter 7.

9.7 Analysis of the Design

The aerodynamic performance of the candidate design can be predicted using the vaneless annular passage and vaned diffuser performance analyses of Chapter 5. Then the vaned diffuser blade loading can be evaluated using one of the blade-to-blade flow analyses of Chapters 12 through 14. The blade loading levels acceptable for vaned diffusers are considerably lower than for impellers. From analysis of several good performing vaned diffusers, the magnitude of the blade loading should generally be limited by

$$|C_s - C_p|/C_{ave} < 0.4 \qquad \text{(9-32)}$$

where C_s and C_p = vane suction and pressure surface velocities, and C_{ave} = average of these two values. Equation (9-32) should be viewed as a maximum limit. A limit of about 0.35 is preferred if possible. Comparison with Eq. (7-60) shows that recommended blade loading limits for vaned diffusers is less than half that recommended for impellers. This is primarily due to the fact that the overall

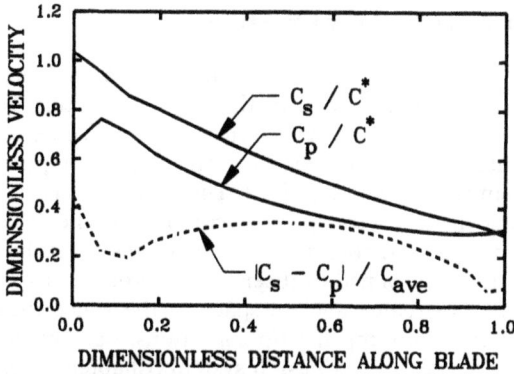

FIGURE 9-10. Typical Vaned Diffuser Blade Loading

(streamwise) diffusion in a vaned diffuser is substantially greater than for an impeller. Figure 9-10 shows a typical blade loading diagram for a vaned diffuser. As usual, smooth velocity profiles that minimize velocity gradients as much as possible should be the designer's goal. The blade surface boundary layers can be analyzed with the two-dimensional boundary layer analysis presented in Chapter 15, but the results provide limited guidance for vaned diffusers. Figure 9-11 shows the boundary layer shape factors (turbulent portion only) from a two-dimensional boundary layer analysis of the blade loading in Fig. 9-10. Noting that separation is expected for shape factors around 2.2–2.4, it can be seen that the boundary layers are predicted to be separated over a large portion of the vane length. This is typical of any effective vaned diffuser design. Basically, to achieve the best static pressure recovery, a vaned diffuser will always be designed for a diffusion level that shows substantial separation in an analysis of the blade surface boundary layers.

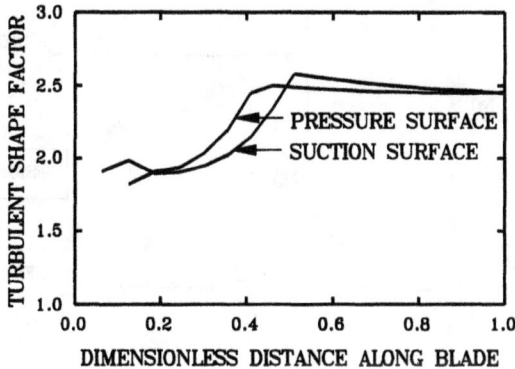

FIGURE 9-11. Vane Surface Boundary Layer Predictions

9.8 A Computerized Design System

The design procedure described in this chapter can be implemented as an interactive, computerized design system. In this form, an optimized vaned diffuser design can be accomplished in a very brief period of time, including all relevant fluid dynamics analysis. A basic flow chart of a computerized vaned diffuser design system is shown in Fig. 9-12. The process starts with the specification of the impeller tip geometry and flow conditions. The designer can use the vaneless diffuser design system to design the vaneless space, but since b_3 and r_3 are selected as part of the vaned diffuser design process, it is better to include construction of the upstream vaneless space in the vaned diffuser design system and work directly from impeller tip conditions. Usually, the process will be started by having the preliminary design system (Chapter 6) set up an input file for its selection. It is also useful to make provision for the stage performance analysis (Chapter 5) to update the impeller tip flow data and geometry to supply the results relevant to the final impeller design. The vaned diffuser inlet and component sizing techniques described in this chapter will then be processed in an interactive fashion, followed by the vane design. The vaneless passage and vaned diffuser performance analyses from the overall stage performance analysis are easily included in this design system to provide direct and immediate access to a performance evaluation. Similarly, the linearized blade-to-blade flow analysis of Chapter 12 is easily included to provide direct and immediate access to a blade loading evaluation. In this way, the vaned diffuser design and analysis can be accomplished in a single design system in an iterative fashion, as indicated by the flow chart. The design system should be capable of updating the diffuser geometry in the input file for the stage performance analysis to permit an overall

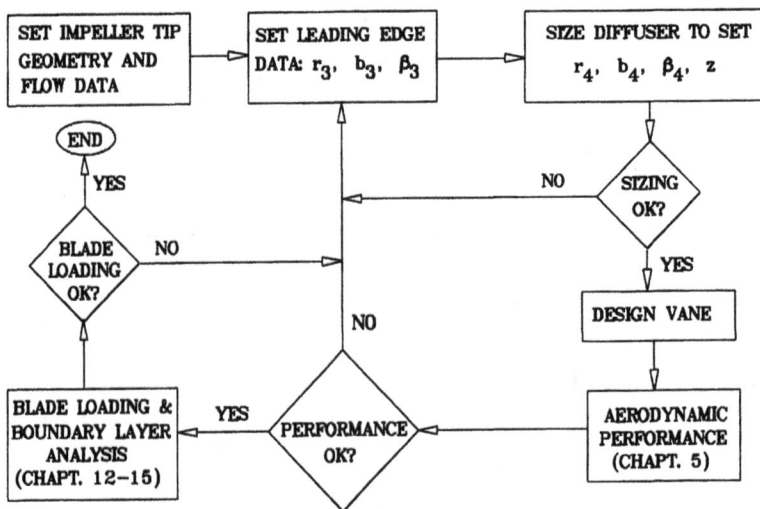

FIGURE 9-12. Flow Chart of the Design Procedure

analysis with the new geometry. It is also useful to provide for the capability to create an input file for one of the more exact blade-to-blade flow analyses presented in Chapters 13 and 14, for a final blade loading evaluation.

EXERCISES

9.1 One consideration in designing any centrifugal compressor component is matching it with the downstream component. In the case of a diffuser, this relates to achieving a good incidence angle for a return channel or a sizing parameter, SP, of Eq. (5-106), close to unity for a volute. Why is this particularly significant in vaned diffuser design?

9.2 Identify two different performance benefits that can be achieved with a vaned diffuser, and state the approximate stage inlet flow coefficient range (e.g., high, medium or low) to which each applies.

9.3 You have been asked if you can improve the efficiency of an existing vaneless diffuser stage by retrofitting it with a vaned diffuser. Identify at least three factors about the existing stage that you should consider before you give an opinion on that question.

9.4 For the retrofit in Exercise 9.3, you are told that a wide stable operating flow range is also needed. Identify at least three additional factors about the existing stage that you need to consider.

Chapter 10

RETURN SYSTEM DESIGN

The return system is probably the least understood component of the centrifugal compressor stage, rivaled only, perhaps, by the volute. The flow through the 180° annular crossover bend, and its interaction with the downstream vane row, presents an extremely complex fluid dynamics problem. The literature offers little guidance relative to design procedures or performance prediction models for this component. Aungier (1993a) is about the only available source offering a detailed aerodynamic design and analysis procedure for the return system. The aerodynamic performance analysis procedure has already been described in Chapter 5. Here, the expanded description of that design procedure is provided. Other than Aungier (1993a), most published experimental and theoretical investigations are attempts to develop a better understanding of these complex flows (e.g., Aungier, 1993a, 1988a; Davis, 1976; Fister et al., 1982; Japikse and Osborne, 1982; Aungier, 1988b; Nykorowytsch, 1983). Where guidance for a return system design is provided, it usually consists of applying basic engineering judgment to control diffusion rates, passage curvatures and other design parameters, supported by empirical categorization of performance from past development activity (e.g., Hohlweg, 1987). While much remains to be learned about these complex return system flows, the design system of Aungier (1993a) has been used with considerable success. For example, the performance data shown in Figs. 5-13 and 5-19 in Chapter 5 relate to a return system designed with this procedure.

NOMENCLATURE

A = area and ellipse axial semiaxis
A_R = area ratio
B = ellipse radial semiaxis
b = passage hub-to-shroud width
C = velocity
C_m = meridional velocity component
C_U = tangential velocity component
c = vane chord length
d = minimum metal thickness of return channel
i = vane incidence angle
K = blade loading parameters
m = meridional coordinate
p = pressure

R = mean streamline radius of curvature
R_C = crossover hub wall radius of curvature
R_O = exit turn shroud radius of curvature
r = radial coordinate
t_b = vane thickness
x = distance along vane chord line
z = number of vanes and axial coordinate
α = flow angle with respect to tangent
β = blade angle with respect to tangent
ϵ = angle between linear portion of return channel contours and the radial direction
θ = polar angle
$2\theta_C$ = equivalent divergence angle
θ_7 = vane trailing edge angle with axial direction
$\overline{\omega}$ = total pressure loss coefficient

Subscripts

av = average value or value at midchord
C = crossover bend parameter
h = parameter on the hub contour
m = a maximum condition
O = exit turn bend parameter
s = parameter on the shroud contour
5 = crossover inlet parameter
6 = crossover exit parameter
$6'$ = return channel vane inlet parameter
7 = return channel vane exit parameter
8 = return channel discharge parameter

10.1   Return System Gas Path Construction

The design system employs the fairly systematic construction shown in Fig. 10-1. The crossover bend construction uses a circular-arc hub contour and an elliptical-arc shroud contour. This requires that r_5, b_5, R_C and r_m be specified (noting that $r_6 = r_5$). Defining $z = 0$ at the shroud wall at the crossover entrance, the center point of the circular arc is located at $z = b_5 + R_C$ and $r = r_5$. The axial and radial semiaxes of the shroud elliptical contour are

$$A_C = R_C + (b_5 + b_6)/2$$
$$B_C = r_m - r_5 \qquad (10\text{-}1)$$

The ellipse center point is at $z = A_C$ and $r = r_5$ and the shroud contour is given by

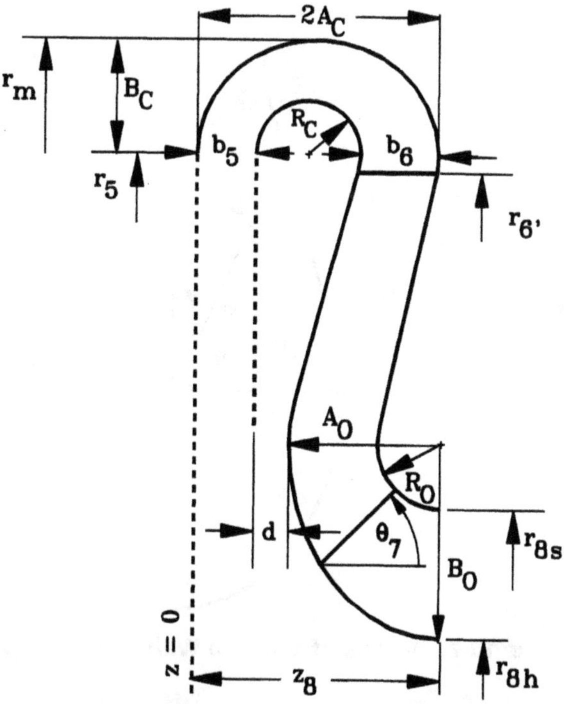

FIGURE 10-1. Return System Passage Construction

$$r = r_5 + B_C\sqrt{1 - (z - A_C)^2/A_C^2} \tag{10-2}$$

The 90° exit turn construction is similar to the crossover bend, except that the arc types are reversed for the two contours. The shroud circular-arc radius, R_O, and the hub elliptical-arc axial semiaxis, A_O, must be specified along with r_{8h}, r_{8s}, z_8. The center point for both the shroud circular arc and the hub elliptical arc is at $z = z_8$ and $r = R_O + r_{8s}$. The hub elliptical-arc radial semiaxis is given by $B_O = r_{8s} - r_{8h} + R_O$ and the hub contour is given by

$$r = r_{8s} + R_O - B_O\sqrt{1 - (z - z_8)^2/A_O^2} \tag{10-3}$$

The contours between r_6 and $R_O + r_{8s}$ are yet to be defined. For this section, straight line contours are used, blended to the crossover and exit turn contours by circular arcs, with arc radii equal to the local bend contour elliptical-arc axial semiaxis or circular-arc radius. The hub contour construction is illustrated in Fig. 10-2. The angle, ϵ, must be calculated to construct this portion of the gas path

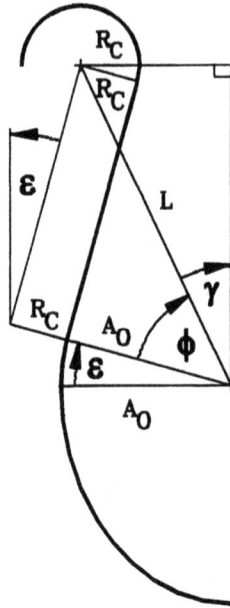

FIGURE 10-2. Details of the Hub Wall Construction

boundary. Then circular-arc extensions can be added to the two bend contours through the angle, ϵ, using arc radii of R_C at the crossover exit and A_O at the exit turn inlet. By connecting the new arc end points with a linear segment, the contour construction is complete. From Fig. 10-2, it can be seen that the linear section will match the slopes at the end points of both circular-arc extensions. Using the construction and nomenclature shown in Fig. 10-2, it is seen that

$$L^2 = (b_5 + R_C - z_8)^2 + (r_5 - R_O - r_{8s})^2 \qquad (10\text{-}4)$$
$$\cos\phi = (R_C + A_O)/L \qquad (10\text{-}5)$$
$$\cos\gamma = (r_6 - R_O - r_{8s})/L \qquad (10\text{-}6)$$
$$\epsilon = 90° - \phi - \gamma \qquad (10\text{-}7)$$

The shroud contour is handled using a construction similar to that of Fig. 10-2, but with L now drawn from the center of the crossover elliptical contour. Here, only the result will be given, i.e.,

$$L^2 = (z_8 - A_O)^2 + (r_5 - R_O - r_{8s})^2 \qquad (10\text{-}8)$$
$$\cos\phi = (A_C + R_O)/L \qquad (10\text{-}9)$$
$$\cos\gamma = (r_6 - R_O - r_{8s})/L \qquad (10\text{-}10)$$
$$\epsilon = 90° - \phi - \gamma \qquad (10\text{-}11)$$

10.2 Return Channel Vane Construction

Figure 10-3 shows the basic return channel vane construction used. The vane camberline used is an adaptation of the vaned diffuser camberline of Aungier 1988a, Chapter 9. The vane leading edge radius, $r_{6'}$, is always less than or equal to r_6. Leading and trailing edge blade loading parameters, $K_{6'}$, and K_7, are specified to define the vane camberline by

$$y = r/r_{6'} \tag{10-12}$$

$$\cot \beta = -A - By - 2Cy^2 - 3Dy^3 \tag{10-13}$$

$$\theta = A \ln(y) + B(y - 1) + C(y^2 - 1) + D(y^3 - 1) \tag{10-14}$$

$$Y = r_{7s}/r_{6'} \tag{10-15}$$

$$D = \frac{(\cot \beta_{6'} - \cot \beta_7)(K_{6'} + K_7 - 2)}{3(Y - 1)^3} \tag{10-16}$$

$$C = \frac{(\cot \beta_{6'} - \cot \beta_7)(K_7 - K_{6'})}{4(Y - 1)^2} - \frac{9}{4} D(Y + 1) \tag{10-17}$$

$$B = K_{6'}(\cot \beta_{6'} - \cot \beta_7)/(Y - 1) - 4C - 9D \tag{10-18}$$

$$A = -\cot \beta_{6'} - B - 2C - 3D \tag{10-19}$$

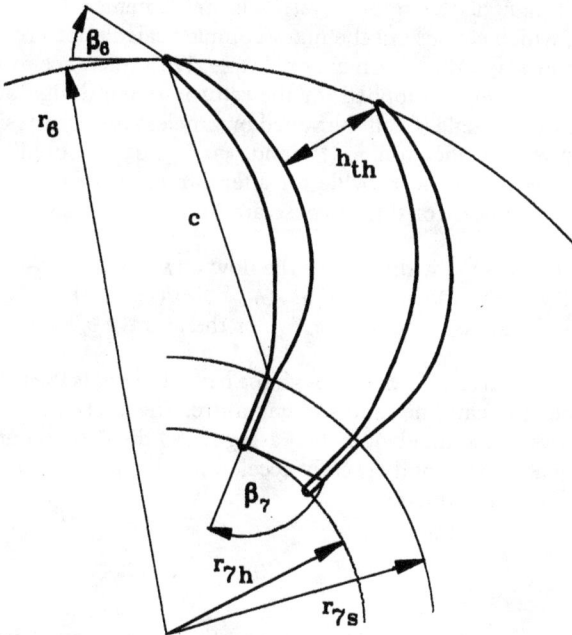

FIGURE 10-3. Return Channel Vane Construction

These equations are applied for radii greater than the vane discharge shroud radius, r_{7S}. Below r_{7S}, a constant blade angle camberline is used. The vane thickness distribution used is given as a function of the distance, x, along the chord line between r_6 and r_{7S}. Below r_{7S}, the vane thickness is held constant. The thickness distribution is given by

$$t_b = t_{b0} + (t_{bm} - t_{b0})y^e \tag{10-20}$$

$$t_{b0} = t_{b6'} + (t_{b7} - t_{b6'})x/c \tag{10-21}$$

$$y = x/x_m; \text{ for } x \le x_m \tag{10-22}$$

$$y = (1 - x)/(1 - x_m); \text{ for } x > x_m \tag{10-23}$$

$$e = \sqrt{\frac{0.4x_m}{c}}\left[0.95\left(1 + \frac{x}{c}\right)(1 - y) + 0.05\right] \tag{10-24}$$

The blade maximum thickness, t_{bm}, and its location x_m/c, are specified, as are $t_{b6'}$ and t_{b7}. Figure 10-3 shows a typical return channel vane designed with these equations.

10.3 A Computerized Interactive Design System

The detailed design of the return system is implemented in a computerized design system, which carries out the many complex calculations required to construct the contours and the return channel vane. With reference to Fig. 10-1, the known design parameters supplied for the return system design are r_m, b_5 and the diffuser exit flow angle (from the vaned or vaneless diffuser design). To complete the crossover specification, b_6, R_C and $r_5 = r_6$ must be supplied. For added flexibility, it is convenient to provide for alternate methods of specifying these data. Convenient choices for this purpose are to have the designer specify

1. Either the passage width, b_6, or the flow angle at the crossover exit.
2. Either the inlet radius, r_5, or $F_A = (A_{av} - A_5)/(A_6 - A_5)$.
3. Either hub radius of curvature, R_C, or the average b/R over the bend.

A_{av} is the passage area midway through the bend and b/R is the ratio of passage width to the mean streamline radius of curvature. The alternate choices provide direct control over vane incidence, the passage area distribution and the distribution of b/R. Their availability greatly accelerates the design process. When F_A is specified, A_{av} is approximated by

$$A_{av} = \pi[r_m^2 - (R_C + r_5)^2] \tag{10-25}$$

When the average b/R is specified, an iterative process must be used where the resident guess for R_C is corrected directly by multiplying it by the calculated average b/R over the desired average b/R until convergence is achieved. With reasonable care in setting up the numerical scheme, convergence is rather rapid and reliable.

With reference to Fig. 10-1, the design specifications for the return channel gas path contours are $r_{6'}$, r_{8h}, r_{8s}, b_8/R_O, d, and the area ratio across the 90° exit turn. The parameter, d, provides control over the minimum metal thickness for the diffuser-return channel wall. The minimum wall thickness sets the minimum axial length achievable with the other specifications. Design specifications for the vane are: the number of vanes, the vane trailing edge angle (θ_7 of Fig. 10-1), $t_{b6'}$, t_{b7}, t_{bm}, $K_{6'}$, K_7, x_m/c, $\beta_{6'}$ and β_7. The entire gas path displayed in Fig. 10-1 is defined by only 12 parameters, which provide direct control over the important passage area and curvature distributions. Specification of the vane requires 10 more parameters, which control both the vane passage area distribution and the blade loading style. Since the vane and all contours are defined analytically, the problem is well suited to a computerized interactive design system. This computerized design system should allow the designer to interactively modify the design parameters, while supplying the following support functions for guidance:

1. Graphical displays of the gas path and vane geometry (Figs. 10-1 and 10-3).
2. Graphical and tabular displays of geometry as well as key area and curvature distributions for evaluation.
3. Display results from the aerodynamic performance analysis (Chapter 5).
4. Display results from the linearized blade-to-blade flow analysis (Chapter 12).

Like the vaned diffuser design system of Chapter 9, the aerodynamic performance analysis and linearized blade-to-blade flow analysis are easily incorporated into this design system to allow direct and immediate access to the evaluation they provide. Similarly, the preliminary design system presented in Chapter 6 should generate the first input file for this design system with its preliminary return system design. Provision for this design system to modify the input file for the overall stage performance analysis should be included to permit a stage performance analysis with the new geometry. Also, provision to create an input file for one of the more exact blade-to-blade flow analyses (Chapters 13 and 14) should be incorporated to allow a final, more precise blade loading evaluation. By including these features, this approach can reduce the design time—to a few hours at most—for even the most complicated return system.

10.4 Return System Design Recommendations

Crossover bend and exit turn design should control the ratio of passage width-to-mean streamline radius of curvature, b/R. Local values should be less than 1, and an average over the bend of less than 0.8 is recommended. Based on experimental results from many return channels, losses appear to increase significantly when the passage equivalent divergence analyses, $2\theta_C$, exceed 9°, where an analogy with simple exhaust diffusers is used (Reneau et al., 1967)

$$2\theta_C = 2\tan^{-1}[b_5(A/A_5 - 1)/(m - m_5)/2] \qquad (10\text{-}26)$$

Simple exhaust diffusers show increased losses for $2\theta_C$ greater than about $11°$. The lower threshold value for swirling flow in a $180°$ bend is not unexpected.

Blade loading for the return channel vanes should feature higher loading near the leading edge than at the trailing edge to minimize flow deviation at the discharge. The moderate inlet Mach numbers normally encountered make this front-loaded vane style a reasonable choice. The vane loading parameters, $K_{6'}$ and K_7, control the loading style in the same way as for vaned diffusers (see discussion in Chapter 9). Values of $K_{6'} = 1.6$ and $K_7 = 0.4$ are typical choices. The vane maximum thickness, and its location along the chord line, can be used to control the vane passage area distribution. They should be used to minimize the local gradients of the mean velocity through the vane passage. Assuming the crossover bend is reasonably well designed in terms of b/R and $2\theta_C$, the return channel vane should be designed for an incidence angle, $i = \beta_{6'} - \alpha_{6'}$, of $2°$ to $4°$. Poor crossover bend designs generally require higher design incidence angles. Periodic evaluations of performance during the design process should guide the designer in selecting this and other design parameters.

EXERCISES

10.1 By substituting y for η and Y for R in Eqs. (9-22) through (9-29), derive Eqs. (10-12) through (10-19). Note that for the radial inward flow in return channels, Eq. (9-23) requires a change in sign for the usual definition of blade angle, i.e.,

$$\cot\beta = -\frac{y\partial\theta}{\partial y}$$

Write expressions for the general blade loading terms analogous to Eqs. (9-24) and (9-25).

10.2 The preliminary design system of Chapter 6 employs a simplified return channel passage using a radial surface as the shroud straight section, as shown in Fig. 6-8. What advantage is derived from the more general formulation used in this chapter as shown in Fig. 10-1?

10.3 When selecting a return channel inlet blade angle or flow angle, how will your choice be likely to differ between a high-flow coefficient stage and a low-flow coefficient stage?

10.4 Section 10.3 suggests three sets of alternative parameters for specifying the crossover bend's basic geometry. For each alternative in each set, qualitatively specify the flow coefficient range (low or high) at which it will be most effective in helping you minimize the total pressure loss level in the return channel, and explain why.

Chapter 11

VOLUTE DESIGN

The aerodynamic design of the volute or scroll remains a rather qualitative and somewhat subjective process. The underlying technology is extremely limited as is the published literature on this subject. Weber and Koronowski (1986), Brown and Bradshaw (1949), Whitfield and Roberts (1983), Japikse (1982), Ferguson (1963), Whitfield and Baines (1990), Cumpsty (1989) are representative of the available information. The situation is most clearly demonstrated by the rather minor differences in the volute technology offered between the earliest and most recent of these references. The best available guidance is the basic definition of loss mechanisms, such as those provided in Weber and Koronowski (1986; Chapter 5), Japikse (1982) Cumpsty (1989). The design process primarily consists of a few basic fluid dynamics principles to control and minimize those loss sources. The volute design process is often subject to many constraints and compromises. Constraints, such as maximum casing size, location of piping to which the volute connects, avoiding necessary auxiliary equipment, etc., often limit the designer's options.

NOMENCLATURE

A = passage area
b = hub-to-shroud passage width
C = absolute velocity
H = characteristic height of passage
L = exit cone length
$\dot{m}$ = mass flow
p = pressure
r = radius
t = wall thickness
W = characteristic width of passage
α = flow angle with respect to tangent
θ = polar angle
$2\theta_C$ = diffuser divergence angle
ρ = gas density

Subscripts

c = mean parameter of the passage cross-section
h = parameter on the hub contour
m = meridional component
s = shroud contour parameter
t = total thermodynamic condition
U = tangential component
5 = volute inlet condition
6 = volute full-collection plane condition
7 = exit cone discharge condition

11.1 Geometrical Construction

Figure 11-1 illustrates the basic configuration of a volute or scroll. It consists of a passage located circumferentially around the diffuser exit, with a progressively increasing cross-sectional area to admit the increasing mass flow as the flow collection process proceeds. The location where all of the mass flow from the diffuser has been collected by the volute is referred to as the full-collection plane. A volute "tongue" divides the collected and uncollected flow at this location. From there, the flow passes through an "exit cone" to the piping flange served by the compressor.

The cross-sectional shape of the passage is usually chosen for convenience rather than from fluid dynamics principles. Circular or elliptical cross-sections are commonly used for cast volutes, such as illustrated in Fig. 11-2. Square or rectangular cross-sectional shapes are often used for fabricated volutes, as shown

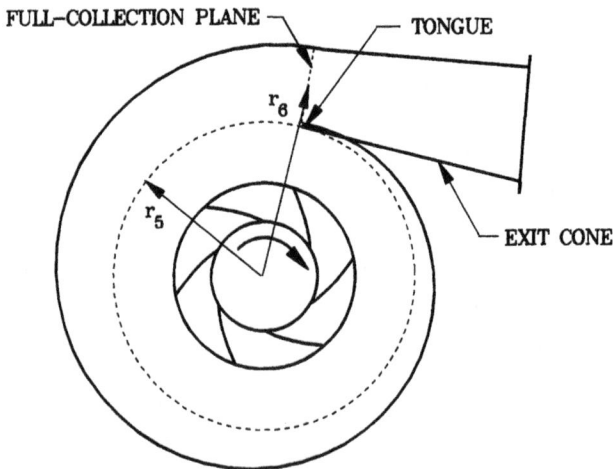

FIGURE 11-1. Volute Nomenclature

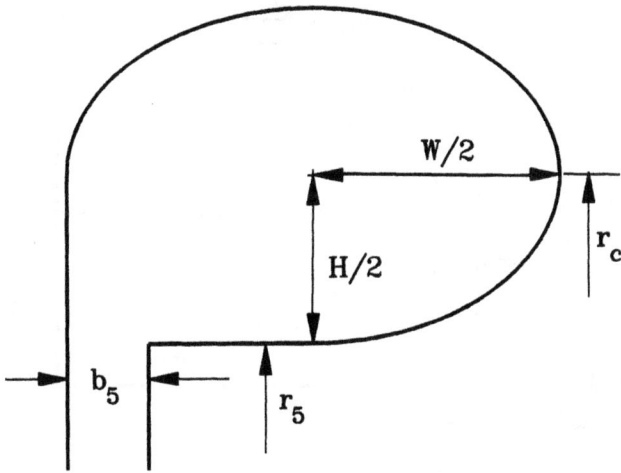

FIGURE 11-2. An Elliptical External Volute

in Fig. 11-3, to simplify manufacturing. Many variations in cross-sectional shape are used, but these two styles will serve to illustrate the design process.

Volutes are also often classified as external, internal and intermediate. Figures 11-2 and 11-3 illustrate external volutes, where the entire passage area is outside the diffuser exit radius. By contrast, an internal volute will locate as much of the

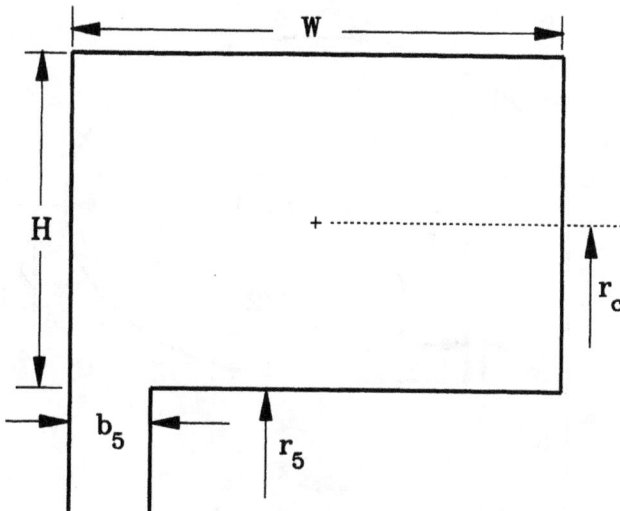

FIGURE 11-3. A Rectangular External Volute

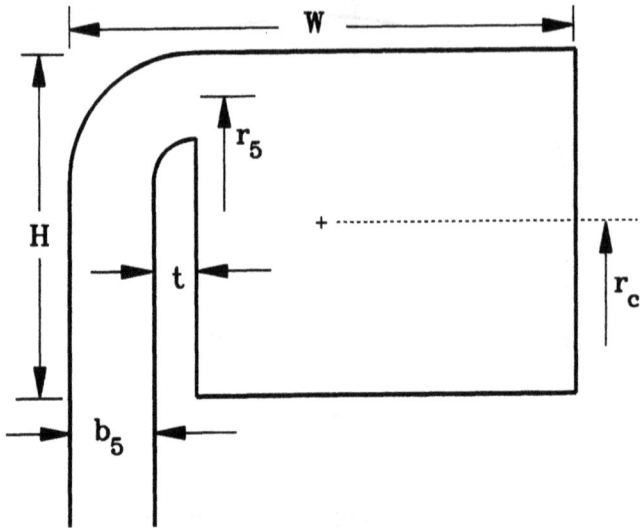

FIGURE 11-4. A Rectangular Internal Volute

passage area as possible inside the diffuser exit radius, to minimize the compressor casing diameter required. Here, a bend is required at the diffuser exit as illustrated in Fig. 11-4. Figure 11-5 shows an intermediate volute. This illustration is a special type of intermediate volute, which can be referred to as semiexternal, since the mean radius, r_c, of the cross-section has the same radius as the diffuser

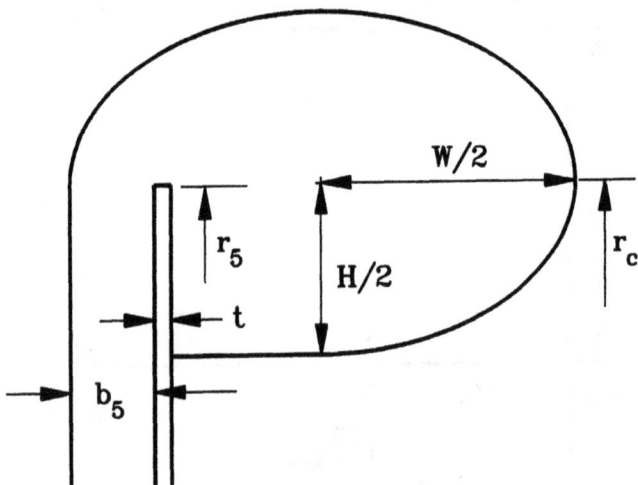

FIGURE 11-5. An Elliptical semiexternal Volute

exit. At first glance, this volute may appear to differ from that of Fig. 11-1 only by the fact that the diffuser hub wall is extended into the volute. But note that r_c for the external volute will continually increase as the area increases.

11.2 Fundamental Design Concepts

The main objective of a design procedure is to define the circumferential variation of r_c and the cross-sectional area. There are two basic approaches used to define the area variation: The simpler method is referred to as the simple area schedule (SAS), which is less commonly used today. In this approach, the desired flow velocity at Station 6 (Fig. 11-1) is used to set the area at that full-collection plane. Then, the volute area is allowed to vary linearly with the polar angle, θ. A more precise method is to establish the area distribution from the principles of conservation of mass and angular momentum (CAM designs), i.e.,

$$(rC)_c = (rC_U)_5 \tag{11-1}$$

$$(\rho C)_c A = \theta(\rho r b C_m)_5 \tag{11-2}$$

which defines the area distribution from the mean passage radius distribution, $r_c(\theta)$. For generality, the gas density is included in Eq. (11-2), but normally it can be regarded as a constant due to the very low Mach number levels typical of volutes. Hence, the local passage area at any circumferential location can normally be approximated by

$$A = \theta r_c b_5 C_{m5}/C_{U5} = \theta r_c b_5 \tan \alpha_5 \tag{11-3}$$

A variant on the CAM design concept is to correct the area for wall friction effects. But due to the very low velocity levels involved, friction losses in volutes are generally quite small. Hence, there seems to be little purpose to that degree of refinement. Departures from satisfying conservation of angular momentum lead to increased volute losses, as seen from Eq. (5-107) in Chapter 5. Consequently, the CAM design procedure must be considered the method of choice in most cases.

The choice for the $r_c(\theta)$ distribution has sometimes been treated as arbitrary (e.g., Cumpsty, 1989). To some degree, it is, but it may be subject to specific limits if the cross-sectional area shape is defined in some systematic fashion. For example, consider the external volute in Fig. 11-3. Here, Eq. (11-3) requires

$$A = HW = 2(r_c - r_5)W = \theta r_c b_5 \tan \alpha_5 \tag{11-4}$$

which defines r_c as well as A if W, its distribution, the aspect ratio (W/H) or some other constraint on W is specified. Hence, the external volute defines an upper limit for $r_c(\theta)$ in this case. A similar result is obtained for the fully internal volute in Fig. 11-4, but here the minimum values will be obtained for the specific cross-sectional shape constraint imposed and the area must be corrected for the portion blocked by the internal diffuser and wall thickness, t. Any distribution between these two extremes can, of course, be used with the desired cross-sectional shape. Similar limitations can be defined for the elliptical volute or any other systematic cross-sectional area definition the designer may choose to employ.

The somewhat ideal one-dimensional CAM design process represented in Eq. (11-3) neglects a number of secondary factors. Flow in the volute will certainly not be uniform over the cross-sectional area. The distorted profiles, viscous effects and secondary flows will tend to produce an actual mean velocity somewhat higher than that calculated from the ideal model. Thus, it is common practice to generalize the basic design equation to the following form:

$$A = \theta S P r_c b_5 \tan \alpha_5 \qquad (11\text{-}5)$$

where SP is called the sizing parameter—it expresses the ratio of the actual volute area to the value satisfying the ideal conservation of angular momentum principle expressed by Eq. (11-3). Since an undersized volute will usually impose a substantial loss penalty, it is common practice to use $SP > 1$ to provide a safety margin. Sometimes, quite large values of SP may be used for vaneless diffuser stages to accommodate off-design performance requirements. But for vaned diffuser stages or where design point performance is the primary interest, SP is usually selected in the 1.0–1.2 range.

11.3   Aerodynamic Design Considerations

It must be recognized that volute design is strongly dependent upon the specific stage type and application involved. The volute and exit cone can only collect the flow from the diffuser and deliver it to the exit flange as efficiently as possible. To a very large degree, the diffuser exit and exit flange flow conditions are imposed constraints, which largely dictate the efficiency achievable by the volute.

It is reasonably clear that the volute cannot recover the meridional velocity head at the diffuser exit. As illustrated in Fig. 11-6, this component of velocity becomes a swirl component in the volute, which will eventually be dissipated, either in the volute or somewhere downstream. The volute has no mechanism to recover any significant portion of this component. Therefore, the first—and usually the largest—volute loss is actually predetermined by the diffuser design, i.e., Eq. (5-105) in Chapter 5. Hence the designer's only effective control over this loss is to minimize C_{m5} when the impeller and diffuser are designed.

The diffuser type is also an important consideration. The flow angle at a vaned diffuser exit will be essentially constant for all flows, meaning the volute can be sized to match all operating conditions. By contrast, vaneless diffuser exit flow angles will vary widely over the operating range, so the volute can be sized to match only a single operating point. Consequently, off-design operating requirements will strongly influence the volute design process. Designers need to be aware of some of the misconceptions reported in the literature, to avoid poor design decisions. Cumpsty (1989) suggests that volute performance will be worse

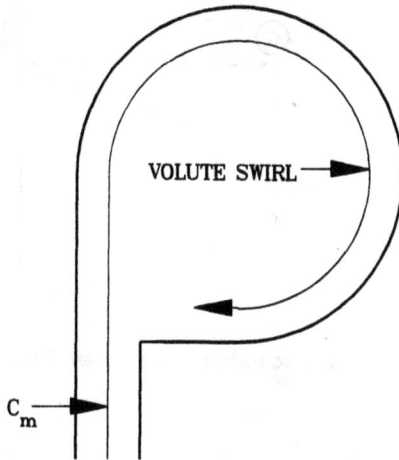

FIGURE 11-6. Secondary Flow in the Volute

for vaned diffusers than for vaneless diffusers. The premise is that since the vaned diffuser turns the flow more radial, the unrecoverable meridional velocity head will be larger. Actually, turning the flow reduces the tangential velocity, which has no direct impact on the meridional velocity head. In fact, a vaned diffuser can normally employ a wider passage than a vaneless diffuser, so the meridional velocity head loss will usually be less for a vaned diffuser. Indeed, that is precisely why vaned diffusers are often used in single-stage, volute discharge designs.

Usually, the discharge flange area, A_7, to be matched by the exit cone is an external constraint. This has definite implications for the volute design. The exit cone can be viewed as an exhaust diffuser. There are definite limits to the amount of diffusion that can be accomplished over a given exit cone length. The highly distorted flow involved as well as the fact that the natural path for the flow leaving the full-collection plane is a log spiral (to conserve angular momentum) suggests the exit cone will be less effective than an equivalent exhaust diffuser. The designer will need to consider this in designing the volute. In general, the full-collection plane area, A_6, should be sized to provide a reasonable diffusion load for the exit cone. As a general rule, classical exhaust diffusers achieve the best static pressure recovery for a diffuser divergence angle ($2\theta_C$ in Fig. 11-7) of about $10°$ to $11°$ (Reneau et al., 1967). Values larger than this result in higher losses, which preclude additional diffusion. Knowing the basic length available for the exit cone (L in Fig. 11-7) and A_7, this observation permits the designer to estimate the optimum value of A_6. Hence, for larger values of A_7, the designer will prefer larger values of r_6 (and therefore, larger A_6). Here again, the basic performance levels achievable may be largely predetermined by the design of the upstream components. Once the diffuser exit flow and geometry is established, the designer's flexibility to control the diffusion level in the exit cone may be quite limited. To maintain a low loss in the volute, the CAM style design is really required, which greatly limits the available choices for C_6 through Eq. (11-5).

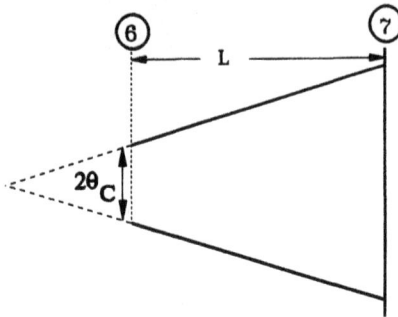

FIGURE 11-7. Basic Exhaust Diffuser Parameters

Consequently, the design of volute style stages should always consider the final disposition of the flow at Station 7, during the stage design process. By the time the volute is designed, very little flexibility is left due to the constraints imposed.

The fully internal volute should be used only when absolutely necessary. The 90° bend required at the diffuser exit is an additional loss source (which can be evaluated using the vaneless passage performance analysis of Chapter 5). In addition, smaller values of A_6 will result from a CAM design, probably leading to excessive diffusion in the exit cone. Indeed, the designer should evaluate the trade-off between accepting the extra bend loss or using a shorter diffuser (when possible) to avoid it. Without evaluation, it is by no means certain which approach will yield the best performance. A significant consideration in volute design is the circumferential pressure distortion imposed on the stage by the volute. This distortion can be substantial, with its influence extending all the way back to the compressor inlet. Such distortions are inevitable when the volute operates off-design with vaneless diffusers. For vaned diffusers, it should be possible to minimize this by proper volute sizing. Indeed, this consideration really favors the semiexternal volute style of Fig. 11-5, since C_c, and therefore pressure, will be essentially uniform around the volute. Where physical constraints permit, it is good practice to select the best choice for r_6 and A_6, and set $r_5 = r_6$. For vaned diffuser stages, this simply involves adding a vaneless annular passage after the vaned diffuser. This provides all of the advantages of the fully external volute, plus minimizing of the circumferential pressure variation.

EXERCISES

11.1 Consider an external volute that has its cross-sections defined by circular arcs—with arc radius, R—in three quadrants and a square in the fourth quadrant with side lengths, R (i.e., Fig. 11-2 with $H = W = 2R$). Derive an equation that can be solved for R in terms of the angular position ($0 \leq \theta \leq 2\pi$) and the known geometry and flow conditions at the volute inlet, b_5, r_5, α_5. Assume the gas density, ρ, is constant and $SP = 1$. Note that this completely defines the cross-sectional area at any angular location.

11.2   Repeat Exercise 11.1 for a semiexternal volute of the same cross-section type, but with the diffuser passage extending into the volute passage with wall thickness, t (i.e., Fig. 11.5 with $H = W = 2R$).

11.3   Consider an external volute with rectangular cross-sections, with the axial width, W, constant (i.e., Fig. 11-3). Derive an equation that can be solved for H in terms of the angular position ($0 \leq \theta \leq 2\pi$) and the known geometry and flow conditions at the volute inlet, b_5, r_5, α_5. Assume the gas density, ρ, is constant and $SP = 1$. Note that this completely defines the cross-sectional area at any angular location.

11.4   Repeat Exercise 11.3 for a semiexternal volute with the same type of cross-section. As with Exercise 11.2, note that the diffuser passage, with wall thickness t, blocks part of the volute cross-sectional area.

11.5   Consider the external and semiexternal volutes of Figs. 11-2 and 11-5, with each having the same cross-sectional shape as well as an identical area and radius at the full-collection plane (Station 6). Which volute will achieve the higher static pressure recovery between Stations 5 and 6? Is this "internal" static pressure recovery an important consideration in selecting the volute to be used?

11.6   You are designing a stage with a vaned diffuser having a specified maximum (casing) diameter. The vaned diffuser discharge diameter is sufficiently less than the maximum diameter that you can choose either of the volutes in Exercise 11.5. Assume that for both volutes the gas density, ρ, is constant and $SP = 1$. Based on the performance analysis models of Chapter 5, Section 5.7, which volute is likely to achieve the lower overall loss? The higher static pressure at station 6?

Chapter 12

QUASI-THREE-DIMENSIONAL
FLOW ANALYSIS

This chapter, and the following three chapters, address techniques for internal flow field analysis that are useful in supporting the aerodynamic design of centrifugal compressors. Here, a method for quasi-three-dimensional inviscid flow analysis in vaned and vaneless annular passages will be reviewed. As discussed in Chapter 3, Section 3.3., quasi-three-dimensional flow analysis employs two-dimensional flow analysis, with interaction between them, to approximate the three-dimensional flow. The concept is generally credited to Wu (1952) and has been used extensively for centrifugal compressor internal flow analysis (e.g., Senoo and Nakase, 1972; Novak and Hearsey, 1976; Colwill, 1979; Northern Research, 1981). The present analysis technique is basically unpublished, although it is mentioned in Aungier (1988b), where its linearized blade-to-blade analysis is described in a form limited to stationary vanes.

The role of quasi-three-dimensional flow analysis in the aerodynamic design process has been described in Chapter 7, Section 7.9, relative to the detailed design of impellers. While that is its most common application, it can be employed in stationary vaned and vaneless components, where a more precise definition of the internal flow field is needed. The linearized blade-to-blade flow analysis used has also been recommended for inclusion in detailed aerodynamic design systems for vaned diffusers (Chapter 9) and return channels (Chapter 10) as a means of rapid blade loading evaluation. Indeed, computational speed is the key feature that keeps the quasi-three-dimensional Euler codes firmly entrenched in virtually all design systems, even though more exact viscous computational fluid dynamics (CFD) codes are now available to many, if not most, designers. In the detailed component design process, the designer may easily process dozens of Euler code solutions in a few hours while accomplishing the gas path and vane design process. The procedure described in this chapter can generate a solution in a matter of seconds, even on a personal computer of modest capability. Currently, this is the only type of analysis method that can provide the type of immediate internal flow analysis a designer needs for efficient, detailed component design activity.

NOMENCLATURE

A = an area inside the blade passage
a = sound speed and meridional portion of the stream function

B = fractional area blockage
b = stream sheet width
C = absolute velocity
C_U = absolute tangential velocity
$\vec{e}_n$ = unit vector normal to a stream surface
$\vec{e}_\theta$ = unit vector in the θ direction
h = enthalpy
M = Mach number
m = meridional coordinate along stream surfaces
$\dot{m}$ = mass flow
n = coordinate normal to stream surfaces
p = pressure
R = rothalpy
r = radius
s = entropy
T = temperature
U = blade speed, ωr
W = relative velocity
W_m = meridional velocity
W_n = relative velocity normal to constant ξ lines
W_U = relative tangential velocity
W_ξ = relative velocity in the ξ direction
y = coordinate along a quasi-normal
Z = total number of blades, $Z_{FB} + Z_{SB}$
Z_{FB} = number of full-length blades
Z_{SB} = number of splitter blades
z = axial coordinate
α = flow angle with respect to tangent
α_C = streamline slope angle with axis
β = blade angle with respect to tangent
ϵ = angle between quasi-normal and true normal
θ = polar angle coordinate
κ_m = streamline curvature
ξ = parameter defined in Eq. (12-42)
ρ = gas density
σ = slip factor, $C_{U2}/C_{U2\,ideal}$
ϕ = stage flow coefficient, $\dot{m}/(\pi \rho_{t0} r_2^2 U_2)$
Φ = stream function
ω = rotation speed

Subscripts

B = a blade parameter
h = parameter on the hub contour
p = blade pressure surface parameter
s = shroud contour or blade suction surface parameter

t = total thermodynamic condition
0 = parameter on blade surface θ_0
1 = parameter on blade surface θ_1

Superscripts

* = sonic condition
$'$ = value relative to the blade's frame of reference and first derivative with respect to m
$''$ = second derivative with respect to m
$^-$ = value at $\eta = \frac{1}{2}$

12.1   Fluid Dynamics Models

The present quasi-three-dimensional flow analysis employs two-dimensional flow analyses on the meridional (hub-to-shroud) plane and on several blade-to-blade stream surfaces. The analyses are made to interact until they are consistent with each other to yield the approximate three-dimensional flow description. Figure 12-1 shows a view of the hub-to-shroud plane while illustrating a typical blade-to-blade surface. In the interest of generality, the analysis will be developed in a rotating coordinate system, since it will be completely valid for stationary components if the rotation speed, ω, is set to zero. The principal simplifying assumption to be used is that the blade-to-blade stream surfaces are axisymmetric, i.e.,

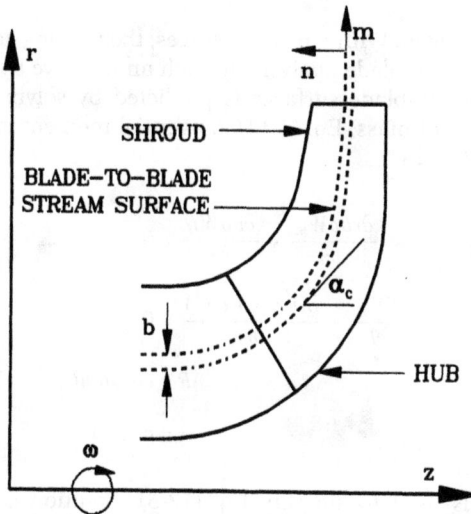

FIGURE 12-1. Hub-to-Shroud Plane Geometry

the effect of stream sheet "twisting" will be neglected. This assumption is not unreasonable, and it greatly simplifies the analysis to be performed.

The hub-to-shroud analysis will solve conservation of mass, Eq. (3-14), normal momentum, Eq. (3-22), and energy, Eq. (3-18)—all simplified to their time-steady, axisymmetric form.

$$\frac{\partial r \rho W_m}{\partial m} + \kappa_n r \rho W_m = 0 \tag{12-1}$$

$$\kappa_m W_m^2 + \frac{W_U}{r} \frac{\partial (r W_U + \omega r^2)}{\partial n} + W_m \frac{\partial W_m}{\partial n} = \frac{\partial R}{\partial n} - T \frac{\partial s}{\partial n} \tag{12-2}$$

$$\frac{\partial R}{\partial m} = 0 \tag{12-3}$$

This model is strictly correct only for vaneless passages. Within a blade passage, this model predicts the hub-to-shroud flow, averaged over the blade passage width. The solution will treat flows with entropy and enthalpy gradients across the passage, but these parameters are assumed constant with respect to the tangential direction. Solution of these equations requires that W_U or β' be specified by some external means. For portions of the passage where no vanes are present, integration of the time-steady, axisymmetric form of Eq. (3-21) with rothalpy, R, and entropy, s, independent of θ, yields the well-known requirement that angular momentum must be conserved along stream surfaces

$$\frac{\partial (r W_U + \omega r^2)}{\partial m} = \frac{\partial r C_U}{\partial m} = 0 \tag{12-4}$$

to supply the data needed. Within blade passages, the missing data will have to be supplied by the blade-to-blade analysis through an iterative solution procedure.

The flow in blade-to-blade surfaces is predicted by solving the time-steady form of conservation of mass, Eq. (3-14), tangential momentum, Eq. (3-21), and the constraint on R and s

$$\frac{\partial r b \rho W_m}{\partial m} + \frac{\partial b \rho W_U}{\partial \theta} = 0 \tag{12-5}$$

$$\frac{\partial W_m}{\partial \theta} - \frac{\partial (r W_U + \omega r^2)}{\partial m} = 0 \tag{12-6}$$

$$R = constant \tag{12-7}$$

$$s = constant \tag{12-8}$$

where Eq. (3-19) is used to simplify Eq. (12-5). Solution of these equations requires specification of the stream surfaces (i.e., r and b). This information must be supplied by the hub-to-shroud flow analysis.

12.2 Gas Path Geometry

The gas path geometry will be specified in the form already described in Chapter 7. As illustrated in Fig. 12-2, a series of linear quasi-normals through the passage are used, defined by the (z, r) coordinates of the end points on the hub and shroud. Note that the blade leading and trailing edges (and splitter blade leading edge, if present) are defined by specific quasi-normals. The blade types and blade construction procedures of Chapter 7 can be used directly for the present analysis. For new designs, the logical approach is to have the gas path design system supply all of the geometrical data in an input file for the present analysis and be capable of updating that information as the design progresses (leaving fluid dynamics input data unchanged). In this way, the quasi-three-dimensional flow analysis can be used to evaluate design changes, with no input data required from the designer after the first run. For analysis of existing designs, all geometry will have to be specified by the designer, requiring significant data preparation activity. This is one reason why Chapter 7, Section 7.8, recommends creating a parallel interactive geometry analysis system while creating the basic gas path design system. A good interactive gas path geometry system can greatly reduce the time required and the risk of error when establishing geometry input for an existing design.

The quasi-normal coordinate system is convenient for conducting the hub-to-shroud flow analysis. To this point, the governing equations are expressed in natural coordinates (m, n, θ), two of which must be determined as part of the solution. Use of the coordinate m is beneficial, since several important parameters are conserved on stream surfaces. This makes the well-known streamline curvature technique (Novak, 1967) an attractive numerical method; it will be used here. But the need to generate normal surfaces as part of the solution is an unnecessary

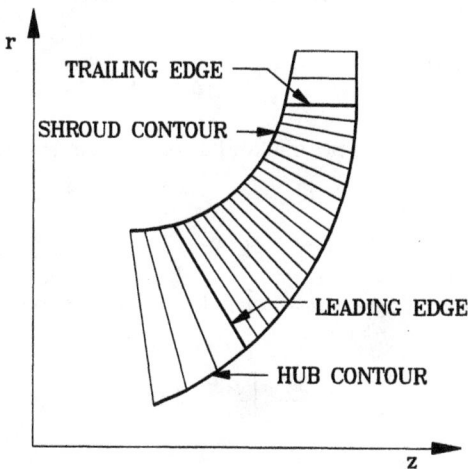

FIGURE 12-2. Quasi-Normal Coordinates

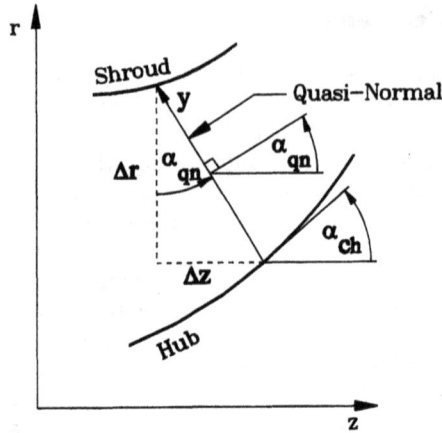

FIGURE 12-3. Quasi-Normal Geometry

complication that can be avoided by using the fixed quasi-normal coordinate, y, in place of n. Figure 12-3 illustrates the primary geometrical data required. On any stream surface, the local stream sheet angle with respect to the axial direction and the streamline curvature are given by

$$\sin \alpha_C = \frac{\partial r}{\partial m} \qquad (12\text{-}9)$$

$$\kappa_m = -\frac{\partial \alpha_C}{\partial m} \qquad (12\text{-}10)$$

and from Fig. 12-3, the quasi-normal angle is given by

$$\tan \alpha_{qn} = \frac{\Delta z}{\Delta r} \qquad (12\text{-}11)$$

The angle between a quasi-normal and a true stream surface normal is given by

$$\epsilon = \alpha_C - \alpha_{qn} \qquad (12\text{-}12)$$

If y is the distance along a quasi-normal, it is easily shown that

$$\frac{\partial}{\partial n} = \frac{1}{\cos \epsilon} \left[\frac{\partial}{\partial y} - \sin \epsilon \, \frac{\partial}{\partial m} \right] \qquad (12\text{-}13)$$

which can be used to eliminate derivatives with respect to n in the governing equations.

12.3   The Hub-to-Shroud Flow Governing Equations

Conservation of mass for the hub-to-shroud flow is conveniently solved in an integral form, i.e.,

$$\dot{m} = 2\pi \int_0^{y_s} K_B r \rho W_m \cos \epsilon \, dy \tag{12-14}$$

where K_B = a blockage factor expressing the fraction of the local stream surface area that is open for flow (with the blade metal blockage subtracted), i.e.,

$$K_B = 1 - Z t_b / (2\pi r \sin \beta) \tag{12-15}$$

where Z = number of blades; t_b = blade thickness; and β = blade angle with respect to tangent. The normal-momentum equation is obtained by combining Eqs. (12-2) and (12-13). Two different forms of this equation are required. When no blades are present at the quasi-normal being considered, Eq. (12-4) provides the angular momentum, rC_U, on each stream surface from the values upstream (values must be specified for the first quasi-normal). Then, the normal-momentum equation is solved in the following form:

$$\frac{\partial W_m}{\partial y} + \left[\kappa_m \cos \epsilon - \frac{\sin \epsilon}{W_m} \frac{\partial W_m}{\partial m} \right] W_m + \frac{W_U}{r W_m} \frac{\partial (rC_U)}{\partial y} = \frac{1}{W_m} \left[\frac{\partial R}{\partial y} - T \frac{\partial s}{\partial y} \right] \tag{12-16}$$

When blades are present, the angular momentum is no longer conserved along stream surfaces, but it must be supplied by the blade-to-blade flow analysis. It is convenient to consider the relative flow angle, α', as the known quantity

$$\cot \alpha' = W_U / W_m \tag{12-17}$$

from which the appropriate form of the normal-momentum equation can be shown to be given by

$$\frac{1}{\sin^2 \alpha'} \frac{\partial W_m}{\partial y} + \left[\kappa_m \cos \epsilon - \frac{\sin \epsilon}{W_m \sin^2 \alpha'} \frac{\partial W_m}{\partial m} + \frac{\cot \alpha'}{r} \left(\frac{\partial r \cot \alpha'}{\partial y} - \sin \epsilon \frac{\partial r \cot \alpha'}{\partial m} \right) \right]$$
$$\cdot W_m + 2\omega \cot \alpha' (\cos \alpha_{qn} - \sin \alpha_C \sin \epsilon) = \frac{1}{W_m} \left[\frac{\partial R}{\partial y} - T \frac{\partial s}{\partial y} \right] \tag{12-18}$$

12.4   Conservation of Mass and Momentum

The conservation of mass and normal-momentum equations can be solved numerically for a specified stream surface geometry and flow angle distribution (in the blade passage). For the purpose of this mass and momentum equation solution, consider those data to be known. From Eqs. (12-16) and (12-18) it is seen that the normal-momentum equation on any quasi-normal can be written in the form

$$\frac{\partial W_m}{\partial y} = f_1(y)W_m + f_2(y) + f_3(y)/W_m \tag{12-19}$$

where f_1, f_2 and f_3 are treated as known functions of y. For quasi-normals within the blade passage,

$$f_1(y) = \sin^2 \alpha' \left[-\kappa_m \cos \epsilon - \frac{\cot \alpha'}{r} \left(\frac{\partial r \cot \alpha'}{\partial y} - \sin \epsilon \frac{\partial r \cot \alpha'}{\partial m} \right) \right] + \frac{\sin \epsilon}{W_m} \frac{\partial W_m}{\partial m}$$

$$\tag{12-20}$$

$$f_2(y) = -2\omega \cos \alpha' \sin \alpha' [\cos \alpha_{qn} - \sin \alpha_C \sin \epsilon] \tag{12-21}$$

$$f_3(y) = \sin^2 \alpha' \left[\frac{\partial R}{\partial y} - T \frac{\partial s}{\partial y} \right] \tag{12-22}$$

while for quasi-normals outside the blade passage

$$f_1(y) = -\kappa_m \cos \epsilon + \frac{\sin \epsilon}{W_m} \frac{\partial W_m}{\partial m} \tag{12-23}$$

$$f_2(y) = 0 \tag{12-24}$$

$$f_3(y) = \frac{\partial R}{\partial y} - T \frac{\partial s}{\partial y} - \frac{W_U}{r} \frac{\partial r C_U}{\partial y} \tag{12-25}$$

Equation (12-25) can also be expressed in the stationary frame of reference if preferred, simply by substituting h_t for R and C_U for W_U. One other case should be considered: At the passage inlet, it may be convenient to specify the inlet flow angle distribution rather than the C_U distribution. Equations (12-20) through (12-22) can be used for this purpose if they are expressed in the stationary frame of reference, while requiring Eq. (12-4) be satisfied. This yields

$$f_1(y) = \sin^2 \alpha \left[-\kappa_m \cos \epsilon - \frac{\cot \alpha}{r} \frac{\partial r \cot \alpha}{\partial y} + \frac{\sin \epsilon}{W_m} \frac{\partial W_m}{\partial m} \right] \tag{12-26}$$

$$f_2(y) = 0 \tag{12-27}$$

$$f_3(y) = \sin^2 \alpha \left[\frac{\partial h_t}{\partial y} - T \frac{\partial s}{\partial y} \right] \tag{12-28}$$

where $\cot \alpha = C_U/W_m$ in Eqs. (12-26) and (12-28). Since the solution is accomplished for specified stream surfaces, these functions can be evaluated for all grid points in the flow field before integrating the momentum equation. The only questionable portion of the solution is the meridional gradient of W_m in $f_1(y)$, which is the one parameter not properly treated from known data since it depends on the solution. Conventional practice has been to use values of W_m from the previous integration to compute this derivative. That approach occasionally contributes to instability in the iteration process, since W_m may change substantially between iterations, particularly during the earlier iterations. A better approach is to observe that the mass flow within a stream tube between successive stream surfaces is constant, i.e.,

$$\Delta \dot{m} = \rho W_m \Delta A \tag{12-29}$$

where ΔA = stream tube area. Hence, the troublesome meridional gradient term can be replaced by

$$\frac{1}{w_m} \frac{\partial W_m}{\partial m} = -\frac{1}{\rho \Delta A} \frac{\partial \rho \Delta A}{\partial m} \tag{12-30}$$

which relates the desired gradient to stream tube geometry (which is constant during the integration of the mass and momentum equations) and gas density. Of course, density still depends on the mass and momentum equations solution, but usually it will not change greatly between major iterations; thus, use of density data from the previous mass and momentum solution iteration to estimate this gradient has been quite beneficial for numerical stability. Also, note that Eq. (12-19) is singular for $W_m = 0$. Equation (12-29) can be used to eliminate that singularity by defining

$$f_4(y) = f_2(y) + f_3(y) \frac{\rho \Delta A}{\Delta \dot{m}} \tag{12-31}$$

and the momentum equation becomes

$$\frac{\partial W_m}{\partial y} = f_1(y) W_m + f_4(y) \tag{12-32}$$

The solution of this linear differential equation is well known, i.e.,

$$W_m = W_{mh} F(y) + F(y) \int_0^y \frac{f_4(y)}{F(y)} \, dy \tag{12-33}$$

where W_{mh} = hub meridional velocity (at $y = 0$) and

$$F(y) = \exp\left[\int_0^y f_1(y)dy\right] \tag{12-34}$$

W_{mh} must be determined from conservation of mass, Eq. (12-14). The procedure used is to integrate the momentum equation for the W_m distribution, integrate Eq. (12-14) for the mass flow and iteratively update W_{mh} until the calculated mass flow agrees with the actual mass flow.

Solution of the conservation of mass equation requires calculation of the gas density through an appropriate equation of state, e.g., one of those discussed in Chapter 2. With the very large number of state calculations required, the pseudo-perfect gas model described in Chapter 2, Section 2.7, can yield a substantial reduction in computation time relative to real gas models, or even relative to ideal gas models with temperature-dependent specific heats. The pseudo-perfect gas model is almost always adequate for this type of analysis of compressor components. To support the state calculations, the conservation of energy, Eq. (12-7), and entropy, Eq. (12-8), must be employed. The passage inlet profiles of the total thermodynamic conditions, and either C_U or α are specified. Then, the appropriate data for each stream surface at the inlet are obtained by interpolation. If α is specified, C_U is calculated during the momentum equation solution from W_m. As shown in Chapter 3, the rothalpy on each stream surface is given by

$$R = h_t - \omega r C_U \tag{12-35}$$

The relative total enthalpy at any point on the stream surface is then given by

$$h_t' = R + \tfrac{1}{2}(\omega r)^2 \tag{12-36}$$

and the static enthalpy by

$$h = h_t' - \tfrac{1}{2}W^2 = h_t - \tfrac{1}{2}C^2 \tag{12-37}$$

Since entropy is known on all stream surfaces from the inlet conditions, all required state calculations are straightforward, following the procedures of Chapter 2. For quasi-normals downstream of the blade row, the appropriate values of angular momentum, rC_U, are obtained from the solution at the blade trailing edge quasi-normal.

12.5   Repositioning Stream Surfaces

Once the mass and momentum equations have been solved for the resident stream surface geometry, the new flow data will not, in general, be consistent with that stream surface pattern. It is necessary to recompute the stream surfaces using the new flow field solution. Then the mass and momentum equation solutions are repeated until the process converges. In principle, repositioning of streamlines is straightforward. Equation (12-14) is integrated for all quasi-normals and new stream surface positions are computed by interpolation such that all stream tubes contain equal mass flows. In practice, it is necessary to employ fairly sophisticated numerical damping procedures to rapidly achieve convergence while avoiding numerical instability. The damping scheme suggested in Novak and Hearsey (1973) has been generalized to centrifugal compressors with excellent success. The stream surfaces are moved a fraction, F, of the distance between the old and new positions. For quasi-normals outside the blade row, F is given by

$$\frac{1}{F} = 1 + \frac{(1 - M_m^2)(\Delta x)^2}{(\Delta m)^2 B^*} \qquad (12\text{-}38)$$

where Δx = hub-to-shroud quasi-normal length; Δm = minimum spacing with the adjacent quasi-normals; M_m = meridional Mach number; and B^* = an empirical constant. For quasi-normals inside the blade passage

$$\frac{1}{F} = 1 + \frac{(1 - M'^2)(\cos \alpha' \Delta x)^2}{(\Delta m)^2 B^*} \qquad (12\text{-}39)$$

where M' = relative Mach number; and α' = relative flow angle. This damping procedure works very well for centrifugal compressors if M_m and M' are limited to values no greater than 0.9 in these equations. B^* is typically about 8 to 16, but the numerical analysis should track the convergence and update B^* to optimize the rate of convergence, reducing B^* when errors increase and increasing it when progress toward convergence is unusually slow but stable.

12.6   The First Iteration

The hub-to-shroud flow analysis must start with an initial guess of the flow field, since no blade-to-blade flow results are available for the first iteration. Initial stream surfaces can be assigned by requiring each stream tube to have equal areas at each quasi-normal. The rothalpy, entropy and inlet angular momentum for all stream surfaces is computed from the specified inlet profiles. The initial values for W_m can be initialized from the stream tube areas, mass flow and the assumption that the static density is equal to the local relative total density. For vaneless passages and quasi-normals upstream of the blade, conservation of angular momentum supplies W_U.

If blades are present, the flow angle at the leading edge obtained by the above procedure will generally be different from the leading edge blade angle. The initial guess for the relative flow angles in the first 15% of the blade passage length can be estimated by

$$\cot \alpha' = \cot \alpha'_{LE} + (\cot \beta - \cot \alpha'_{LE})[(m - m_1)/(m_2 - m_1)/0.15]^2 \qquad (12\text{-}40)$$

and the relative flow angle is set equal to the local blade angle following this "inlet slip" region. A similar procedure can be used to impose a slip factor or an assigned trailing edge flow or deviation angle if desired, but this really isn't necessary in most cases. Once the trailing edge flow is initialized, conservation of angular momentum provides W_U for quasi-normals downstream of the blade.

These initialization procedures are relatively conservative, in the sense that they are almost always sufficient to successfully start the analysis. But, completion of the first blade-to-blade flow analysis to estimate the relative flow angles in a blade row has, consistently, greatly accelerated the rate of convergence of the hub-to-shroud flow analysis. Hence, the number of hub-to-shroud flow iterations performed prior to the first blade-to-blade flow analysis should be limited to, say, 12–15 (i.e., ignore the convergence requirement, if necessary). This consistently leads to faster overall solution convergence.

12.7 Choked Flow

The most common convergence problem encountered is the attempt to analyze the flow for a specified mass flow that is beyond the choke limit for the passage. This situation can be recognized in the mass and momentum conservation solution. The flow is locally beyond the choke limit if

$$\frac{\partial \dot{m}_c}{\partial W_{mh}} \leq 0 \qquad (12\text{-}41)$$

where $\dot{m}_c$ = calculated mass flow. This may be a true choke, or it may be caused by numerical errors on the early iterations. The mass and momentum solution procedure should provide for computing the maximum mass flow the particular quasi-normal can pass when this situation is encountered. The maximum mass flow corresponds to the gradient in Eq. (12-41) equal to zero. The solution should continue, solving for the largest mass flow acceptable (but, of course, not exceeding the mass flow specified in case the "choke" is due to numerical errors). This process will lead to a solution for the choking mass flow if the specified mass flow cannot be passed. In the case of temporary choke caused by numerical error, the mass flow will gradually increase until the specified mass flow is reached, as the numerical errors are reduced.

12.8 The Blade-to-Blade Flow Governing Equations

The blade-to-blade flow analysis solves the two-dimensional flow in the stream surfaces to define the complete flow field. The hub-to-shroud flow analysis supplies the stream surface geometry, and the stream sheet thicknesses, wherein the blade-to-blade flow will be solved. That specified data reduces the problem to a

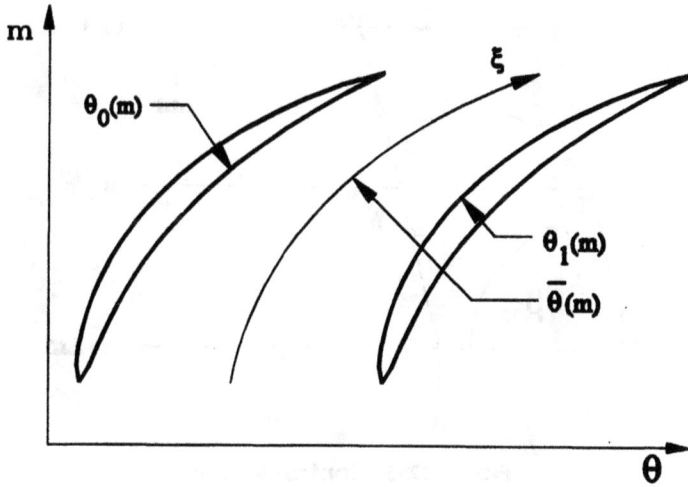

FIGURE 12-4. Blade-to-Blade Plane Geometry

two-dimensional flow problem. Figure 12-4 shows the basic geometry to be considered. Within the blade passage, the flow field is bounded by the two blade surfaces represented in functional form as $\theta_0(m)$ and $\theta_1(m)$. It is convenient to introduce the alternate coordinates (ξ, η)

$$\xi = \int_0^m \frac{dm}{\sin\beta} \tag{12-42}$$

$$\eta = [\theta - \theta_0]/[\theta_1 - \theta_0] \tag{12-43}$$

where β = angle of a constant η line with respect to the tangential direction

$$\cot\beta = \left[\frac{r\partial\theta}{\partial m}\right]_\eta = \cot\beta_0 + \eta[\cot\beta_1 - \cot\beta_0] \tag{12-44}$$

Note that η varies from 0 to 1 between the two blade surfaces. The blade-to-blade passage width is

$$S = r(\theta_1 - \theta_0) \tag{12-45}$$

Define velocity components, W_ξ and W_n, parallel and normal to constant η lines, respectively, by

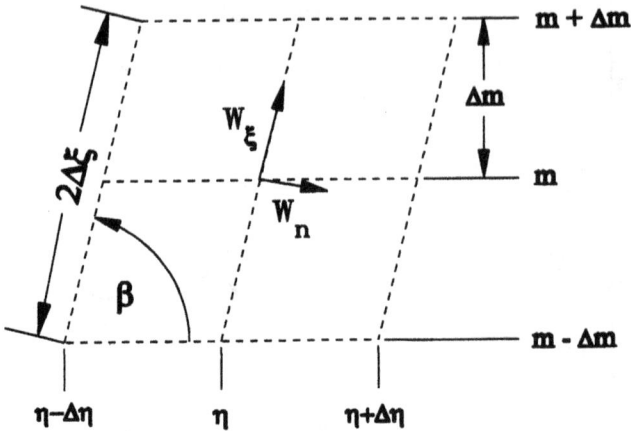

FIGURE 12-5. Control Volume

$$W_\xi = W_m \sin \beta + W_U \cos \beta \qquad (12\text{-}46)$$

$$W_n = W_U \sin \beta - W_m \cos \beta \qquad (12\text{-}47)$$

$$W_m = W_\xi \sin \beta - W_n \cos \beta \qquad (12\text{-}48)$$

$$W_U = W_n \sin \beta + W_\xi \cos \beta \qquad (12\text{-}49)$$

The governing equations for the blade-to-blade flow analysis are conservation of mass, Eq. (3-14), the requirement that the absolute flow is irrotational, Eq. (3-21), and the requirement that rothalpy and entropy are constant on each stream surface. The basic equations can be developed in the (ξ, η) space by relating the derivatives with respect to m and θ to derivatives with respect to ξ and η through Eqs. (12-42) and (12-43), and substituting into the basic equations. But a more accurate numerical solution will result from a more fundamental development from the control volume shown in Fig. 12-5. Conservation of mass requires

$$2\Delta m \left[\left(\frac{\rho b W_n}{\sin \beta} \right)_{(\xi, \eta - \Delta \eta)} - \left(\frac{\rho b W_n}{\sin \beta} \right)_{(\xi, \eta + \Delta \eta)} \right]$$

$$+ 2\Delta \eta [(Sb\rho W_m)_{(m - \Delta, m, \eta)} - (Sb\rho W_m)_{(m + \Delta m, \xi)}] = 0 \qquad (12\text{-}50)$$

where the specific grid point on the control volume is identified with subscripts. Taking the limit as Δm and $\Delta \eta$ approach zero yields the continuity equation of interest

$$\frac{\partial}{\partial \eta} \left[\frac{\rho b W_n}{\sin \beta} \right] + \frac{\partial (S \rho b W_m)}{\partial m} = 0 \tag{12-51}$$

The condition of irrotational absolute flow requires

$$\vec{\nabla} \times (\vec{W} + r\omega \vec{e}_\theta) = \vec{\nabla} \times \vec{C} = 0 \tag{12-52}$$

Application of Stokes' theorem to the control volume in Fig. 12-5 is a convenient method to develop the irrotationality condition. Stokes' theorem relates the vector line integral of velocity around any closed path to the integral of the normal vorticity over the area included in the path

$$\oint_C \vec{W} \cdot \vec{dr} = \int_A [\vec{e}_n \cdot (\vec{\nabla} \times \vec{W})] da = - \int_A [\vec{e}_n \cdot (\vec{\nabla} \times r\omega \vec{e}_\theta)] da \tag{12-53}$$

where Eq. (12-52) was used to evaluate $(\vec{\nabla} \times \vec{W})$. In the case of the present control volume, this can be expressed as

$$\left[\left(\frac{W_\xi}{\sin \beta} \right)_{(\xi, \eta + \Delta \eta)} - \left(\frac{W_\xi}{\sin \beta} \right)_{(\xi, \eta - \Delta \eta)} \right] (2\Delta m)$$

$$+ [(S W_U)_{(m - \Delta m, \eta)} - (S W_U)_{(m + \Delta m, \eta)}](2\Delta \eta)$$

$$= 4 \Delta \eta \Delta m \frac{S}{r} \frac{\partial r^2 \omega}{\partial r} \tag{12-54}$$

where the curl operator of Eq. (3-56) has been used. In the limit as Δm and $\Delta \eta$ approach zero, this yields

$$\frac{\partial}{\partial \eta} \left[\frac{W_\xi}{\sin \beta} \right] = \frac{\partial S W_U}{\partial m} + 2 S \omega \sin \alpha_C \tag{12-55}$$

The continuity equation can be identically satisfied by defining a stream function, Ψ, by

$$\dot{m} \frac{\partial \Psi}{\partial m} = -\rho b (W_U - W_m \cot \beta) \tag{12-56}$$

$$\dot{m} \frac{\partial \Psi}{\partial \eta} = S \rho b W_m \tag{12-57}$$

where $\dot{m}$ = mass flow in the stream sheet. Substitution of Eqs. (12-56) and (12-57) into Eq. (12-51) with the aid of Eqs. (12-46) through (12-49) will confirm that the continuity equation is identically satisfied by this stream function. Note that Ψ varies from 0 to 1 as θ varies from θ_0 to θ_1. From the stream function definition, it is easily shown that

$$W_m = \frac{\dot{m}}{S\rho b} \frac{\partial \Psi}{\partial \eta} \tag{12-58}$$

$$W_U = \frac{\dot{m}}{\rho b} \left[\frac{\cot \beta}{S} \frac{\partial \Psi}{\partial \eta} - \frac{\partial \Psi}{\partial m} \right] \tag{12-59}$$

12.9   Linearized Blade-to-Blade Flow

Up to this point, the blade-to-blade flow equations are completely general. Now, the equations will be linearized, following the approach described in Aungier (1988b) by defining the stream function as

$$\Psi(m, \eta) = a(m)[\eta - \eta^2] + \eta^2 \tag{12-60}$$

where $a(m)$ is a function of m, only. Note that Ψ varies from 0 to 1 as required. Equation (12-57) shows that this definition requires the quantity $\rho b W_m$ to vary linearly with η. Since the stream function satisfies the continuity equation, it remains to solve Eq. (12-55). It is convenient to solve this equation in an integral form. Integrating through the equation with respect to η, from 0 to 1, yields

$$\frac{W_1}{\sin \beta_1} - \frac{W_0}{\sin \beta_0} = \int_0^1 \frac{\partial S W_U}{\partial m} d\eta + 2S\omega \sin \alpha_C \tag{12-61}$$

where Eq. (12-61) uses the fact that $W = W_\xi$ on the blade surfaces. Noting that on the blade surfaces

$$W = \frac{W_m}{\sin \beta} = \frac{1}{\sin \beta} \frac{\dot{m}}{S\rho b} \frac{\partial \Psi}{\partial \eta} \tag{12-62}$$

equation (12-60) can be used to solve for the left-hand side of Eq. (12-61)

$$\frac{W_1}{\sin \beta_1} = \frac{\dot{m}(2 - a)}{S\rho \sin^2 \beta_1} \tag{12-63}$$

$$\frac{W_0}{\sin \beta_0} = \frac{\dot{m} a}{S\rho b \sin^2 \beta_0} \tag{12-64}$$

The integral term in Eq. (12-61) is more complex. From Eqs. (12-58) through (12-60), it is easily shown that

$$SW_U = \dot{m}[\cot \beta(a - 2a\eta + 2\eta) - a' S(\eta - \eta^2)]/(\rho b) \tag{12-65}$$

where the prime denotes the total derivative with respect to m. For convenience, define

$$f(m, \eta) = \dot{m} \cot \beta/(\rho b); \quad g(m, \eta) = \dot{m} S/(\rho b) \tag{12-66}$$

to yield

$$\frac{\partial S W_U}{\partial m} = \frac{\partial f}{\partial m}(a - 2a\eta + 2\eta) + (1 - 2\eta)fa'$$

$$- (ga'' + \frac{\partial g}{\partial m} a')(\eta - \eta^2) \tag{12-67}$$

Integrals with respect to η will be approximated by

$$\int_0^1 F(\eta)d\eta = [F_0 + 4\bar{F} + F_1]/6 \tag{12-68}$$

where the overbar designates a value at $\eta = \frac{1}{2}$. Equation (12-68) is derived from a truncated Taylor series expansion with values of F known at $\eta = 0, \frac{1}{2}$ and 1. After some tedious algebra, the above equations can be combined to yield

$$\int_0^1 \frac{\partial S W_U}{\partial m} d\eta = [af_0' + f_0a' + 4\bar{f}' - \bar{g}a'' - \bar{g}'a' + f_1'(2 - a) - f_1a']/6$$

$$\tag{12-69}$$

Then, combining Eqs. (12-61), (12-63), (12-64) and (12-69), the equation governing the potential flow in the blade-to-blade stream surface is reduced to the following simple linear differential equation:

$$a'' + Aa' + Ba = C \tag{12-70}$$

where

$$A(m) = [\bar{g}' - f_0 + f_1]/\bar{g} \tag{12-71}$$

$$B(m) = \frac{f_1' - f_0'}{\bar{g}} - \frac{6}{\bar{g}S^2} \left[\frac{g_1}{\sin^2 \beta_1} + \frac{g_0}{\sin^2 \beta_0} \right] \tag{12-72}$$

$$C(m) = \frac{12\omega \sin \alpha_C + 4\bar{f}' + 2f_1'}{\bar{g}} - \frac{12g_1}{\bar{g}S^2 \sin^2 \beta_1} \tag{12-73}$$

If f and g are known at $\eta = 0, \frac{1}{2}$ and 1, Eq. (12-70) can be solved for $a(m)$, subject to suitable boundary conditions. That provides a complete definition of the blade-to-blade flow; suitable leading and trailing edge boundary conditions still have to be selected. At the leading edge, the overall (integrated) angular momentum is equal to the known passage inlet angular momentum, i.e., the average relative tangential velocity is known. Using the integral approximation of Eq. (12-68), this requires

$$6S\overline{W_{ULE}} = f_0 a + 4\bar{f} - \bar{g}a' + f_1(2 - a) \tag{12-74}$$

Hence, the leading edge boundary condition can be expressed as

$$a' + Ba = C \tag{12-75}$$

where

$$B = (f_1 - f_0)/\bar{g} \tag{12-76}$$
$$C = (4\bar{f} + 2f_1 - 6S\overline{W_{ULE}})/\bar{g} \tag{12-77}$$

At the trailing edge, the well-known Kutta condition requires $W_0 = W_1$, i.e., the flow must leave the blade trailing edge with the same pressure on each side of the blade, since there is no longer a blade force to sustain a difference in pressure. From Eqs. (12-63) and (12-64), this requires

$$(2 - a)\rho_0 \sin \beta_0 = a\rho_1 \sin \beta_1 \tag{12-78}$$

and noting that if the velocities are equal, the densities must also be equal, the trailing edge boundary condition is

$$a = \frac{2 \sin \beta_0}{\sin \beta_0 + \sin \beta_1} \tag{12-79}$$

The solution is carried out entirely inside the blade passage; thus, normally the angles on the two blade surfaces will not be equal.

12.10 Numerical Solution for the Stream Function

Equation (12-70) and its boundary conditions can be easily solved via a finite-difference numerical scheme. Distribute N grid points, equally spaced along m. Designating the grid point number by a subscript, the meridional derivatives are approximated with three-point differences. For any interior point i

$$F_i' = (F_{i+1} - F_{i-1})/(2\Delta m) \tag{12-80}$$
$$F_i'' = (F_{i+1} - 2F_i + F_{i-1})/(\Delta m)^2 \tag{12-81}$$

and at the end points

$$F_1' = (4F_2 - 3F_1 - F_3)/(2\Delta m) \tag{12-82}$$

$$F_1'' = (F_3 - 2F_2 + F_1)/(\Delta m)^2 \tag{12-83}$$

$$F_N' = (3F_N - 4F_{N-1} + F_{N-2})/(2\Delta m) \tag{12-84}$$

$$F_N'' = (F_N - 2F_{N-1} + F_{N-2})/(\Delta m)^2 \tag{12-85}$$

These difference approximations are substituted into the governing equations to yield N algebraic equations for the N grid points. When expressed in matrix form, this yields a tridiagonal matrix, except for the leading edge grid point, which has one extra term, i.e.,

$$
\begin{bmatrix}
E_1 & F_1 & H_1 & 0 & 0 & \cdots & 0 & 0 & 0 \\
D_2 & E_2 & F_2 & 0 & 0 & \cdots & 0 & 0 & 0 \\
0 & D_3 & E_3 & F_3 & 0 & \cdots & 0 & 0 & 0 \\
& & & \cdot & & & & & \\
& & & \cdot & & & & & \\
& & & \cdot & & & & & \\
0 & 0 & 0 & 0 & \cdots & 0 & D_{N-1} & E_{N-1} & F_{N-1} \\
0 & 0 & 0 & 0 & \cdots & 0 & 0 & D_N & E_N
\end{bmatrix}
\begin{bmatrix}
a_1 \\ a_2 \\ a_3 \\ \cdot \\ \cdot \\ \cdot \\ a_{N-1} \\ a_N
\end{bmatrix}
=
\begin{bmatrix}
G_1 \\ G_2 \\ G_3 \\ \cdot \\ \cdot \\ \cdot \\ G_{N-1} \\ G_N
\end{bmatrix}
$$

The algebra is tedious, but it is easily shown that the parameters in the matrix are

Leading edge: $H_1 = -1$, $E_1 = 2B_1\Delta m - 3$, $F_1 = 4$, $G_1 = 2C_1\Delta m$

Interior points: $D_i = 1 - \frac{1}{2}A_i\Delta m$, $E_i = B_i(\Delta m)^2 - 2$, $F_i = 1 + \frac{1}{2}A_i\Delta m$, $G_i = C_i(\Delta m)^2$

Trailing edge: $D_N = 0$, $E_N = 1$, $G_N = 2\sin\beta_0/(\sin\beta_0 + \sin\beta_1)$

Inversion of this matrix to solve for the a_i terms is rather simple. The basic steps to be performed (in sequence) involve updating the matrix parameters (D_i, E_i, F_i, G_i and H_i) to achieve all $E_i = 1$, and all other elements of the left matrix as zero. The process is as follows:

1. At the leading edge ($i = 1$):

$$F_1 \rightarrow F_1/E_1$$
$$H_1 \rightarrow H_1/E_1$$
$$G_1 \rightarrow G_1/E_1$$
$$E_1 \rightarrow 1$$

2. For interior points in sequence $i = 2$ to $(N - 1)$:

$$W = E_i - D_i F_{i-1}$$
$$F_i \rightarrow (F_i - D_i H_{i-1})/W$$
$$G_i \rightarrow (G_i - D_i G_{i-1})/W$$
$$D_i \rightarrow 0$$
$$H_i \rightarrow 0$$
$$E_i \rightarrow 1$$

3. At the trailing edge $(i = N)$:

$$W = E_N - D_N F_{N-1} = 1$$
$$E_N \rightarrow (E_N - D_N F_{N-1})/W = 1$$
$$G_N \rightarrow (G_N - D_N G_{N-1})/W = G_N$$
$$D_N \rightarrow 0$$
$$E_N \rightarrow 1$$
$$a_N = G_N$$

4. For interior points in sequence $i = (N-1)$ to 2:

$$a_i = G_i - F_i a_{i+1}$$

5. At the leading edge $(i = 1)$:

$$a_1 = G_1 - F_1 a_2 - H_1 a_3$$

This stream function solution can be extended to include a set of splitter blades, such that each splitter blade lies midway between the full blades and is identical to them except that its leading edge lies inside the full blade passage on a specified quasi-normal. The procedure is to first process the solution without splitters, viewing that solution as valid up to the splitter blade leading edge. Then, set the flow at the splitter leading edge equal to that predicted by this full blade solution, using it as an upstream boundary condition for the two splitter blade passages within the full blade passage. The prediction process for the splitter passages is the same as outlined above, except for the upstream boundary conditions. Indeed, the same computing logic can be used if η and Ψ are scaled to vary from 0 to 1 across each passage. For the passage adjacent to surface θ_0, it is easily shown that the scaled data and the splitter passage leading edge boundary condition are related to the local full blade passage data as follows:

$$\tilde{\eta} = 2\eta$$

$$\tilde{\Psi} = \frac{\Psi(\eta)}{\Psi(0.5)} = \frac{4\Psi}{a+1}$$

$$\tilde{a}_{LE} = \frac{2a}{a+1} \tag{12-86}$$

and similarly, data for the other passage are

$$\tilde{\eta} = 2\eta - 1$$

$$\tilde{\Psi} = \frac{4\Psi - a - 1}{3 - a}$$

$$\tilde{a}_{LE} = \frac{2}{3 - a} \qquad\qquad (12\text{-}87)$$

Hence, to treat splitter blades, it is only necessary to apply the same prediction procedure to three different passage analyses instead of just one.

12.11   Iteration for Gas Density

It should be noted that the stream function solution described above requires that the gas density be known at all N meridional stations at $\eta = 0$, $\frac{1}{2}$ and 1, since f and g in Eqs. (12-66) contain density. Clearly, an iterative procedure is required to progressively refine the estimates of ρ and Ψ until they are self-consistent. The process can be started by setting the density equal to the local relative total density as an initial guess. Then, the stream function solution yields a new estimate of the velocity field through Eqs. (12-58) and (12-59). Gas density can then be re-estimated using conservation of energy and the equation of state as described earlier for the hub-to-shroud solution, via Eqs. (12-36) and (12-37). This iterative process will be repeated until successive estimates of density and velocity have achieved acceptable convergence. Note that the gas density calculations always lag the stream function calculation by one iteration. This procedure yields a very stable numerical scheme with rapid convergence, so long as the flow is subsonic throughout. Since the blade-to-blade analysis is a potential flow method, the governing equations are elliptic in mathematical form. But when Mach numbers exceed unity, the fluid flow problem becomes hyperbolic in form. It is well known that potential flow methods will diverge when supersonic velocities are encountered.

There is a fairly simple unpublished procedure that this author has used to extend the analysis, and its capability, into the supersonic range. The process is quite simple and proceeds as follows: The flow velocities are monitored during the analysis for the presence of supersonic values. When encountered, the inlet total temperature is increased enough to hold all velocities just under sonic conditions. The stream sheet thicknesses are also adjusted such that the predicted subsonic velocities will correctly conserve mass when the actual inlet total temperature and actual stream sheet thicknesses are employed. After convergence is achieved, the predicted velocities are accepted as the correct values, but the inlet total temperature and stream sheet thicknesses are returned to their correct values, and all flow field thermodynamic data are recomputed. Since the velocity field is considered correct, it is always possible to compute how much the gas density will change when the correct inlet total temperature is imposed on the final solution. Hence for each meridional station, the stream sheet thickness is adjusted to compensate such that the final solution will satisfy conservation of mass. Since we are treating axisymmetric stream surfaces, only overall mass conservation at each meridional station can be achieved. If $\bar{\rho}(m)$ is the average density that will exist when the total temperatures are readjusted, and $\bar{\rho}_c(m)$ is the calculated average density with the modified temperatures, the solution should use a corrected stream sheet thickness given by

$$b_c(m) = b(m)\bar{\rho}(m)/\bar{\rho}_c \qquad (12\text{-}88)$$

The stream sheet thicknesses are readjusted on each density iteration until the solution converges (Chapter 13 describes this procedure in detail). As a result, the analysis yields a solution that satisfies both the irrotationality condition and mass conservation, i.e., it satisfies all of the governing equations. But, of course, once supersonic velocities are present, shocks may form and the flow actually may not be irrotational. Thus the procedure really only serves to provide a solution at Mach numbers greater than unity, which would otherwise not be possible. So long as Mach numbers are not too much greater than unity, the result should be reasonably valid.

12.12 Quasi-Three-Dimensional Flow

It remains only to describe the coupling and related numerical damping procedures. Upon completion of the hub-to-shroud analysis, the stream surface geometries are supplied to the blade-to-blade analysis. The stream sheet thicknesses and mass flow are also required; these are interrelated, thus having relative significance rather than absolute values. If the stream sheet mass flow is defined as $\Delta\dot{m}$, the stream sheet thicknesses follow directly from the hub-to-shroud flow field data, i.e.,

$$b = \Delta\dot{m}/(\rho r W_m) \qquad (12\text{-}89)$$

Since the two analyses will normally use different calculation grid spacings, interpolation is required.

After the blade-to-blade flow analyses are completed on all stream sheets, the mass-averaged meridional and tangential velocity components must be computed at all meridional stations. Equations (12-58), (12-59) and (12-60) supply those data at $\eta = 0$, $\frac{1}{2}$ and 1 for all meridional computing stations. The mass-averaged data are given by

$$\tilde{W}_m = \frac{\displaystyle\int_0^1 \rho b W_m^2\, d\eta}{\displaystyle\int_0^1 \rho b W_m\, d\eta} \qquad (12\text{-}90)$$

$$\tilde{W}_U = \frac{\int_0^1 \rho b W_m W_U d\eta}{\int_0^1 \rho b W_m d\eta} \qquad (12\text{-}91)$$

These are evaluated with the three-point integral approximation of Eq. (12-68). Then, the revised average flow angles to be used in the next hub-to-shroud flow iteration are given by

$$\cot \alpha' = \tilde{W}_U / \tilde{W}_m \qquad (12\text{-}92)$$

Again, interpolation is required to obtain values on the grid points used by the hub-to-shroud analysis.

Convergence criteria required to ensure adequate solution convergence should include the following:

1. Hub-to-shroud flow stream surface positions.
2. Hub-to-shroud flow gas densities and velocities.
3. Blade-to-blade flow gas densities and velocities.
4. Blade-to-blade flow relative flow angles.

Items 1 to 3 confirm convergence of the individual analyses, while Item 4 checks for convergence on successive blade-to-blade analyses to ensure overall solution convergence.

With the correction procedure for $M' > 1$ as described above, and in Chapter 13, the blade-to-blade flow analysis is extremely stable and reliable. Virtually no numerical damping is required for that portion of the analysis. Numerical damping procedures for the stream surface repositioning has also been described and is the primary damping technique for the hub-to-shroud analysis. It is good practice to provide for some additional damping in this portion of the analysis. The streamline curvature technique is quite sensitive to large changes in stream surface curvature and slope. This author's quasi-three-dimensional flow analysis imposes damping in the form

$$\kappa_m = (\kappa_{m_i} + D\kappa_{m_{i-1}})/(1 + D) \qquad (12\text{-}93)$$

where subscripts i and $i-1$ = iteration numbers; and D = a damping factor. Stream surface slopes are damped by the same method, also using D. Normally, $D = 1$ is used, but the numerical analysis will increase it if successive hub-to-shroud iterations show increasing errors. While usually not necessary, including this numerical damping logic greatly improves reliability, to the point that this author's analysis almost never fails to converge.

The flow chart shown in Fig. 12-6 illustrates the basic logic flow required to perform this type of analysis. Figures 12-7 through 12-11 illustrate typical results obtained for centrifugal compressor impellers. The case presented is a fictitious design processed through the preliminary design procedure and one

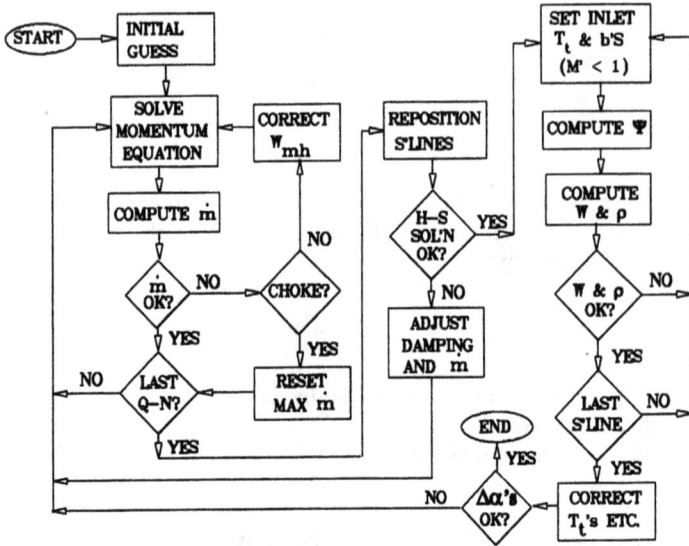

FIGURE 12-6. Flow Chart of the Analysis

pass through the gas path design system. Clearly, reasonably good blade load-ings were obtained, but the rate of diffusion of W_m along the shroud should be improved. It is simply not good design practice to diffuse W_m on the shroud below its discharge value. Adjustment of the hub-and-shroud contours and the passage

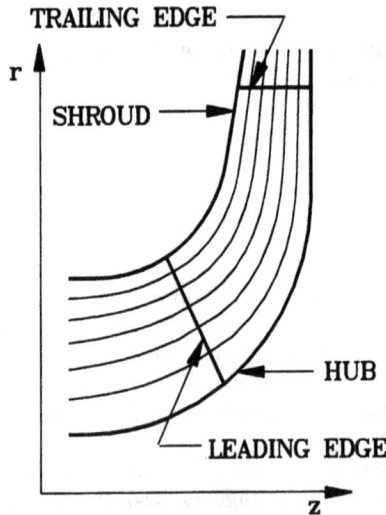

FIGURE 12-7. Stream Surface Pattern

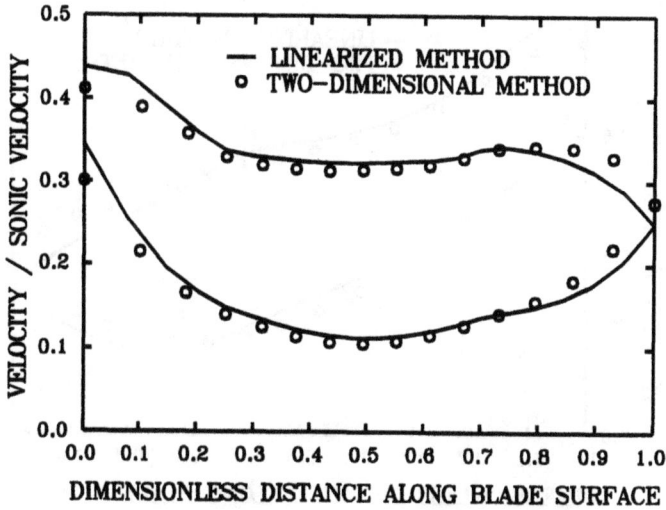

FIGURE 12-8. Hub Surface Blade Loading

area should correct this problem fairly easily. Considering the preliminary status of this example, the quasi-three-dimensional flow analysis results would be quite encouraging to the designer. Figures 12-8 through 12-10 also show predictions of the blade surface velocities supplied by the more exact two-dimensional blade-to-blade flow analysis described in Chapter 13. It can be seen that

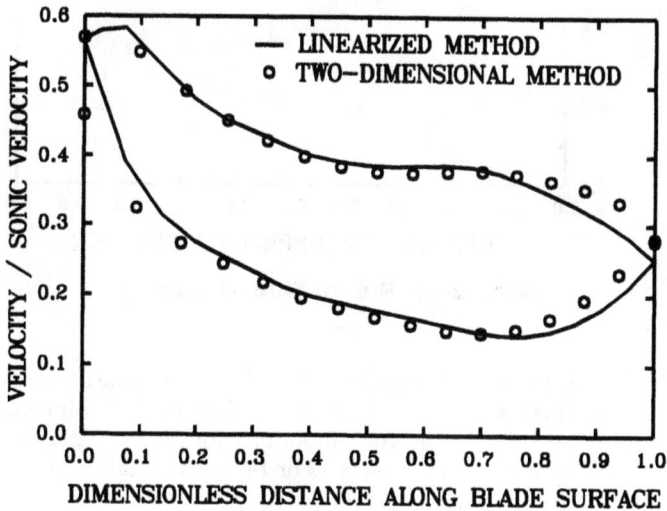

FIGURE 12-9. Mean Surface Blade Loading

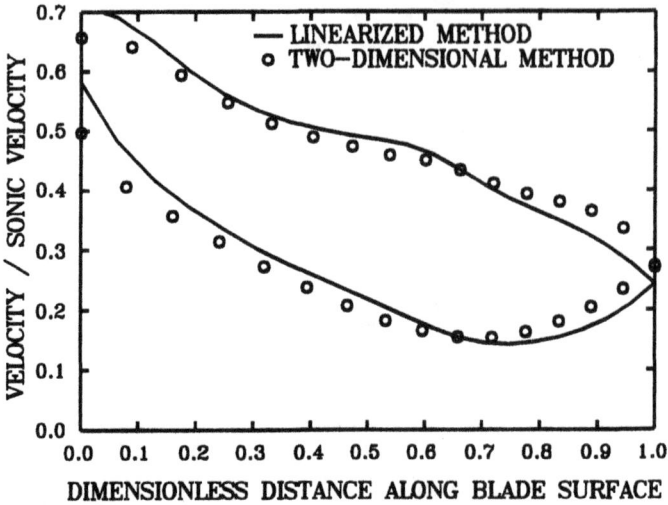

FIGURE 12-10. Shroud Surface Blade Loading

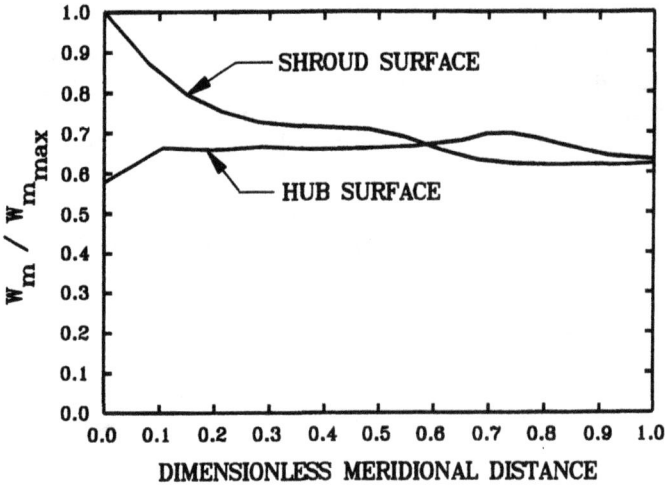

FIGURE 12-11. Hub-to-Shroud Loading

the linearized blade-to-blade flow analysis provides remarkable accuracy for such a simple method. It is good practice to provide the quasi-three-dimensional flow analysis with the capability to generate input files for the more precise blade-to-blade flow analysis methods of Chapters 13 or 14 for this final check on the blade loading distributions.

Chapter 13

POTENTIAL FLOW ANALYSIS IN THE BLADE-TO-BLADE PLANE

In Chapter 12, the flow in the blade-to-blade stream surfaces was analyzed with a linearized method, where the stream function is assumed to vary linearly with η. This assumption results in a substantial reduction in computation time to produce a very efficient quasi-three-dimensional flow analysis. This linearized method has also been recommended for a fast blade loading evaluation for the vaned diffuser and return system detailed design procedures of Chapters 9 and 10. Comparison of the linearized method with the more exact method in this chapter shown in Figs. 12-8 through 12-10 has shown that the linearized method yields excellent prediction accuracy. But it also shows that a more exact analysis yields slightly different results, particularly near the blade leading and trailing edges. Consequently, a more exact blade-to-blade flow analysis should be part of any modern aerodynamic design system to provide a final blade loading evaluation. If the design system is properly structured, this final blade-to-blade flow analysis will require almost no effort on the part of the designer. Any design or analysis method using the linearized method (i.e., Chapters 9, 10 and 12) will have all data required to run the more exact method. By enabling those analyses to create the input file for the more exact blade-to-blade flow analysis, the designer can easily process the final analysis without any input data preparation.

This chapter describes a numerical method to solve the two-dimensional potential flow equations in a blade-to-blade stream surface. Potential flow (i.e., irrotational absolute flow) analysis is, in itself, a simplifying assumption. This is a classical boundary value problem, where the boundary conditions imposed on all boundaries of the solution domain completely determine the solution. For practical purposes, this can be considered a time-steady, subsonic flow analysis, although the technique introduced in Chapter 12 to extend it into the transonic regime will also be used. This analysis is sufficient to treat most centrifugal compressor analysis problems. For cases involving Mach numbers too high for this model, a general time-marching blade-to-blade flow analysis will be presented in Chapter 14. Potential flow analysis can be conducted through a variety of numerical methods. The blades can be replaced with distributed singularities (e.g., Senoo and Nakase, 1972). The streamline curvature technique used in Chapter 12 for the hub-to-shroud flow has also been used extensively (e.g., Novak and Hearsey, 1976). Definition of a potential function or stream function casts the problem in a form well suited to solution by relaxation (Katsanis, 1968; 1969) or matrix (Smith and Frost, 1969) methods. The present method is a matrix method based on a stream function. Indeed, it is really a generalization of the linearized matrix method presented in Chapter 12.

NOMENCLATURE

a = sound speed
b = stream sheet width
C = absolute velocity
C_U = absolute tangential velocity
F = matrix or matrix elements of coefficients for stream function equations
$\hat{F}$ = temporary storage matrix or matrix elements
h = enthalpy
I = the number of grid points in the meridional direction
J = the number of grid points in the tangential direction covering the solution domain
M = Mach number
m = meridional coordinate along stream surfaces
$\dot{m}$ = mass flow in a single blade passage
N = total number of tangential grid points, $J + 1$
p = pressure
Q = matrix or matrix elements for right-hand side of stream function equations
R = rothalpy
r = radius
S = $r(\theta_1 - \theta_0)$
s = entropy
T = temperature
U = blade speed, ωr
W = relative velocity
W_m = meridional velocity
W_n = relative velocity normal to constant ξ lines
W_U = relative tangential velocity
W_ξ = relative velocity in the ξ direction
Z = total number of blades, $Z_{FB} + Z_{SB}$
Z_{FB} = number of full-length blades
Z_{SB} = number of splitter blades
z = axial coordinate
α = flow angle with respect to tangent
α_C = streamline slope angle with axis
β = blade angle or angle of constant η curve with respect to tangent
θ = polar angle coordinate
ρ = gas density
Ψ = stream function
ω = rotation speed

Subscripts

ch = choke condition
i = meridional grid point number
j = tangential grid point number
k = row number in matrices for stream function
m = column number in matrix $\hat{F}$
n = column number in matrix F
p = blade pressure surface parameter
s = blade suction surface parameter
t = total thermodynamic condition
0 = parameter on blade surface θ_0
1 = parameter on blade surface θ_1

Superscripts

$'$ = value relative to the blade row's frame of reference
$*$ = sonic condition

13.1 Definition of the Problem

To a very large extent, the problem to be solved has already been defined in Chapter 12. The flow is analyzed in a thin stream sheet bounded by stream surfaces as illustrated in Fig. 13-1. The stream surfaces are assumed to be axisymmetric between the blade surfaces, and the absolute flow is assumed to be irrotational and isentropic in the blade-to-blade plane. The (ξ, η) coordinates of Chapter 12, Eqs. (12-42) and (12-43), and the velocity components, W_ξ and W_n, of Eqs. (12-46) through (12-49) will also be used here. The basic governing equations were derived in Chapter 12, and are repeated here for easy reference. They are conservation of mass or the continuity equation

$$\frac{\partial}{\partial \eta} \left[\frac{\rho b W_n}{\sin \beta} \right] + \frac{\partial (S \rho b W_m)}{\partial m} = 0 \tag{13-1}$$

and the requirement that the absolute flow is irrotational

$$\frac{\partial}{\partial \eta} \left[\frac{W_\xi}{\sin \beta} \right] = \frac{\partial S W_U}{\partial m} + 2 S \omega \sin \alpha_C \tag{13-2}$$

where β is the angle of a constant η curve with the tangential direction; the coordinates (ξ, η) are given by

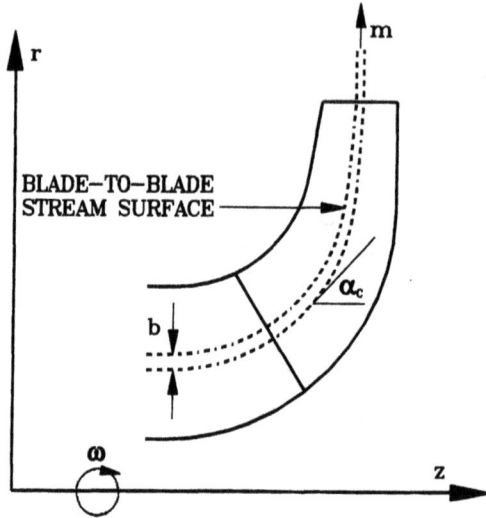

FIGURE 13-1. Blade-to-Blade Stream Sheet

$$\xi = \int_0^m \frac{dm}{\sin\beta} \qquad (13\text{-}3)$$

$$\eta = \frac{\theta - \theta_0}{\theta_1 - \theta_0} \qquad (13\text{-}4)$$

and the velocity components are related by

$$W_\xi = W_m \sin\beta + W_U \cos\beta \qquad (13\text{-}5)$$

$$W_n = W_U \sin\beta - W_m \cos\beta \qquad (13\text{-}6)$$

$$W_m = W_\xi \sin\beta - W_n \cos\beta \qquad (13\text{-}7)$$

$$W_U = W_n \sin\beta + W_\xi \cos\beta \qquad (13\text{-}8)$$

The continuity equation can be identically satisfied by defining a stream function, Ψ, by

$$\dot{m}\,\frac{\partial\Psi}{\partial m} = -\rho b(W_U - W_m \cot\beta) \qquad (13\text{-}9)$$

$$\dot{m}\,\frac{\partial\Psi}{\partial\eta} = S\rho b W_m \qquad (13\text{-}10)$$

where $\dot{m}$ = mass flow in the stream sheet. Note that Ψ varies from 0 to 1 as θ varies from θ_0 to θ_1. From the stream function definition, it is easily shown that

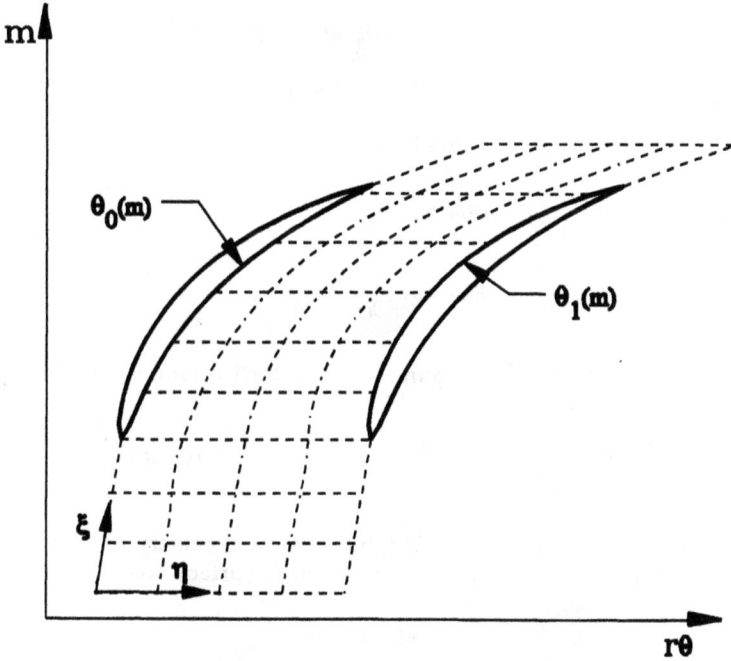

FIGURE 13-2. Blade-to-Blade Flow Grid

$$W_m = \frac{\dot{m}}{S\rho b} \frac{\partial \Psi}{\partial \eta} \tag{13-11}$$

$$W_U = \frac{\dot{m}}{\rho b} \left[\frac{\cot \beta}{S} \frac{\partial \Psi}{\partial \eta} - \frac{\partial \Psi}{\partial m} \right] \tag{13-12}$$

Since Ψ is no longer assumed linear with η, a general finite-difference method is used here to solve for the flow on a finite-difference grid structure, such as that illustrated in Fig. 13-2. Introducing the stream function into Eq. 13-2 yields

$$\frac{\partial}{\partial \eta} \left[\frac{\dot{m}(1 + \cot^2 \beta)}{S\rho b} \frac{\partial \Psi}{\partial \eta} - \frac{\dot{m}\cot \beta}{\rho b} \frac{\partial \Psi}{\partial m} \right]$$

$$= \frac{\partial}{\partial m} \left[\frac{\dot{m}}{\rho b} \left(\cot \beta \frac{\partial \Psi}{\partial \eta} - S \frac{\partial \Psi}{\partial m} \right) \right] + 2S\omega \sin \alpha_C \tag{13-13}$$

This equation is simplified by defining the following functions

$$A(m,\eta) = \dot{m}/(S\rho b \, \sin^2 \beta) \qquad (13\text{-}14)$$

$$B(m,\eta) = \dot{m} \, \cot \beta/(\rho b) \qquad (13\text{-}15)$$

$$C(m,\eta) = \dot{m} \, S/(\rho b) \qquad (13\text{-}16)$$

$$D(m,\eta) = \frac{\partial A}{\partial \eta} - \frac{\partial B}{\partial m} \qquad (13\text{-}17)$$

$$E(m,\eta) = \frac{\partial C}{\partial m} - \frac{\partial B}{\partial \eta} \qquad (13\text{-}18)$$

to yield the following basic governing equation for irrotational absolute flow:

$$A \frac{\partial^2 \Psi}{\partial \eta^2} - 2B \frac{\partial^2 \Psi}{\partial \eta \partial m} + C \frac{\partial^2 \Psi}{\partial m^2} + D \frac{\partial \Psi}{\partial \eta} + E \frac{\partial \Psi}{\partial m} = 2S\omega \, \sin \alpha_C \qquad (13\text{-}19)$$

A potential flow problem of this type requires boundary conditions for all boundaries of the solution domain. For the blade surface boundaries

$$\Psi(m,0) = 0$$
$$\Psi(m,1) = 1 \qquad (13\text{-}20)$$

For all points upstream and downstream of the blade passage, the flow must be periodic with η, i.e., the flow in all blade passages must be identical. Hence points on these side boundaries can be treated in the same way as points internal to the solution boundary by extending the flow field into adjacent passages using the periodicity condition and requiring

$$\Psi(m,1) = \Psi(m,0) + 1$$
$$W_m(m,0) = W_m(m,1)$$
$$W_U(m,0) = W_U(m,1)$$
$$\rho(m,0) = \rho(m,1) \qquad (13\text{-}21)$$

At the upstream and downstream boundaries, it is common practice to require the flow field to be uniform with η, which is suitable if the boundaries are far enough away from the blades. A less stringent condition is to require that the flow angle be uniform with η. This boundary condition is easily imposed by setting the boundary values of β equal to the flow angle and requiring

$$\frac{\partial \Psi}{\partial m} = 0 \qquad (13\text{-}22)$$

This approach requires assigning $\theta_0(m)$ and $\theta_1(m)$ upstream and downstream of the blades such that β is equal to the flow angle on the boundary of interest.

This is easily accomplished by assigning a linear variation of β with m between the (average) blade leading or trailing edge value and the appropriate flow angle. Since the definition of β can be specified as

$$\cot \beta = \left[\frac{r\partial\theta}{\partial m} \right]_\eta \tag{13-23}$$

simple integration yields the values $\theta_0(m)$ and $\theta_1(m)$ upstream and downstream of the blades. This set of upstream and downstream boundary conditions is well suited to a matrix type solution procedure for the stream function equation, and imposes a fairly mild constraint to reduce the solution's sensitivity to the distance between these boundaries and the blade passage. But for centrifugal compressor applications, the flow angles are usually not known in advance. At the upstream boundary, the mass-averaged angular momentum, $(rC_U)_{in}$, is usually the known quantity. Hence, the most appropriate upstream boundary condition is

$$\int_0^1 \rho W_m^2 \cot \beta d\eta + r^2 \omega = \dot{m}(rC_U)_{in} \tag{13-24}$$

This will not complicate the analysis significantly. As was done in the linearized blade-to-blade flow analysis of Chapter 12, the solution for Ψ will be generated with the gas density field treated as known. The density field will be iteratively updated using successive Ψ solutions until the process converges. As the density field is updated, Eq. (13-24) can be used to update the upstream flow angle as well. As was done in Chapter 12, it is generally preferable to impose a trailing edge Kutta condition rather than assign the discharge flow angle. This requires that the pressures and velocities on the two sides of the blade be equal at the trailing edge, i.e.,

$$(W_0)_{te} = (W_1)_{te} \tag{13-25}$$

As will be shown later in this chapter, it is fairly simple to incorporate an iterative numerical scheme to adjust the downstream flow angle to satisfy this constraint while converging on the density field.

13.2   The Stream Function Solution

The solution for the stream function involves solving Eq. (13-19) subject to the above boundary conditions while treating the gas density field as known. Consequently, all functions and parameters in Eq. (13-19), except for the stream function, are known. The solution domain is divided into a grid structure as illustrated in Fig. 13-2, where the spacings between each grid point, Δm and $\Delta \eta$, are constant. The partial derivatives are approximated with three-point finite difference approximations derived from appropriate Taylor series expansions. For example, at any grid point (m, η), the truncated Taylor series expansions in the m direction

$$\Psi(m + \Delta m, \eta) = \Psi(m, \eta) + \frac{\partial \Psi}{\partial m} \Delta m + \frac{\partial^2 \Psi}{\partial m^2} \frac{(\Delta m)^2}{2}$$

$$+ \frac{\partial^3 \Psi}{\partial m^3} \frac{(\Delta m)^3}{6} + \cdots \tag{13-26}$$

$$\Psi(m - \Delta m, \eta) = \Psi(m, \eta) - \frac{\partial \Psi}{\partial m} \Delta m + \frac{\partial^2 \Psi}{\partial m^2} \frac{(\Delta m)^2}{2}$$

$$- \frac{\partial^3 \Psi}{\partial m^3} \frac{(\Delta m)^3}{6} + \cdots \tag{13-27}$$

can be solved for the finite difference approximations for the first and second partial derivatives with respect to m. Similar expansions yield approximations for the other partial derivatives required. If we designate the (m, η) grid point numbers as subscripts, the finite difference approximations at the interior grid point (i, j) are easily shown to be

$$\frac{\partial \Psi}{\partial m} = \frac{\Psi_{i+1,j} - \Psi_{i-1,j}}{2\Delta m} \tag{13-28}$$

$$\frac{\partial \Psi}{\partial \eta} = \frac{\Psi_{i,j+1} - \Psi_{i,j-1}}{2\Delta \eta} \tag{13-29}$$

$$\frac{\partial^2 \Psi}{\partial m^2} = \frac{\Psi_{i+1,j} - 2\Psi_{i,j} + \Psi_{i-1,j}}{(\Delta m)^2} \tag{13-30}$$

$$\frac{\partial^2 \Psi}{\partial \eta^2} = \frac{\Psi_{i,j+1} - 2\Psi_{i,j} + \Psi_{i,j-1}}{(\Delta \eta)^2} \tag{13-31}$$

$$\frac{\partial^2 \Psi}{\partial m \partial \eta} = \frac{\Psi_{i+1,j+1} - \Psi_{i-1,j+1} - \Psi_{i+1,j-1} + \Psi_{i-1,j-1}}{4\Delta m \Delta \eta} \tag{13-32}$$

These approximations can also be used to compute the functions D and E in Eqs. (13-17) and (13-18). This requires first partial derivatives on the boundaries of the solution domain not provided by these centered difference approximations. Again, truncated Taylor series can be used to define one-sided difference approximations, i.e.,

$$\frac{\partial \Psi}{\partial m} = \frac{4\Psi_{i+1,j} - 3\Psi_{i,j} - \Psi_{i+2,j}}{2\Delta m} \tag{13-33}$$

$$\frac{\partial \Psi}{\partial \eta} = \frac{4\Psi_{i,j+1} - 3\Psi_{i,j} - \Psi_{i,j+2}}{2\Delta \eta} \tag{13-34}$$

$$\frac{\partial \Psi}{\partial m} = \frac{3\Psi_{i,j} - 4\Psi_{i-1,j} + \Psi_{i-2,j}}{2\Delta m} \tag{13-35}$$

$$\frac{\partial \Psi}{\partial \eta} = \frac{3\Psi_{i,j} - 4\Psi_{i,j-1} + \Psi_{i,j-2}}{2\Delta \eta} \tag{13-36}$$

After some tedious algebra, substitution of the above finite difference approximations into Eq. (13-19) yields the desired approximation to compute Ψ at any interior point, (i,j) as

$$\Psi_{i,j} + \hat{A}_{i,j}\Psi_{i-1,j} + \hat{B}_{i,j}\Psi_{i+1,j} + \hat{C}_{i,j}\Psi_{i,j-1} + \hat{D}_{i,j}\Psi_{i,j+1}$$
$$+ \hat{E}_{i,j}[\Psi_{i+1,j+1} - \Psi_{i+1,j-1} - \Psi_{i-1,j+1} + \Psi_{i-1,j-1}] = \hat{Q}_{i,j} \tag{13-37}$$

where the coefficients are defined by

$$\hat{A}_{i,j} = -\left[\frac{C_{i,j}}{(\Delta m)^2} - \frac{E_{i,j}}{2\Delta m}\right] \bigg/ \left[\frac{2A_{i,j}}{(\Delta \eta)^2} + \frac{2C_{i,j}}{(\Delta m)^2}\right] \tag{13-38}$$

$$\hat{B}_{i,j} = -\left[\frac{C_{i,j}}{(\Delta m)^2} + \frac{E_{i,j}}{2\Delta m}\right] \bigg/ \left[\frac{2A_{i,j}}{(\Delta \eta)^2} + \frac{2C_{i,j}}{(\Delta m)^2}\right] \tag{13-39}$$

$$\hat{C}_{i,j} = -\left[\frac{A_{i,j}}{(\Delta \eta)^2} - \frac{D_{i,j}}{2\Delta \eta}\right] \bigg/ \left[\frac{2A_{i,j}}{(\Delta \eta)^2} + \frac{2C_{i,j}}{(\Delta m)^2}\right] \tag{13-40}$$

$$\hat{D}_{i,j} = -\left[\frac{A_{i,j}}{(\Delta \eta)^2} + \frac{D_{i,j}}{2\Delta \eta}\right] \bigg/ \left[\frac{2A_{i,j}}{(\Delta \eta)^2} + \frac{2C_{i,j}}{(\Delta m)^2}\right] \tag{13-41}$$

$$\hat{E}_{i,j} = \frac{B_{i,j}}{2\Delta m \Delta \eta} \bigg/ \left[\frac{2A_{i,j}}{(\Delta \eta)^2} + \frac{2C_{i,j}}{(\Delta m)^2}\right] \tag{13-42}$$

$$\hat{Q}_{i,j} = -[2S\omega \sin \alpha_C]_{i,j} \bigg/ \left[\frac{2A_{i,j}}{(\Delta \eta)^2} + \frac{2C_{i,j}}{(\Delta m)^2}\right] \tag{13-43}$$

The governing equation for Ψ must be solved for a specified number of grid points, I, in the meridional direction and, J, in the tangential direction. To permit application of the periodicity condition outside of the blade passage, an extra grid point in the tangential direction will be used for a total of $N = J + 1$. This extra tangential grid will also be used to handle splitter blades, when they are present. Now let us define a dummy index, k, given by

$$k = (i - 1)N + j \tag{13-44}$$

which maps our grid points into a form suitable for a matrix solution. Note that this mapping relates the grid point indexes, i and j, for the points of interest to the new index k by

$$
\begin{array}{lll}
i-1, j-1 \rightarrow k-N-1 & i, j-1 \rightarrow k-1 & i+1, j-1 \rightarrow k+N-1 \\
i-1, j+1 \rightarrow k-N+1 & i, j \rightarrow k & i+1, j \rightarrow k+N \\
i-1, j \rightarrow k-N & i, j+1 \rightarrow k+1 & i+1, j+1 \rightarrow k+N+1
\end{array}
$$

Hence, Eq. (13-37) at any interior grid point, k, can be written

$$
\hat{E}_k \Psi_{k-N-1} + \hat{A}_k \Psi_{k-N} - \hat{E}_k \Psi_{k-N+1} + \hat{C}_k \Psi_{k-1} + \Psi_k - \hat{D}_k \Psi_{k+1}
$$
$$
- \hat{E}_k \Psi_{k+N-1} + \hat{B}_k \Psi_{k+N} + \hat{E}_k \Psi_{k+N+1} = \hat{Q}_k \tag{13-45}
$$

which can be viewed as a matrix equation of the form

$$
[F_{k,n}][\Psi_k] = [Q_k] \tag{13-46}
$$

where k and n can vary from 1 to $(I \times N)$. The remainder of this section describes some useful techniques for obtaining a very efficient numerical solution to this matrix equation. Readers not interested in those details may want to simply accept the fact that solution is possible and skip to Section 13.3.

The square matrix, $[F_{k,n}]$ is a very sparse matrix with nonzero values limited to the band $(k-N-1) \leq n \leq (k+N+1)$. Hence, there are only $(2N+3)$ possible nonzero values in any row of this matrix; this fact can be used to greatly reduce the computer storage required for this matrix, since only the nonzero entries need be stored in the computer's memory. Indeed, it will be seen that it is necessary to store only nine nonzero matrix elements for each row of the matrix $[F_{k,n}]$. For all interior points, including those on the right-hand side boundary ($j = N - 1$) outside the blade row, the relevant (nonzero) elements of $[F_{k,n}]$ are

$$
\begin{array}{lll}
F_{k,k-N-1} = \hat{E}_k & F_{k,k-N} = \hat{A}_k & F_{k,k-N+1} = -\hat{E}_k \\
F_{k,k-1} = \hat{C}_k & F_{k,k} = 1 & F_{k,k+1} = \hat{D}_k \\
F_{k,k+N-1} = -\hat{E}_k & F_{k,k+N} = \hat{B}_k & F_{k,k+N+1} = \hat{E}_k \\
Q_k = \hat{Q}_k
\end{array}
$$

For points on the left-hand side boundary, i.e., $j = 1$ or $k = (i-1)N+1$, the relevant (nonzero) matrix elements are given by Eq. (13-20), i.e., $\Psi_{i,1} = \Psi_{i,N-1} - 1$

$$
F_{k,k} = 1 \qquad F_{k,k+N-2} = -1 \qquad Q_k = -1
$$

For points on the right-hand side inside the blade row: $\Psi_{i,N-1} = 1$, so the relevant (nonzero) matrix elements are

$$
F_{k,k} = 1 \qquad Q_k = 1
$$

Similarly, outside the blade row the periodicity condition requires $\Psi_{i,N} = \Psi_{i,2} + 1$, or

$$F_{k,k} = 1 \qquad F_{k,k-N+2} = -1 \qquad Q_k = 1$$

At the upstream boundary, the requirement that the meridional derivative of Ψ be zero can be imposed using Eq. (13-33), i.e.,

$$\Psi_{1,j} = \tfrac{4}{3}\Psi_{2,j} - \tfrac{1}{3}\Psi_{3,j} \tag{13-47}$$

This condition appears to complicate the problem by substantially increasing the band width of the possible nonzero elements in the array $[F]$, thus substantially increasing the computer storage requirements for this array. But that problem is easily avoided by introducing Eq. (13-47) directly into the governing equation for $i = 2$ to decouple the matrix inversion from data at $i = 1$. Then the matrix inversion can start at $i = 2$. When Ψ is determined from the matrix inversion, Eq. (13-47) can be used directly to compute Ψ for points on the upstream boundary (i.e., $i = 1$). Introducing Eq. (13-47) into Eq. (13-37) for $i = 2$ yields

$$\Psi_{2,j}[3 + 4\hat{A}_{2,j}] + \Psi_{3,j}[3\hat{B}_{2,j} - \hat{A}_{2,j}] + \Psi_{2,j-1}[3\hat{C}_{2,j} + 4\hat{E}_{2,j}]$$
$$+ \Psi_{2,j+1}[3\hat{D}_{2,j} - 4\hat{E}_{2,j}] + 4\hat{E}_{2,j}[\Psi_{3,j+1} - \Psi_{3,j-1}] = 3\hat{Q}_{2,j} \tag{13-48}$$

and in the form of Eq. (13-45) with $k = N + j$

$$\Psi_k + \Psi_{k+N}\,\frac{3\hat{B}_k - \hat{A}_k}{3 + 4\hat{A}_k} + \Psi_{k-1}\,\frac{3\hat{C}_k + 4\hat{E}_k}{3 + 4\hat{A}_k} + \Psi_{k+1}\,\frac{3\hat{D}_k - 4\hat{E}_k}{3 + 4\hat{A}_k}$$

$$+ \frac{4\hat{E}_k}{3 + 4\hat{A}_k}\,[\Psi_{k+N+1} - \Psi_{k+N-1}] = \frac{3\hat{Q}_k}{3 + 4\hat{A}_k} \tag{13-49}$$

Hence, for $i = 2$ (or $k = N + j$) with $1 < j < N - 1$, the relevant (nonzero) matrix elements are

$$F_{k,k-1} = [3\hat{C}_k + 4\hat{E}_k]/[3 + 4\hat{A}_k] \qquad F_{k,k} = 1$$
$$F_{k,k+1} = [3\hat{D}_k - 4\hat{E}_k]/[3 + 4\hat{A}_k] \qquad F_{k,k+N-1} = -4\hat{E}_k]/[3 + 4\hat{A}_k]$$
$$F_{k,k+N} = [3\hat{B}_k - \hat{A}_k]/[3 + 4\hat{A}_k] \qquad F_{k,k+N+1} = 4\hat{E}_k/[3 + 4\hat{A}_k]$$
$$Q_k = 3\hat{Q}_k/[3 + 4\hat{A}_k]$$

At the downstream boundary, where $i = I$, the requirement that the meridional derivative of Ψ be zero can be imposed using Eq. (13-35), i.e.,

$$\Psi_{I,j} = \tfrac{4}{3}\Psi_{I-1,j} - \tfrac{1}{3}\Psi_{I-2,j} \tag{13-50}$$

and, like the upstream boundary condition, Eq. (13-50) will be introduced directly into Eq. (13-37) for $i = I - 1$ to decouple the matrix inversion from data at $i = I$. Then, after the matrix is inverted, Ψ can be computed on the downstream boundary directly from Eq. (13-50). Thus, at $i = I - 1$

$$\Psi_{i,j} + \Psi_{i-1,j}\frac{3\hat{A}_{i,j} - \hat{B}_{i,j}}{3 + 4\hat{B}_{i,j}} + \Psi_{i,j-1}\frac{3\hat{C}_{i,j} - 4\hat{E}_{i,j}}{3 + 4\hat{B}_{i,j}} + \Psi_{i,j+1}\frac{3\hat{D}_{i,j} + 4\hat{E}_{i,j}}{3 + 4\hat{B}_{i,j}}$$

$$+ \frac{4\hat{E}_{i,j}}{3 + 4\hat{B}_{i,j}}[\Psi_{i-1,j-1} - \Psi_{i-1,j+1}] = \frac{3\hat{Q}_{i,j}}{3 + 4\hat{B}_{i,j}} \qquad (13\text{-}51)$$

It can be seen that Eq. (13-51) contains no terms involving $\Psi_{i+1,j}$, so the matrix inversion need proceed only to $i = I - 1$. In the form of Eq. (13-45) with $k = (I - 2)N + j$, Eq. (13-51) can be written as

$$\Psi_k + \Psi_{k-N}\frac{3\hat{A}_k - \hat{B}_k}{3 + 4\hat{B}_k} + \Psi_{k-1}\frac{3\hat{C}_k - 4\hat{E}_k}{3 + 4\hat{B}_k} + \Psi_{k+1}\frac{3\hat{D}_k + 4\hat{E}_k}{3 + 4\hat{B}_k}$$

$$+ \frac{4\hat{E}_k}{3 + 4\hat{B}_k}[\Psi_{k-N-1} - \Psi_{k-N+1}] = \frac{3\hat{Q}_k}{3 + 4\hat{B}_k} \qquad (13\text{-}52)$$

For $i = I - 1$ [or $k = (I - 2)N + j$] with $1 < j < N - 1$, the relevant (nonzero) matrix elements are

$$\begin{aligned}
F_{k,k-N-1} &= 4\hat{E}_k/[3 + 4\hat{B}_k] & F_{k,k-N} &= [3\hat{A}_k - 4\hat{B}_k]/[3 + 4\hat{B}_k] \\
F_{k,k-N+1} &= -4\hat{E}_k/[3 + 4\hat{B}_k] & F_{k,k-1} &= [3\hat{C}_k - 4\hat{E}_k]/[3 + 4\hat{B}_k] \\
F_{k,k} &= 1 & F_{k,k+1} &= [3\hat{D}_k + 4\hat{E}_k]/[3 + 4\hat{B}_k] \\
Q_k &= 3\hat{Q}_k/[3 + 4\hat{B}_k]
\end{aligned}$$

With specific, but reasonable restrictions, it is rather simple to include the option to treat a set of splitter blades, i.e., a set of short blades starting somewhere inside the passage. The following restrictions are required:

- An odd number of grid points must be used between the side boundaries, i.e., N must be an even integer.
- The splitter blades must be identical to the full blades.
- The splitter blades must be located midway between the full blades.

An additional tangential grid point has already been included to impose the periodicity condition. Within the blade passage, that additional node is not being used, so it can be employed to store data for one side of the splitter blade. Hence, grid point $j = N/2$ will contain data for the first splitter blade surface, and $j = N$ for the second surface. The parameters S and β must be modified when splitter blades are present. In the absence of splitters, $S = r(\theta_1 - \theta_0)$. When a splitter blade is present, its blade metal blockage needs to be included. Since the splitter blades and full blades are assumed to be identical, the modified value is easily shown to be $S = 2r(\theta_1 - \theta_0 - \pi/Z_{FB})$, where Z_{FB} is the number of full blades. Similarly, in the absence of splitter blades, β is given by

$$\cot \beta = \cot \beta_0 + \eta[\cot \beta_1 - \cot \beta_0] \qquad (13\text{-}53)$$

which follows directly from Eq. (13-4). When splitters are present, it is easily shown that

$$\cot \beta = \cot \beta_0 + 2\eta[\cot \beta_1 - \cot \beta_0]; \text{if} \quad j \leq N/2$$

$$\cot \beta = \cot \beta_0 + (2\eta - 1)[\cot \beta_1 - \cot \beta_0]; \text{if} \quad j > N/2 \qquad (13\text{-}54)$$

For any grid point on the splitter blade surface, the relevant equation for the stream function is

$$\Psi_{i,j} = \Psi_{i-1,j}$$
$$\Psi_k = \Psi_{k-N} \qquad (13\text{-}55)$$

Hence, the relevant (nonzero) matrix elements for splitter blade surface points ($j = N/2$) are

$$F_{k,k} = 1 \qquad F_{k,k-N} = -1$$

A review of the above tabulations of the nonzero matrix elements in the array $[F_{k,n}]$ shows that there are only 11 distinct locations in any row. Since $F_{k,k} = 1$ in every case, it need not be stored in memory. Similarly, special handling of all equations corresponding to $j = 1$ will avoid the need for storing the F_{k+N-2} coefficients for those points; therefore, there are actually only nine distinct values to be stored for each row in the matrix. Consequently, computer storage is greatly reduced if a storage matrix is defined $[\hat{F}_{k,m}]$, where k is the same row number as in $[F_{k,n}]$, but m varies only from 1 to 9 (or $N + 2$, if that is larger, as will be seen later). Table 13-1 shows a correspondence between m and n that can be used to recover the proper coefficients for any row, k, from the smaller storage matrix. Since the total number of rows in the matrix is equal to $I \times N$, use of a square matrix to store these data can quickly exhaust available computer memory. By reducing the number of columns to be stored to 9 (or $N + 2$), rather than $I \times N$, a dramatic reduction in computer storage requirements is achieved. To set up the matrices for solution $[\hat{F}_{k,m}]$ and $[Q_k]$ are initialized with all elements equal to zero. Then the nonzero matrix elements for all grid points are inserted in accordance with the tabulations presented for the various grid point types, using the m-to-n correspondence shown in Table 13-1.

Table 13-1. Mapping from Main Matrix to Storage Matrix

m	1	2	3	4	5	6	7	8	9
n	$k - N - 1$	$k - N$	$k - N + 1$	$k - N + 2$	$k - 1$	$k + 1$	$k + N - 1$	$k + N$	$k + N + 1$

The inversion of the matrix is accomplished by first eliminating all $F_{k,n}$ terms for $n < k$ and normalizing each row to yield $F_{k,k} = 1$. This will leave only coefficients with $n \geq k$. We will need to store the results in our storage matrix, $[\hat{F}_{k,m}]$. For $n \geq k$, all nonzero elements in our banded matrix lie in $k \leq n \leq k + N + 1$, i.e., a maximum of $N + 2$ coefficients, so the storage matrix must have at least that many columns. Upon completion of this portion of the inversion, the temporary storage matrix will contain the $F_{k,n}$ coefficients for $k \leq n \leq k + N + 1$, which are mapped into the temporary matrix $[\hat{F}_{k,m}]$ by $m = n - k + 1$. Anticipating this result, all equations for $j = 1$ can be treated immediately since they are not coupled to other equations. Hence, for all $j = 1$ equations, set

$$\hat{F}_{k,1} = 1 \qquad \hat{F}_{k,N-1} = -1 \qquad Q_k = 1$$

Now sweep through the matrix row by row, starting with $k = N + 2$ (or $i = 2, j = 2$) to $k = (I)(N) - N + 1$ (or $i = I - 1, j = N$), but skipping all $j = 1$ equations. Recover all nonzero coefficients from $[\hat{F}_{k,m}]$ and eliminate the $[F_{k,n}]$ coefficients for $n < k$. Each of these is known from preceding equations, so simply subtract the appropriate multiple of the preceding equation from the current one. Divide through the equation by $F_{k,k}$ and store the resulting nonzero coefficients in $[\hat{F}_{k,m}]$, using the mapping $m = n - k + 1$. Next, sweep through the matrix in reverse, bottom to top, eliminating all of the $n > k$ coefficients. Each of these coefficients will be known from subsequent rows of the matrix that have already been treated. When complete, all $F_{k,k} = 1$ and all other coefficients in $[F]$ are zero, so the stream function is given by $\Psi_k = Q_k$. Finally, compute Ψ on the upstream and downstream boundaries using Eqs. (13-47) and (13-50), and the stream function solution is complete.

Programming this matrix solution procedure is more complex than other methods (e.g., relaxation techniques), but the dramatic reduction in computation time and its excellent numerical stability makes it well worth the effort.

13.3 The Gas Density Solution

The stream function solution described above is valid for the gas density values used in the solution. When completed, the relative velocity components at all grid points can be computed from Eqs. (13-11) and (13-12). In general, these velocities will not be consistent with the gas density field used. As was done in Chapter 12 with the linearized blade-to-blade flow analysis, the gas density field is updated in an iterative fashion and the stream function solution is repeated until the gas density field is converged. As shown in Chapter 3, if the inlet total enthalpy is constant in the tangential direction, conservation of energy requires that the rothalpy, R, be constant on a stream surface. From Eqs. (3-4) through (3-7), this requires

$$h = h_t' - \tfrac{1}{2}W^2 = R + \tfrac{1}{2}(\omega r)^2 - \tfrac{1}{2}W^2 \qquad (13\text{-}56)$$

and since the flow is assumed to be isentropic, entropy is also known from the specified inlet conditions. This permits calculation of all other thermodynamic conditions using an appropriate equation of state from Chapter 2. For each iteration on the gas density field, the upstream value of β must be readjusted using Eq. (13-24). Similarly, the Kutta condition, Eq. (13-25), must be imposed on each gas density field iteration. A suitable correction to the trailing edge angular momentum—to reduce any difference between the trailing edge velocities on both sides of the full blades—has been found to be

$$\Delta(rC_U) = \frac{2[(\tilde{W}_0)_{te} - (\tilde{W}_1)_{te}]r\Delta m}{S(\cos\beta_0 + \cos\beta_1)} \tag{13-57}$$

$$\tilde{W}_{te} = \frac{1}{2}W_{te} + \frac{(W_m)_{te}}{\cos\beta_0 + \cos\beta_1} \tag{13-58}$$

Equation (13-58) can be viewed as imposing the Kutta condition as an average for the flow inside and outside the blade row at the trailing edge. For centrifugal compressors, this refinement is usually unnecessary, but for blades having large differences between β_0 and β_1 at the trailing edge (e.g., turbine blades), it results in a much more stable solution and is more representative of the Kutta condition's requirement that the flow leave the blades with the same pressure on both sides. Equation (13-57) is an approximation developed using Stokes' theorem, Eq. (12-53), applied to the closed path formed by the blade surfaces and the last two constant m surfaces on the blade. The downstream boundary's angle can now be adjusted by

$$\Delta(\cot\beta) = \frac{\Delta(rC_U)}{r\int_0^1 W_m\,d\eta} \tag{13-59}$$

With the use of suitable numerical damping methods, this procedure will rapidly converge to the desired trailing edge Kutta condition, usually in about the same number of iterations as that required to converge on the gas density field.

This procedure yields a very stable numerical scheme with rapid convergence, as long as the flow is subsonic throughout. Since the analysis is a potential flow method, the governing equations are elliptic in mathematical form. But when Mach numbers exceed unity, the fluid flow problem becomes hyperbolic in form. It is well known that potential flow methods will diverge when supersonic velocities are encountered. A number of approximations have been used to extend potential flow methods into the transonic flow range. These methods typically adjust the inlet mass flow and blade speed to preserve the basic inlet flow angles while reducing the Mach numbers to subsonic values. When a solution is achieved, the stream function is regarded as a known quantity and some type of streamline curvature calculation is used to compute the velocities and densities corresponding to the actual blade speed and inlet conditions. Katsanis (1969) presents a good example of this type of correction procedure. Its weakness lies in the fact that the resulting flow properties and stream function are no longer consistent, i.e., the solution does not satisfy the condition of irrotational absolute flow. A much better unpublished approach has already been described in Chapter 12, Section 12.11. That approach can also be used in the present analysis to extend it to transonic flow problems. For purposes of clarification, the transonic correction procedure will be illustrated for the type of perfect gas (or pseudo-perfect gas) model discussed in Chapter 2. The procedure can also be used for nonideal gases, but that involves numerical methods for the equation of state calculations. For a thermally and calorically perfect gas model, Eq. (2-21) and (2-41)—expressed in a frame of reference fixed to the blade—can be combined to predict the relative total temperature, T'_{tc}, for which the local velocity, W, and sound speed are equal, i.e.,

$$T'_{tc} = W^2 \left[\frac{1}{\gamma R} + \frac{1}{2c_p} \right] \tag{13-60}$$

Using Eq. (13-60) at all grid points being considered, the change in local relative total temperature required to hold $M' \le 1$ is computed as

$$\Delta T'_t = T'_{tc} - T'_t$$
$$\Delta T'_t \ge 0; \quad \text{required} \tag{13-61}$$

If the largest value of $\Delta T'_t$ is added to the inlet total temperature, it is easily shown from the energy equation that all relative velocities in the flow field will be less than or equal to the local sonic velocity. It is convenient to maintain the inlet relative total density unchanged by this correction. Hence, the corrected inlet total pressure is

$$P_{tc} = P_t[1 + (\Delta T'_t)/T_t] \tag{13-62}$$

Now, we want to correct the stream sheet thicknesses, b, such that the predicted velocity field will be valid for both the actual and the corrected inlet conditions. That is, when the solution is complete, we want the ability to reset the inlet conditions and stream sheet thicknesses to their actual values, recompute all thermodynamic data for the same velocity field and continue to satisfy conservation of mass. Since the velocity field is to be invariant with this transonic correction process, on all iterations, the corrected static density, ρ_c, is given by the analysis and conservation of energy, which permits us to calculate the actual gas density, ρ, that will exist when the temporary inlet conditions are replaced by the actual inlet conditions. Thus, to ensure mass conservation at any meridional computing station, the corrected stream sheet thickness to be used in the analysis is given by

$$b_c = b \int_0^1 (\rho/\rho_c)d\eta \qquad (13\text{-}63)$$

This correction process is accomplished on all iterations until a converged solution is obtained. Hence, the analysis solves a subsonic flow problem for which this potential flow model is valid. As a final precaution, the validity of this transonic correction is checked by integrating $b\rho W_m$ across the passage at each meridional computing station for comparison with the known mass flow. These normally show excellent agreement for modest levels of supersonic Mach numbers, but will eventually deteriorate as the Mach number level is increased. Nevertheless, it does permit solution of the many problems where Mach numbers locally exceed unity. If the Mach number levels get too high for this approach, the general time-marching blade-to-blade flow analysis of Chapter 14 must be used.

13.4   Some Useful Features

It is useful to include a check for choked flow and to take corrective action when it is encountered. Not only is it of value in defining the choke limit to prevent additional solution attempts beyond this limit, but it is quite possible to predict a false choke in the early iterations due to numerical errors. For stations within the blade passage, the local choking mass flow corresponds to the average Mach number reaching unity for the station, i.e.

$$\dot{m}_{ch} = \dot{m} \int_0^1 (W^*/W)d\eta \qquad (13\text{-}64)$$

where $W^* =$ local sonic velocity (without the transonic flow correction). Outside the blade passage, choke is easily shown to be associated with sonic values of the meridional velocity, i.e.

$$\dot{m}_{ch} = \dot{m} \int_0^1 (a/W_m)d\eta \qquad (13\text{-}65)$$

If the estimated choking mass flow is less than the true mass flow, this smaller value is used for the next iteration. As the solution proceeds, it is not uncommon for the estimated choking mass flow to increase as the numerical errors are reduced, often returning to the original specified mass flow by the time convergence is achieved. In these cases, the choke flow logic simply prevents numerical failures while the iteration procedure reduces the numerical errors to eliminate the false choked flow indication.

Like the effect of near-sonic and transonic Mach numbers, solution method used at the blade leading and trailing edges can be a source of numerical problems for this type of solution. The fluid dynamicist's first inclination is to locate the first and last grid points on the blade surfaces at the extreme ends, such that they touch the blades but have no blade thickness (see Fig. 13-3). This can

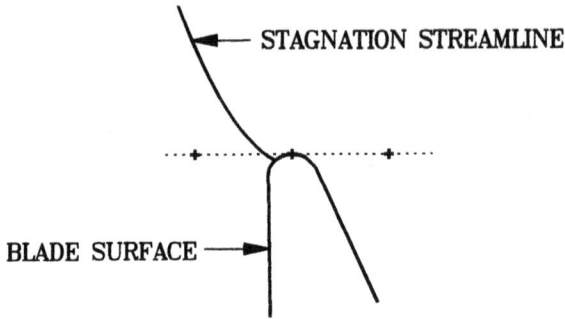

FIGURE 13-3. External Leading-Edge Grid Point

result in considerable ambiguity, since it is not obvious as to where the stagnation streamline impacts with the blade at the leading edge or leaves it at the trailing edge. Unless one uses an extremely fine grid in these regions, the stream function's derivatives needed to determine velocities through Eqs. (13-11) and (13-12) are not well defined. Figure 13-3 illustrates this type of grid structure near the blade leading edge. In the case illustrated, the finite difference approximation to the derivative of Ψ with respect to η at the leading edge grid point and the grid point to its left are quite inaccurate. This follows from the fact that $\Psi = 1$ on both the stagnation streamline and the body surface. Except for the case of one unique approach incidence angle, this situation will always exist for the leading edge grid point and one of the grid points adjacent to it. The one-sided difference approximation of Eq. (13-34) can be used to improve the accuracy, but it requires some complex numerical logic to identify which grid point requires this treatment. And it introduces a destabilizing effect on the numerical analysis, since the stagnation streamline location may alternate from one side of the blade to the other during the early iterations. These problems can be easily eliminated by locating the first and last grid points on the blade surfaces inside the blade passages as illustrated in Fig. 13-4. It can be seen that the finite difference approximations

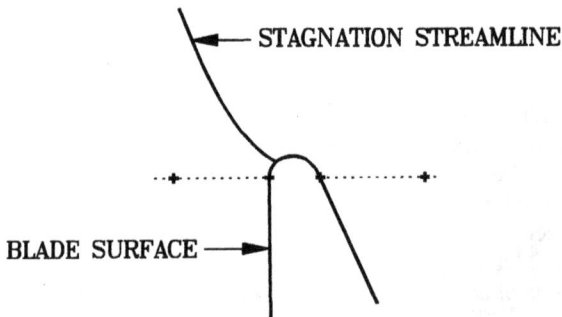

FIGURE 13-4. Internal Leading-Edge Grid Point

are now well defined. Of course, at extremely large approach incidence angles, the stagnation streamline may again move down the blade to produce a situation similar to that shown in Fig. 13-3. But even in these cases, experience has shown that the influence of inaccuracies in the finite difference approximations is localized to one of the blade surface leading or trailing edge grid points. This simple approach to setting the finite-difference grid structure virtually eliminates the numerical inaccuracy and instability problems near the leading and trailing edges induced by the grid structure shown in Fig. 13-3.

13.5 Typical Results

To illustrate results from this two-dimensional blade-to-blade flow analysis, typical axial-flow compressor and turbine blades will be considered. Unlike the centrifugal compressor application, high-quality experimental data are readily available for both axial-flow compressor and axial-flow turbine blade cascade tests—for comparison with the flow predicted by the analysis.

Figure 13-5 compares blade surface velocity distributions predicted by the present blade-to-blade flow analysis technique with the experimental data reported in Dunevant et al. (1955) for a NACA 65-series axial-flow compressor blade. It can be seen that good agreement with the experiment is achieved, except near the tailing edge, where the experimental data shows higher velocities than those predicted by the analysis. The author's blade-to-blade flow analysis incor-

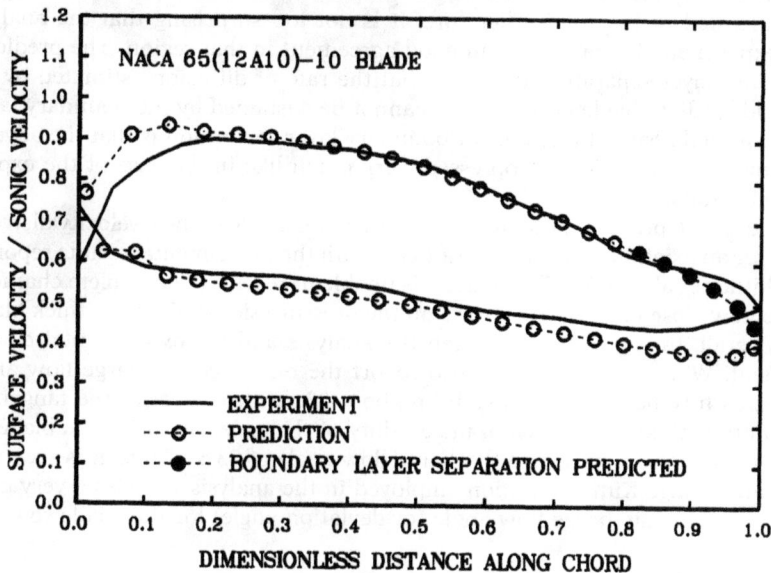

FIGURE 13-5. Axial Compressor Blade Loading

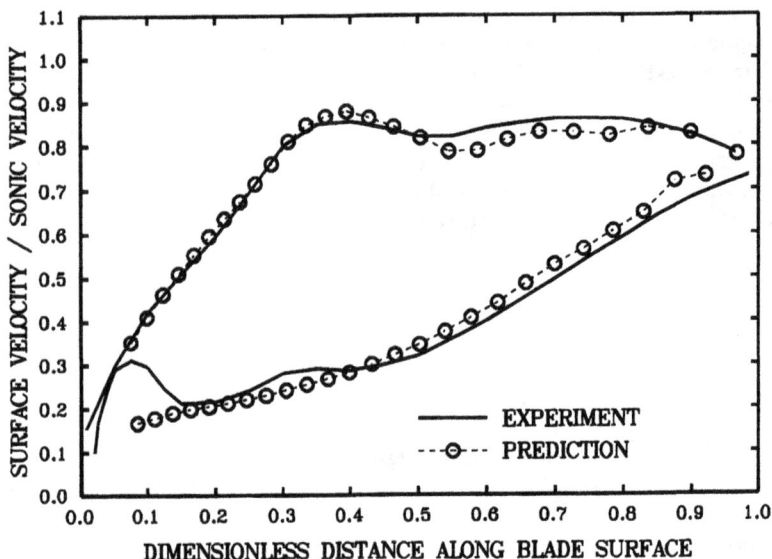

FIGURE 13-6. Axial Turbine Blade Loading

porates the two-dimensional laminar and turbulent boundary layer analysis that is described in Chapter 15. In the blade-to-blade flow analysis of this blade row, the boundary layer analysis indicated that the boundary layer will separate over almost exactly the same region where the predictions and experiment disagree, as illustrated in Fig. 13-5. Therefore, it is not too surprising that the analysis and experimental data are not in good agreement in that region. The predicted boundary layer separation indicates that the rate of diffusion estimated by the inviscid blade-to-blade flow analysis cannot be sustained by the boundary layer. The expected result of this type of boundary layer separation is that the adverse pressure gradient will be suppressed, very much like in the case of the experimental results.

Figure 13-6 presents a comparison of predicted blade surface velocity distributions for an axial-flow turbine stator blade with the experimental results reported in Whitney et al. (1967). The analysis is unable to resolve the complete character of the flow close to the leading edge on the pressure side of this very thick blade; but, overall, the agreement between the analysis and the experimental data is excellent. Whitney et al. (1967) also report the measured discharge flow angle from this turbine stator to have been about 23° with respect to the tangential direction, with an experimental uncertainty of the order of ±0.5°. The blade-to-blade flow analysis predicts a discharge flow angle of 23.5°. Thus, it is seen that the trailing edge Kutta condition employed in the analysis supplies a very good estimate of the discharge flow angle (or deviation angle) for this blade row.

Chapter 14

TIME-MARCHING ANALYSIS OF THE BLADE-TO-BLADE PLANE FLOW

Chapters 12 and 13 have reviewed methods, to predict flow in the blade-to-blade plane, that are suitable for subsonic and slightly supersonic flows. While this covers many centrifugal compressor analysis problems, the design of high pressure ratio compressors can involve a substantial supersonic flow character, beyond the capability of those potential flow methods. A modern design and analysis system should include an analysis capable of considering any Mach number level. The potential flow methods presented in Chapters 12 and 13 are suitable for solving subsonic flow problems where the governing equations are elliptic in mathematical form. There, the flow analysis involves solving a classical boundary value problem. For supersonic flow, the time-steady equations of fluid dynamics become hyperbolic in form. In this case, the solution procedure required is some type of marching solution, such as the method of characteristics. With mixed subsonic-supersonic flow, the problem is highly complex, requiring two types of solution procedures to be matched to each other in some fashion.

A better approach is to employ the time-unsteady form of the governing equations, i.e., Eqs. (3-24) through (3-28). The time-unsteady equations are hyperbolic in time for any Mach number level. Hence, a single numerical technique can be used for subsonic, supersonic or mixed subsonic-supersonic flow problems. Van Neumann and Richtmyer (1950) are generally credited with the first practical application of this technique. The general approach was formerly referred to as the time-dependent method, although today, the term time-marching method is more commonly used. The concept is to solve the time-unsteady equations of motion to advance the solution in time until the solution reaches a steady state. Strictly speaking, a steady state is never reached, since the solution approaches the steady state asymptotically. But for practical purposes, it is only necessary to reach a point where variations in the flow data with time are negligibly small.

One important complication involving this method is that a numerical method explicitly advancing the solution in time is known to be unstable. Von Neumann and Richtmyer (1950) found a means of achieving numerical stability by introducing stabilizing terms similar in form to viscous terms. This is often referred to as "artificial viscosity." Early attempts to use this method often produced less than satisfactory results due to the influence of the stabilizing terms. Alternative methods to stabilize the analysis (e.g., Lax, 1954; Lax and Wendroff, 1964;

Aungier, 1970; 1971) gradually produced accurate numerical representations of the flow fields. Once techniques were developed to control the stabilizing term influence, the time-dependent method became a very powerful way to compute for a wide range of fluid flow problems. This author became involved with the development of this computational method in the late 1960s, while seeking a better analysis technique for hypersonic reentry vehicle flows. That investigation resulted in a time-dependent method that gives the user complete control over the influence of the stabilizing terms on the solution (Aungier, 1970; 1971). It permits the user to always achieve solutions with negligible stabilizing term influence, simply by reducing the time step used to advance the solution in time.

The use of the time-marching method for the blade-to-blade flow application received considerable attention a few years after its successful application to reentry aerodynamics (e.g., Gopalakrishnan and Bozzola, 1973; Denton, 1982). This author has used Aungier's (1970; 1971) method to solve the blade-to-blade flow problem; familiarity with the method played a role in this choice, but it was primarily selected because it offers total control over the influence of stabilizing terms. This time-marching blade-to-blade analysis has been used for radial and axial blades in both compressors and turbines, including an extremely wide range of Mach number levels. This chapter will describe the technique in the context of that application.

NOMENCLATURE

A = area
a = sound speed
B = constant defined in Eq. (14-17)
b = stream sheet width
C = absolute velocity and constant defined in Eq. (14-18)
C_U = absolute tangential velocity
D = constant defined in Eq. (14-19)
E = constant defined in Eq. (14-20)
f = body force term and a general function
g = general function
h = enthalpy
K = constant defined in Eq. (14-31)
M = Mach number
m = meridional coordinate, along stream surface
p = pressure
q_n = velocity defined in Eq. (14-7)
R = rothalpy
r = radius
S = $r(\theta_1 - \theta_0)$
s = entropy
T = temperature
t = time
U = blade speed, ωr
u = velocity component along x coordinate

v = velocity component along y coordinate
W = relative velocity
W_m = meridional velocity
W_n = velocity component normal to ξ coordinate
W_U = relative tangential velocity
W_ξ = velocity component tangent to ξ coordinate
x = coordinate normal to a boundary and spacial coordinate for stability analysis
y = coordinate tangent to a boundary
Z = total number of blades, $Z_{FB} + Z_{SB}$
Z_{FB} = number of full-length blades
Z_{SB} = number of splitter blades
z = axial coordinate
α = flow angle with respect to tangent
α_C = streamline slope angle with axis
β = blade angle or angle of constant η curve with respect to tangent
η = dimensionless tangential coordinate (see Fig. 14-3)
θ = polar angle coordinate
μ = stabilizing term coefficient
ν = stabilizing term coefficient
ξ = streamwise coordinate (see Fig. 14-3)
ρ = gas density
Φ = artificial stabilizing terms
ω = rotation speed

Subscripts

m = partial derivative with respect to m
mm = second partial derivative with respect to m
p = blade pressure surface parameter
s = blade suction surface parameter
t = total thermodynamic condition and partial derivative with respect to time
x = partial derivative with respect to x
xx = second partial derivative with respect to x
y = partial derivative with respect to y
yy = second partial derivative with respect to y
η = partial derivative with respect to η
$\eta\eta$ = second partial derivative with respect to η
0 = parameter on blade surface θ_0
1 = parameter on blade surface θ_1

Superscripts

$'$ = value relative to the blade row's frame of reference
$*$ = sonic condition

14.1 Definition of the Problem

The problem to be solved is quite similar to those defined in Chapters 12 and 13. The flow will be analyzed in a thin stream sheet bounded by stream surfaces as illustrated in Fig. 14-1. The stream surfaces are assumed to be axisymmetric between the blade surfaces. The (ξ, η) coordinates of Chapters 12 and 13, from Eqs. (12-42) and (12-43), will again be used (Fig. 14-2). The flow and thermodynamic conditions will be assumed to be constant on the upstream boundary. From Chapter 3, Eqs. (3-14) through (3-18), the governing equations are

$$\frac{\partial \rho}{\partial t} + \frac{1}{r} \left[\frac{\partial r \rho W_m}{\partial m} + \frac{\partial \rho W_U}{\partial \theta} \right] + \frac{\rho W_m}{b} \frac{\partial b}{\partial m} = 0 \tag{14-1}$$

$$\frac{\partial W_m}{\partial t} + W_m \frac{\partial W_m}{\partial m} + \frac{W_U}{r} \frac{\partial W_m}{\partial \theta} - \frac{\sin \alpha_C}{r} [W_U + \omega r]^2 = -\frac{1}{\rho} \frac{\partial p}{\partial m} \tag{14-2}$$

$$\frac{\partial W_U}{\partial t} + W_m \frac{\partial W_U}{\partial m} + \frac{W_U}{r} \frac{\partial W_U}{\partial \theta} + \frac{W_m \sin \alpha_C}{r} [W_U + 2\omega r] = -\frac{1}{r\rho} \frac{\partial p}{\partial \theta} \tag{14-3}$$

$$\frac{\partial R}{\partial t} - \frac{1}{\rho} \frac{\partial p}{\partial t} + W_m \frac{\partial R}{\partial m} + \frac{W_U}{r} \frac{\partial R}{\partial \theta} = 0 \tag{14-4}$$

One could simply express these equations in the (ξ, η) coordinates, convert them to a finite-difference form and proceed with a numerical solution. But time-marching solutions benefit from a more rigorous approach to the basic conservation relations. Lax (1954) has noted that the equations should be expressed in

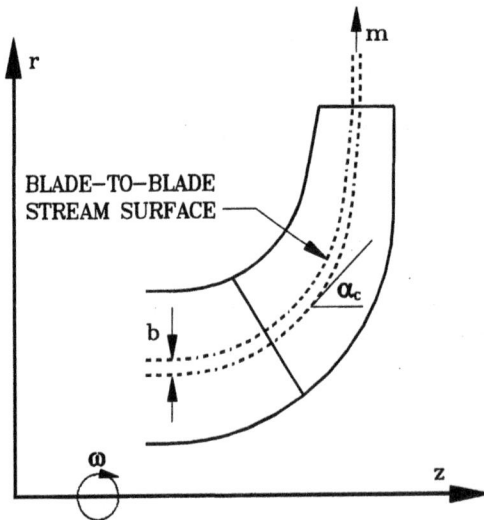

FIGURE 14-1. Blade-to-Blade Stream Sheet

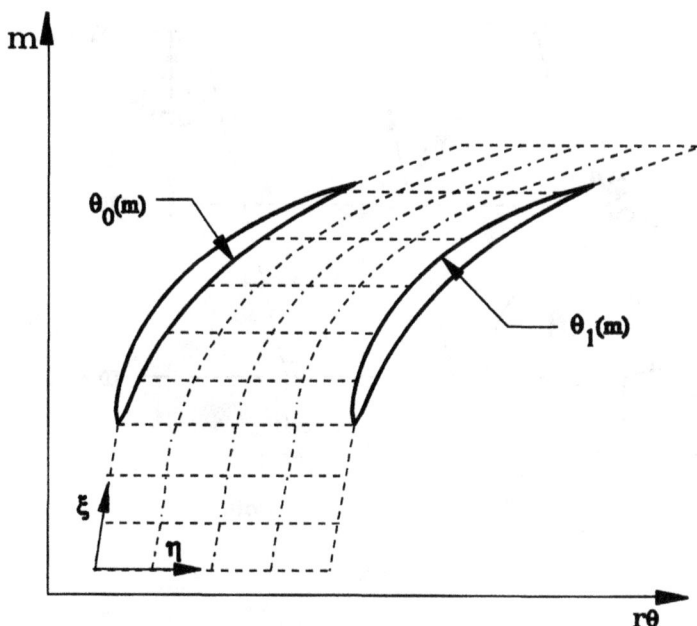

FIGURE 14-2. Blade-to-Blade Flow Grid

conservation form to ensure their validity across shock waves. Shock waves can exist in a time-marching solution, so provision to treat them properly is necessary. Derivation of equations in the conservation form is simplified by introducing the velocity components (W_ξ, W_n) shown in Fig. 14-3 and q_n, defined by

$$W_\xi = W_m \sin \beta + W_U \cos \beta \tag{14-5}$$

$$W_n = W_U \sin \beta - W_m \cos \beta \tag{14-6}$$

$$q_n = W_n / \sin \beta = W_U - W_m \cot \beta \tag{14-7}$$

The governing equations are derived by applying the integral form of the equations of motion to the control cell in Fig. 14-3. The integral form of the continuity equation is

$$\int_V \frac{\partial \rho}{\partial t}\, dV = \int_A \rho \vec{W} \cdot d\vec{A} \tag{14-8}$$

where V and A = volume and area integrals, respectively; and the vector dot product = a component normal to the surface area (with the normal directed outward from the control cell). Integral momentum conservation is given by

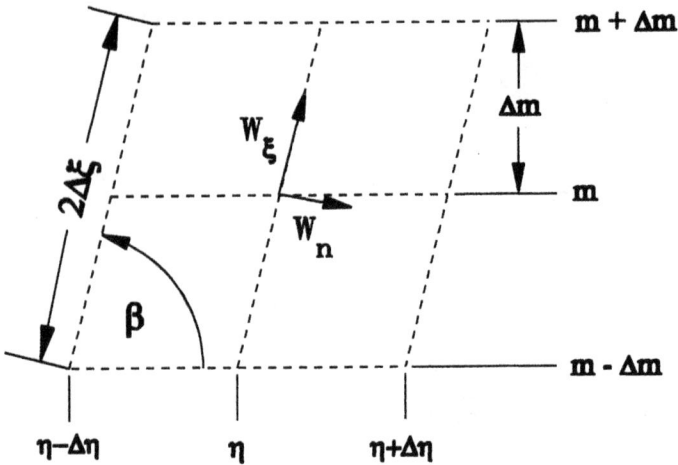

FIGURE 14-3. Control Volume

$$\int_V \frac{\partial \rho \vec{W}}{\partial t}\, dV = \int_A (\rho \vec{W})\vec{W}\cdot d\vec{A} + \int_A p\vec{e}\cdot d\vec{A} = \int_V \vec{f}\,dV \qquad (14\text{-}9)$$

where $\vec{e}$ = a unit vector in the direction of $\vec{W}$ and $\vec{f}$ = a body force. In the case of our rotating curvilinear coordinate system, the body force term will account for the Coriolis and centrifugal acceleration terms. The integral energy equation is

$$\int_V \left[\rho\, \frac{\partial h'_t}{\partial t} - \frac{\partial p}{\partial t} \right] dV + \int_A \rho h'_t \vec{W}\cdot d\vec{A} = \int_V \rho \vec{f}\cdot\vec{W}dV \qquad (14\text{-}10)$$

After some tedious algebra, Eqs. (14-8) through (14-10), applied to the control volume of Fig. 14-3, yield the following set of equations in conservation form:

$$Sb\,\frac{\partial \rho}{\partial t} + \frac{\partial}{\partial m}\, [Sb\rho W_m] + \frac{\partial}{\partial \eta}\, [b\rho q_n] = 0 \qquad (14\text{-}11)$$

$$Sb\,\frac{\partial \rho W_m}{\partial t} + \frac{\partial}{\partial m}\, [Sb(\rho W_m^2 + p)] - \cot\beta\,\frac{\partial bp}{\partial \eta} + \frac{\partial}{\partial \eta}\, [b\rho q_n W_m]$$

$$= \frac{1}{r}\, Sb\rho \sin\alpha_C (W_U + \omega r)^2 + P\,\frac{\partial Sb}{\partial m} \qquad (14\text{-}12)$$

$$Sb \frac{\partial \rho W_U}{\partial t} + \frac{1}{r} \frac{\partial}{\partial m} [rSb\rho W_m(W_U + \omega r)] + \frac{\partial}{\partial \eta} [b(\rho q_n W_U + p)]$$

$$= \omega r \frac{\partial Sb\rho W_m}{\partial m} \tag{14-13}$$

$$Sb \frac{\partial(\rho R - p)}{\partial t} + \frac{\partial}{\partial m} [Sb\rho W_m R] + \frac{\partial}{\partial \eta} [b\rho q_n R] = 0 \tag{14-14}$$

Equations (14-11) through (14-14) are appropriate for solution at all grid points that are not on the blade surface. On the blade surfaces, the velocity normal to the surface must vanish, so only one momentum equation can be solved. Thus, a momentum equation in the ξ direction (see Fig. 14-2) is needed for solution on the blade surfaces. Application of integral momentum conservation, Eq. (14-9), in the ξ direction yields

$$Sb \frac{\partial \rho W_\xi}{\partial t} + \frac{\partial}{\partial m} [Sb(\rho W_m W_\xi + p \sin \beta)] + \frac{\partial \rho b q_n W_\xi}{\partial \eta}$$

$$= p \frac{\partial Sb \sin \beta}{\partial m} + Sb\rho \sin \alpha_C \sin \beta r \omega^2 \tag{14-15}$$

and since the velocity normal to the surface must vanish,

$$W_m = W_\xi \sin \beta; \qquad W_U = W_\xi \cos \beta \tag{14-16}$$

Equations (14-15) and (14-16) replace Eqs. (14-12) and (14-13) for points on the blade surfaces.

14.2 Boundary Conditions

Boundary conditions for the side boundaries of the solution domain illustrated in Fig. 14-2 are basically identical to the potential flow analysis described in Chapter 13. The velocity normal to the blade surfaces, which is imposed by Eq. (14-16), must vanish. For the side boundaries upstream and downstream of the blades, the basic equations of motion can be applied directly, since the flow in all blade passages is assumed to be identical. Hence, a point on these boundaries can be solved just like any other interior point in the solution domain simply by using this "periodicity condition" to extend the solution into the adjacent passages. The boundary conditions imposed on the upstream and downstream boundaries require more care. Specification of valid boundary conditions on these boundaries requires careful consideration of the type of flow involved.

As previously noted, the time-dependent equations of fluid mechanics are hyperbolic in form. The distinguishing property of hyperbolic equations is the existence of certain characteristic directions, usually called characteristics. On a characteristic, derivatives of the dependent variables normal to it may be discontinuous; indeed, the characteristic is a line or surface in the dependent variable space on which such discontinuities can occur. To better understand the physics of this problem, it is useful to present a few basic concepts on unsteady characteristics theory. Consider a simple Cartesian coordinate system with x normal to the boundary and y tangent to it. Conservation of mass, momentum energy can be expressed as

$$\frac{\partial \rho}{\partial t} + \rho \frac{\partial u}{\partial x} + u \frac{\partial \rho}{\partial x} = -B \qquad (14\text{-}17)$$

$$\frac{\partial u}{\partial t} + u \frac{\partial u}{\partial x} + \frac{1}{\rho} \frac{\partial p}{\partial x} = -C \qquad (14\text{-}18)$$

$$\frac{\partial v}{\partial t} + u \frac{\partial v}{\partial x} = -D \qquad (14\text{-}19)$$

$$\frac{\partial R}{\partial t} - \frac{1}{\rho} \frac{\partial p}{\partial t} + u \frac{\partial R}{\partial x} = -E \qquad (14\text{-}20)$$

where B, C, D and E are terms relating to derivatives with respect to y. These can be treated as constants for our purpose of investigating the characteristic directions normal to the boundary. By adding and subtracting Eq. (14-17) from Eq. (14-18), it is easily shown that

$$\frac{\partial u}{\partial t} + (u \pm a) \frac{\partial u}{\partial x} \pm \frac{1}{\rho a} (u \pm a) \frac{\partial p}{\partial x} = -\left(C \pm B \frac{a}{\rho} \right) \qquad (14\text{-}21)$$

where the speed of sound, a, is defined by the isentropic gradient of pressure with respect to density, i.e.,

$$a = \left(\frac{\partial p}{\partial \rho} \right)_S \qquad (14\text{-}22)$$

Now, define directional derivatives by

$$\frac{d}{dt} = \frac{\partial}{\partial t} + (u \pm a) \frac{\partial}{\partial x} \qquad (14\text{-}23)$$

$$\frac{d}{dt} = \frac{\partial}{\partial t} + u \frac{\partial}{\partial x} \qquad (14\text{-}24)$$

and our original set of equations can be written as

$$\frac{du}{dt} \pm \frac{1}{\rho a} \frac{dp}{dt} = -\left(C \pm B \frac{a}{\rho}\right); \qquad \frac{dx}{dt} = u \pm a \qquad (14\text{-}25)$$

$$\frac{dv}{dt} = -D; \qquad \frac{dx}{dt} = u \qquad (14\text{-}26)$$

$$\frac{dR}{dt} - \frac{1}{\rho} \frac{\partial p}{\partial t} = -E; \qquad \frac{dx}{dt} = u \qquad (14\text{-}27)$$

Note that the dependent variables can be completely defined using only information along specific lines in the x-t space, i.e., there are three characteristics associated with our simple problem. The characteristics are defined by

$$\frac{dx}{dt} = u + a \qquad (14\text{-}28)$$

$$\frac{dx}{dt} = u - a \qquad (14\text{-}29)$$

$$\frac{dx}{dt} = u \qquad (14\text{-}30)$$

Equations (14-28) and (14-29) define the characteristics that determine u and p, while Eq. (14-30) defines the characteristic that determines v and R. Figure 14-4 shows an x-t diagram for a point in space, x_0, for subsonic values of u. Two of the characteristics determining the dependent variables at $t + \Delta t$ lie entirely

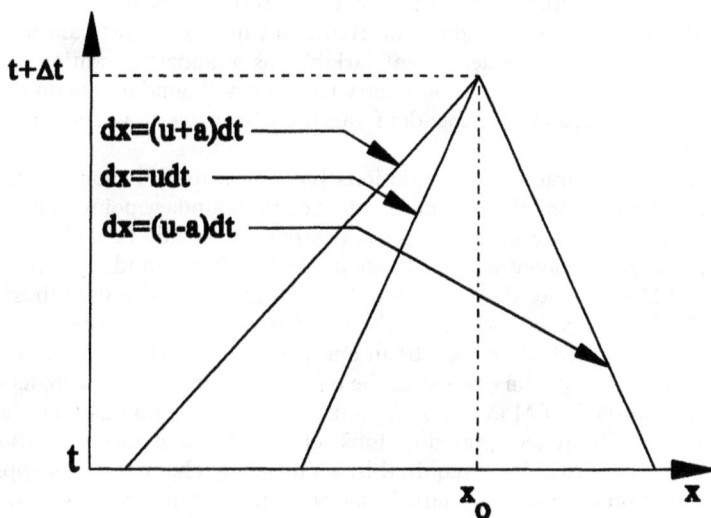

FIGURE 14-4. Subsonic Flow x-t Diagram

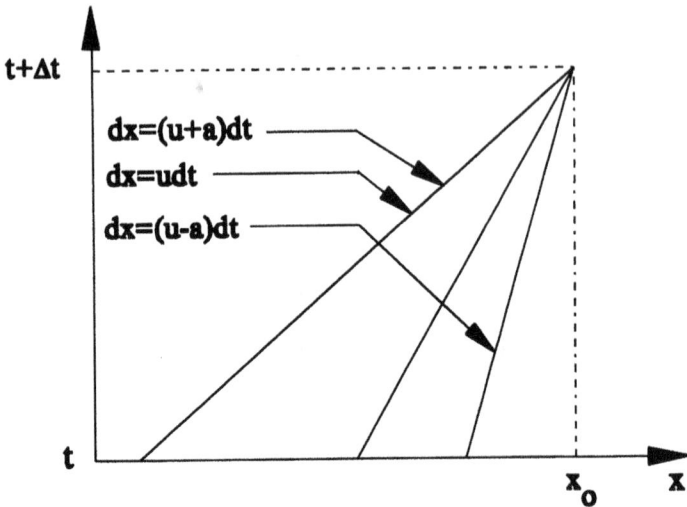

FIGURE 14-5. Supersonic Flow x-t Diagram

upstream of X_0, while the third lies downstream of this point. If x_0 lies on an upstream boundary, this tells us that one dependent variable can be computed from the downstream flow, i.e., as part of the flow field analysis. But the other dependent variables must be assigned as boundary conditions. If x_0 is on a downstream boundary, only one boundary condition is needed, with all other parameters computed from the upstream flow as part of the flow field analysis. Figure 14-5 illustrates the situation for supersonic values of u, showing that all characteristics lie on the upstream side of x_0. Hence, a supersonic upstream boundary requires specification of all dependent variables as boundary conditions. In contrast, a supersonic downstream boundary requires no boundary conditions, i.e., the analysis can predict all dependent variables from the flow field within the solution domain.

The unsteady characteristics equations have been used by some authors as the means of computing the flow parameters on all boundary points. This author used them to advantage for treating the upstream unsteady bow shock boundary on hypersonic reentry vehicles (Aungier, 1970; 1971), an idea adopted from the work of Moretti and Abbett (1966); Moretti and Abbett also used these equations on body surface boundaries, although it is not clear why the added complexity was warranted. Gopalakrishnan and Bozzola (1973) employed unsteady characteristics for boundary points in the blade-to-blade flow problem, again following the approach of Moretti and Abbett. For stationary boundaries, standard one-sided finite difference approximations achieve the same purpose without the cumbersome numerical logic required by an unsteady characteristics approach. The contribution of unsteady characteristics to the present problem is to define the correct handling of boundary points, i.e., the number of dependent parameters that should be specified and the number to be computed.

It can be noted that the hyperbolic form of the governing equations results in quite a different boundary condition logic than that for the elliptic equation solutions in Chapters 12 and 13. For the elliptic equations, all dependent variables have to be assigned on the boundaries (directly or indirectly). As a result, the discharge flow angle is not directly predicted without an additional constraint, such as the Kutta condition. Indeed, elliptic solvers such as those described earlier are valid for any discharge flow angle. In that regard, they cannot truly model the flow field without imposing some constraining assumption. In contrast, there is no such problem for the time-marching analysis. The discharge tangential velocity will be directly predicted, such that no additional (empirical) constraint is required.

The specific choices of dependent variables to be defined as boundary conditions is somewhat arbitrary. Our dependent variables are W_m, W_U, R and ρ (or p). At the upstream boundary, we need to define two boundary conditions when W_m is subsonic. The most logical choices are the tangential velocity, W_U, and the rothalpy, R, since they are parameters always known. Equation (14-11) can be solved for ρ, and W_m and p follow from the equation of state and the definition of rothalpy. For supersonic values of W_m, all dependent variables must be assigned at the upstream boundary. At the downstream boundary, we need one boundary condition for subsonic values of W_m. For supersonic values of W_m, no boundary condition is required but, in general, one is needed. For the blade-to-blade flow problem, it is not particularly meaningful to impose the assumption that the flow on the boundary has a supersonic W_m. The normal practice is to always specify the discharge pressure as the downstream boundary condition. If that assigned pressure requires a supersonic W_m, the solution must ignore the assigned pressure and predict it from the upstream flow. But without the specification of downstream pressure, there is no unique indication of whether W_m should be subsonic or supersonic.

14.3   Fundamental Concepts in Numerical Stability

A critical issue regarding use of the time-dependent method is numerical stability. Aungier (1968) successfully addressed this issue using a stability analysis applied to a simple, linearized momentum equation of the form

$$u_t + K u_x = \mu u_{xx} \qquad (14\text{-}31)$$

where the subscripts denote partial derivatives. Central difference approximations are used for the spatial partial derivatives and a forward difference approximation is used for the partial derivative with respect to time. The term on the right-hand side of Eq. (14-31) is an artificial viscosity or stabilizing term included to obtain a stable numerical scheme

$$u_t \rightarrow [u(x, t + \Delta t) - u(x, t)]/(\Delta t)$$
$$u_x \rightarrow [u(x + \Delta x, t) - u(x - \Delta x, t)]/(2\Delta x)$$
$$u_{xx} \rightarrow [u(x + \Delta x, t) - 2u(x, t) + u(x - \Delta x, t)]/(\Delta x)^2 \qquad (14\text{-}32)$$

Following Von Neumann and Richtmyer (1950), u is replaced by

$$u \rightarrow u_0 + (\delta u) \exp(\alpha t - i\beta x) \qquad (14\text{-}33)$$

where δu is a constant small perturbation on u. Von Neumann's stability criterion is that errors should decay exponentially with time, i.e.,

$$| \exp(\alpha t)| \leq 1 \qquad (14\text{-}34)$$

after substitution of Eqs. (14-32) and (14-33) into Eq. (14-31) and some tedious algebra, Aungier (1968) obtained the result given in Eq. (14-35)

$$\frac{K^2(\Delta t)^2}{(\Delta x)^2} \sin^2(\beta\Delta x) + \left[1 - \frac{2\mu\Delta t}{(\Delta x)^2} (1 - \cos(\beta\Delta x))\right]^2 \leq 1 \qquad (14\text{-}35)$$

Clearly, if μ vanishes, a numerical scheme using a central difference approximation for u_x is always unstable for any positive Δt, i.e., the stability condition cannot be satisfied without the artificial viscosity term. Defining ν as

$$\mu = \nu \frac{(\Delta x)^2}{2\Delta t} \qquad (14\text{-}36)$$

the criterion for stability is

$$\Delta t \leq \frac{\sqrt{\nu}\Delta x}{K} \qquad (14\text{-}37)$$

Indeed the numerical solution will be stable if $0 \leq \nu \leq 1$. From Eqs. (14-23) and (14-25), it can be expected that the linearized momentum equation can be adjusted to the real fluid dynamics problem by substituting the quantity $|u \pm a|$ for K. Hence, the stability criterion becomes

$$\Delta t \leq \frac{\sqrt{\nu}\Delta x}{|u \pm a|} \qquad (14\text{-}38)$$

which reduces to the well-known Courant-Friedricks-Lewy (CFL) stability criterion (Courant et al., 1928) if $\nu = 1$. The CFL limit expresses the maximum time step that can be used in the finite difference solution. Once an unsteady characteristic from an adjacent grid point reaches the point of interest, the information upon which it is based (from the previous time step) ceases to be valid for any further advancement in time. Hence, $\nu \leq 1$ is required.

A similar stability analysis can be conducted for a backward difference approximation for u_x.

$$u_x \rightarrow [u(x,t) - u(x - \Delta x, t)]/\Delta x \tag{14-39}$$

which yields the result

$$\Delta t \leq \frac{\Delta x}{\dfrac{2\mu}{\Delta x} + K} \tag{14-40}$$

which is generalized to the real fluid dynamics problem by referring to Eqs. (14-23) and (14-25) as

$$\Delta t \leq \frac{\Delta x}{\left|\dfrac{2\mu}{\Delta x} + u \pm a\right|} \tag{14-41}$$

This suggests that if $u > a$ it should be possible to achieve a stable numerical solution with no stabilizing terms, which was confirmed to be true by Aungier (1970; 1971). Similarly, if $u = 0$, Eq. (14-41) suggests that stabilizing terms may not be necessary. This situation occurs when u is normal to a solid boundary. Indeed, experience shows that a backward difference approximation normal to a solid boundary does not require artificial stabilizing terms.

Similarly if a forward difference approximation is used for u_x,

$$u_x \rightarrow [u(x + \Delta x, t) - u(x, t)]/\Delta x \tag{14-42}$$

the stability analysis yields the result

$$\Delta t \leq \frac{\Delta x}{\dfrac{2\mu}{\Delta x} - K} \tag{14-43}$$

which is generalized to the real fluid dynamics problem by referring to Eqs. (14-23) and (14-25) as

$$\Delta t \leq \frac{\Delta x}{\left|\dfrac{2\mu}{\Delta x} - u \mp a\right|} \tag{14-44}$$

This suggests that if u is negative and $|u| > a$, it should be possible to achieve a stable numerical solution with no stabilizing terms. Similarly, if $u = 0$, Eq. (14-44) suggests that stabilizing terms may not be necessary. Again, this situation occurs when u is normal to a solid boundary and experience shows that a forward difference approximation normal to a solid boundary does not require artificial stabilizing terms.

Consequently, the simple stability analysis provides excellent guidance with regard to the necessity for artificial stabilizing terms. Aungier (1970; 1971) exploited these results further—the basic stability criterion was inverted to predict the stabilizing coefficient required locally at each grid point in the solution domain, and independently in each coordinate direction. For example, to solve the general equation $g_t = f(x, y, t)$, a stable numerical scheme is developed by introducing stabilizing terms in the form

$$g_t = f(x, y, t) + \mu^{(x)} g_{xx} + \mu^{(y)} g_{yy} \tag{14-45}$$

and inverting Eq. (14-38) to predict the magnitude of the stabilizing coefficients required for stability. Substitution into Eq. (14-36) then yields

$$\mu^{(x)} = (|u| + a)^2 \Delta t / 2; \qquad \mu^{(y)} = (|v| + a)^2 \Delta t / 2 \tag{14-46}$$

When applied to the multidimensional reentry vehicle problem, this consistently yielded a stable numerical scheme. Further, this approach has the unique feature of always employing the minimum magnitude possible for the stabilizing terms, set independently for each coordinate direction. One need only specify Δt to calculate these minimum values. To advance the solution in time, the finite difference form used is

$$g(x, y, t + \Delta t) = g(x, y, t) + [f(x, y, t) + \mu^{(x)} g_{xx} + \mu^{(y)} g_{yy}] \Delta t \tag{14-47}$$

From the form of the stabilizing terms in Eq. (14-46) it can be seen that the stabilizing term influence in Eq. (14-47) is a second-order one with respect to Δt, while the true dynamic terms are first order. As a result, the stabilizing term influence can be made as small as necessary by simply reducing the magnitude of the time step used. Of course, this means that more iterations are needed to reach a quasi-steady-state solution, but the method does ensure that the stabilizing term influence can be negligible if one is willing to perform enough time iterations to achieve the final solution.

14.4 Numerical Stability for the Blade-to-Blade Flow Application

While the procedures described in Section 14.3 proved to be totally adequate for the reentry vehicle application, the highly skewed and distorted grid structure used for the blade-to-blade flow application (see Fig. 14-2) requires some refinement. Direct application of the previous methods discussed yields

$$\mu^{(\xi)} = \tfrac{1}{2}(|W_m| + a)^2 \Delta t \qquad (14\text{-}48)$$

$$\mu^{(\eta)} = \tfrac{1}{2}(|W_U| + a)^2 \Delta t \qquad (14\text{-}49)$$

Direct application of these equations normally leads to a stable solution, but exceptions do occur. For highly skewed grid lines, it is also necessary to check stability requirements along and normal to the ξ coordinate, or occasional numerical instability will be encountered, i.e., one should require

$$\mu^{(\xi)} \geq \tfrac{1}{2}[(|W_\xi| + a)\sin\beta]^2 \Delta t \qquad (14\text{-}50)$$

$$\mu^{(\eta)} \geq \tfrac{1}{2}[(|W_n| + a)/\sin\beta]^2 \Delta t \qquad (14\text{-}51)$$

With these additional constraints, almost all analyses will be numerically stable, but some exceptions have been encountered by the author when the grid is highly skewed and the grid point spacing in the tangential direction is much less than the meridional grid point spacing. Here, the previous stability analyses provide no insight into the problem. From extensive numerical experimentation, an empirical correction was developed to eliminate those problems. If the meridional stabilizing term coefficient from Eq. (14-48) or Eq. (14-50) is increased by

$$\mu^{(\xi)} \to \mu^{(\xi)} + \tfrac{1}{2}[(|W_U| + a)\cos^2\beta\,\Delta m/(S\Delta\eta)]^2 \Delta t \qquad (14\text{-}52)$$

the problem is totally eliminated. Note that this increment is very small unless both β and $S\Delta\eta$ are very small.

For typical centrifugal compressor applications, the above logic is sufficient to provide an accurate numerical method. The author's solution procedure employs one additional feature, motivated primarily for the very high Mach numbers often encountered in turbine blades. The first partial derivatives with respect to m in the governing equations are approximated by a weighted average of forward and backward difference approximations.

$$f_x \to F[f(m + \Delta m, \theta) - f(m, \theta)]/\Delta m + (1 - F)[f(m, \theta) - f(m - \Delta m, \theta)]/\Delta m$$

$$(14\text{-}53)$$

where the weighting factor, F, is given by

$$F = \left[\frac{1}{2} - \frac{2\overline{W}_m|\overline{W}_m|}{(|\overline{W}_m| + \overline{a} + ||\overline{W}_m| - \overline{a}|)^2} \right] \qquad (14\text{-}54)$$

$$\overline{W}_m = \tfrac{1}{4}[W_m(m + \Delta m, \theta) + 2W_m(m, \theta) + W_m(m - \Delta m, \theta)] \qquad (14\text{-}55)$$

$$\overline{a} = \tfrac{1}{4}[a(m + \Delta m, \theta) + 2a(m, \theta) + a(m - \Delta m, \theta)] \qquad (14\text{-}56)$$

and the meridional stabilizing term coefficient is adjusted by

$$\mu^{(\xi)} \to 4\mu^{(\xi)}F(1-F) \qquad (14\text{-}57)$$

This procedure simply extends the stability analysis results presented earlier relative to forward and backward difference approximations. For $W_m = 0$, the meridional derivatives are approximated with standard central differences, using the basic stabilizing term coefficient outlined above. As $W_m \to a$, F and μ both approach 0, i.e., for positive, supersonic values of W_m, the analysis will use backward differences and no stabilizing term with respect to the meridional direction. If W_m is negative, the transition is from central differences to forward differences. Hence, as W_m increases, the analysis progresses from central difference approximations to "upwind" differencing with progressively reduced values of $\mu^{(\xi)}$. For turbine blades with high supersonic discharge Mach numbers, this yields a faster and more stable numerical solution with sharper shock-wave capturing when embedded shocks are present. For typical centrifugal compressor problems, the benefits are less dramatic due to the more modest Mach number levels encountered. In principal, the same process could be used for the first partial derivatives in the tangential direction. It was, in fact, used by this author for a period of time. But occasionally numerical stability problems associated with interactions with the side boundary conditions were encountered. In addition, no obvious benefit could be observed for the high Mach number turbine problems studied. Consequently, the author's analysis now uses standard central difference approximations for the tangential derivatives.

To complete the definition of the stabilizing terms, the time step to be used must be defined and the specific form of the stabilizing terms needs to be specified. The time step is normally set to some fraction of the maximum allowable value given by the CFL limit. During each time iteration, the maximum time step is defined by checking the CFL limit at each grid point, i.e.

$$\Delta t_{\max} \leq \Delta m/(|W_m| + a)$$
$$\Delta t_{\max} \leq S\Delta\eta/(|W_U| + a) \qquad (14\text{-}58)$$

and the time step is set by

$$\Delta t = \mu_0 \Delta t_{\max}$$
$$0 < \mu_0 < 1 \qquad (14\text{-}59)$$

The normal procedure is to start the analysis with a relatively large value of μ_0 (typically about 0.75) to advance the solution in time more rapidly while the solution is far from a steady-state solution. Then, μ_0 is steadily reduced to a value specified by the user of the analysis (typically, about 0.25) to reduce the stabilizing term influence as the solution approaches a steady state. The analysis monitors the rate of change of the dependent variables with time and adjusts μ_0 to approach the lower limit value as the solution approaches its convergence tolerance. This technique provides a good compromise to achieve reasonable computer running times and minimal stabilizing term influence.

The stabilizing term influence can also be reduced by proper selection of the form of the terms used for each equation. It is important to choose a form that is expected to yield minimal magnitudes of the second derivative when the solution approaches a steady state. The following forms have been found to be good choices, where these artificial stabilizing terms are added directly to the right-hand side of the designated equation numbers:

Equation (14 – 11): $\Phi_c = Sb[\mu^{(\xi)}\rho_{mm} + \mu^{(\eta)}\rho_{\eta\eta}]$

Equation (14 – 12): $\Phi_m = [\mu^{(\xi)}(Sb\rho W_m)_{m\dot{m}} + \mu^{(\eta)}(sb\rho W_m)_{\eta\eta}] + \Phi_c W_m$

Equation (14 – 13): $\Phi_U = Sb\rho[\mu^{(\xi)}(rC_U)_{mm} + \mu^{(\eta)}(rC_U)_{\eta\eta}]/r + \Phi_c W_U$

Equation (14 – 14): $\Phi_R = Sb\rho[\mu^{(\xi)}R_{mm} + \mu^{(\eta)}R_{\eta\eta}] + \Phi_c R$

Equation (14 – 15): $\Phi_\xi = \mu^{(\xi)}(Sb\rho W_\xi)_{mm} + \Phi_c W_\xi$

Note that for Eqs. (14-12) through (14-15), a term involving Φ_c has been added. This is a correction to the specific conservation equations, to account for the error in mass conservation caused by the stabilizing term Φ_c. Since each of those equations will be in error by a known amount due to the mass balance error, accuracy is improved by imposing this correction.

14.5   The Solution Procedure

The solution procedure is reasonably straightforward. The user specifies the inlet total thermodynamic conditions, inlet tangential velocity, rotation speed and the discharge static pressure. A finite-difference grid structure is defined over the solution domain as illustrated in Fig. 14-2. Initially, the angle, β, of the grid outside the blade row is set to the mean camberline angle of the blade end point. As the solution proceeds, this angle is continually adjusted to correspond to the flow angle at the point closest to the relevant blade end point. This is done to enable a valid solution of Eq. (14-15) by ensuring that $W_n = 0$ at the side boundary grid point closest to the blade—a necessary condition for a valid finite-difference solution of Eq. (14-15). Typically, initial values for all grid points are assigned by assuming the local static gas density is equal to the local total density (incompressible flow approximation) and that $W_n = 0$ for all grid points in the blade passage. Grid points upstream and downstream of the blade passage are initialized by assuming axisymmetric flow, using conservation of angular momentum to assign W_U and conservation of mass to assign W_m.

The governing equations are then solved for each interior grid point for ρ, W_m, W_U and R. The equation of state and conservation of energy, Eqs. (3-5) and (3-6), yield all other fluid dynamics and thermodynamic data. The upstream and downstream boundary conditions are then imposed and the process is repeated in an iterative fashion until a quasi-steady state is reached, or the user halts the run. The analysis creates a data file, which permits the solution to be restarted from the last time iteration completed. This allows the user to stop the run and complete it later, without starting over; and also to tighten the convergence tolerance and perform additional iterations, if there is any doubt that convergence is adequate. Since the author runs the analysis on personal computers, the analysis is currently restricted to a perfect (or pseudo-perfect) gas equation of state with constant specific heats.

14.6 Typical Results

To illustrate results from this analysis, predictions for the same axial flow compressor and axial flow turbine blades considered in Chapter 13 will be considered. Unlike the centrifugal compressor application, high-quality experimental data are readily available for both axial flow compressor and turbine blade cascade tests for comparison with the flow predicted by the analysis. Figure 14-6 compares predicted blade surface velocity distributions for a NACA 65-series axial flow compressor blade with experimental data reported in Dunevant et al. (1955). For comparison purposes, the predictions from the potential flow analysis of Chapter 13 are also shown. It can be seen that rather good agreement with the experiment is achieved, although the time-marching method tends to predict higher blade surface velocities than either the experimental data or the potential flow analysis. It is interesting that the time-marching solution is in better agreement with experimental data near the leading and trailing edges than is the potential flow analysis. Presumably, the artificial stabilizing terms have an influence similar to viscous effects, which appear to be significant in those regions.

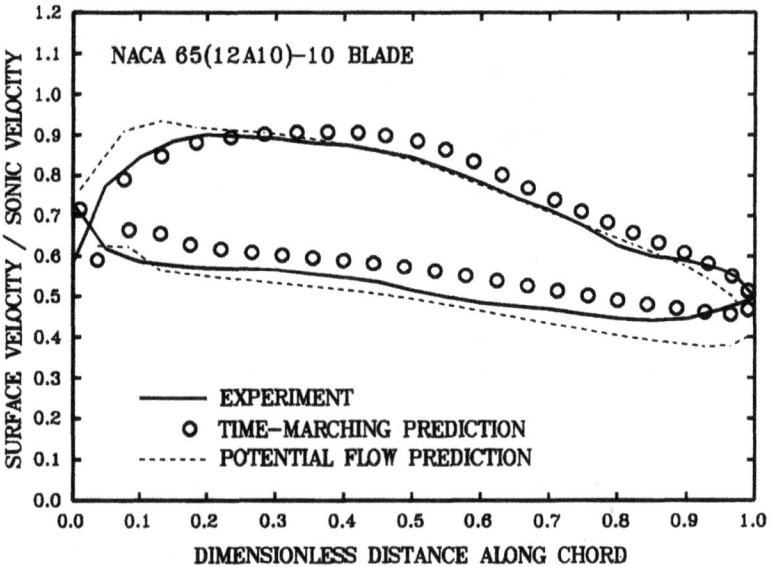

FIGURE 14-6. Axial Compressor Blade Loading

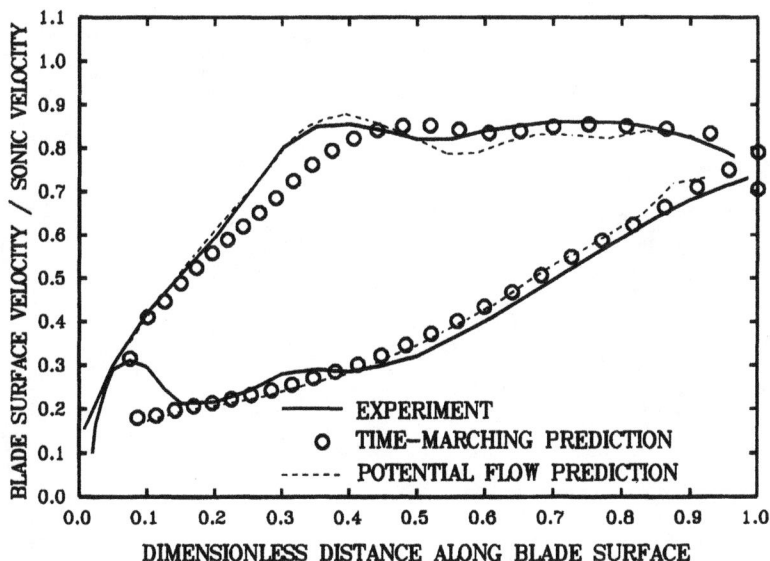

FIGURE 14-7. Axial Turbine Blade Loading

Also, the direct nature of the downstream boundary condition, as opposed to the imposed Kutta condition of Chapter 13, may play a role in this better agreement. Overall, both analytical methods yield rather good predictions. But, for subsonic flow problems such as this, the potential flow analysis would be the preferred method due to the absence of artificial stabilizing term effects, not to mention the much shorter computer running time.

Figure 14-7 presents a similar comparison of predicted blade surface velocity distributions, from the two analyses for an axial flow turbine stator blade, with the experimental results reported in Whitney et al. (1967). In this case, there is less basis for preferring one analysis method over the other. The potential flow analysis appears to resolve the suction surface velocity distribution a little better, but otherwise the two predictions are quite similar. Neither analysis is able to resolve the complete character of the flow close to the leading edge on the pressure side of the blade for this very thick blade, but, overall, the agreement between the analyses and the experimental data is excellent. Whitney et al. (1967) also report the measured discharge flow angle from this turbine stator to have been about 23° with respect to the tangent, with an experimental uncertainty of the order of ±0.5°. The time-marching blade-to-blade flow analysis predicts a discharge flow angle of 23.4° (versus 23.5° for the potential flow analysis). Hence, it is seen that both methods yield a good estimate of the discharge flow angle (or deviation angle) for this blade row.

Chapter 15

BOUNDARY LAYER ANALYSIS

The detailed aerodynamic design of centrifugal compressor stage components relies heavily on the use of internal flow analysis methods, such as the quasi-three-dimensional flow analysis method presented in Chapter 12 and the blade-to-blade flow analyses of Chapters 13 and 14. These so-called Euler methods neglect the effects of viscosity in the interest of obtaining the computational speed and fast response a designer needs for the iterative process of establishing the annulus wall contours and blade geometry. These inviscid flow analyses provide valuable guidance to the designer in the form of blade loading distributions and velocity or pressure distributions along the gas path boundaries. But the omission of the important viscous effects means that the designer's skill and experience in the interpretation of the results is critical to success. Experienced designers develop very definite opinions as to blade loading styles that will be effective and to levels of adverse pressure gradients that can be tolerated. In effect, they apply judgment as to how viscous effects will modify the inviscid flow analyses they use for detailed design.

Boundary layer theory provides a useful supplement to the Euler methods that can provide an approximate evaluation of viscous effects. Analysis of two-dimensional or axisymmetric, three-dimensional boundary layers can be added on to an inviscid flow analysis with no significant increase in computer running time. The basic assumption of boundary layer theory is that viscous effects are confined to a thin layer close to the physical surfaces. In many cases, this is far from true in a centrifugal compressor. In addition, boundary layers in centrifugal compressors are almost always highly three-dimensional in nature. So, addition of these approximate boundary layer analyses yields a simplified or idealized model of viscous effects, but it does bring the Euler methods used in detailed design one step closer to the real flow problem.

In contrast, the general three-dimensional boundary layer problem is very complex, resulting in computer running times too great for it to be used in the iterative component design process. In principle, general three-dimensional boundary layer analysis could be used for final evaluation of a component design. But today it is reasonably common for designers to have access to a viscous flow computational fluid dynamics (CFD) code for this purpose. Availability of the more exact viscous CFD codes has largely eliminated efforts to use general three-dimensional boundary layer analysis in centrifugal compressor design and analysis.

A general overview of the relevant boundary layer approximations and derivation of the basic governing equations is supplied in Chapter 3, Section 3.4. Here, specific useful boundary layer analysis techniques will be described. Two-dimensional boundary layer theory provides an approximate analysis for blade surface boundary layers. And axisymmetric, three-dimensional boundary layer theory presents an approximate analysis for boundary layers along the hub-and-shroud walls. The author's design system employs both these models for the analysis described in Chapter 12, and the two-dimensional model for the analyses described in Chapters 13 and 14.

NOMENCLATURE

b = stream sheet thickness
C = absolute velocity
c_f = skin friction coefficient
E = boundary layer entrainment function
f = blade force per unit mass
H = shape factor, δ^*/θ
H_k = shape factor defined by Eq. (15-25)
H_1 = shape factor, δ_1^*/θ_{11}
H_2 = shape factor, δ_2^*/θ_{22}
K = parameter defined by Eq. (15-10)
m = meridional coordinate and tangential velocity profile power-law exponent
n = meridional velocity profile power-law exponent
p = pressure
R = rothalpy
Re_θ = momentum thickness Reynolds number, Eq. (15-23)
r = radius
S = shape factor defined in Eq. (15-26) and the spacing between blades
u = velocity in two-dimensional boundary layer analysis (equivalent to W)
V_U = tangential velocity component relative to the wall, $W_U + r(\omega - \omega_w)$
W = velocity relative to a blade row
x = distance along the wall
y = distance normal to the wall
z = axial coordinate
α = flow angle with respect to the tangential direction
α_C = wall slope angle with axial direction
δ = boundary layer thickness
δ_E = energy thickness
δ_h = enthalpy thickness
δ_u = velocity thickness
δ' = density thickness
δ^* = displacement thickness
δ_1^* = meridional displacement thickness
δ_2^* = tangential displacement thickness
η = dimensionless distance across boundary layer
θ = tangential coordinate (polar angle) and momentum thickness
θ_{11} = meridional momentum thickness

θ_{12} = meridional momentum flux thickness
θ_{22} = tangential momentum thickness
Λ = shape factor defined by Eq. (15-8)
μ = gas viscosity
ρ = gas density
τ = boundary layer shear stress
v_1 = meridional blade force defect thickness
v_2 = tangential blade force defect thickness
ω = rotation speed

Subscripts

e = boundary layer edge parameter
m = meridional component
s = parameter along boundary layer edge streamline direction
t = total thermodynamic condition
U = tangential component
w = parameter at a wall

Superscripts

$'$ = value relative to rotating frame of reference

15.1 Two-Dimensional Laminar Boundary Layer Analysis

While boundary layers in centrifugal compressors are almost always turbulent, the early portion of blade surface boundary layers is laminar. Hence, laminar boundary layers must be included in the two-dimensional boundary layer analysis. The author prefers the method of Gruschwitz (1950), which is reviewed in some detail by Schlichting (1968). Basically, this is a generalization of the classical Karmen-Pohlhausen solution (Pohlhausen, 1921) to treat compressible flows. The basic equation to be solved is the momentum-integral equation

$$\frac{1}{b} \frac{\partial b \rho_e u_e^2 \theta}{\partial x} + \rho_e u_e \delta^* \frac{\partial u_e}{\partial x} = \tau_w \tag{15-1}$$

which was derived earlier as Eq. (3-38). Here, b = stream sheet thickness; ρ_e and u_e = the boundary layer edge density and velocity, respectively; and τ_w = wall shear stress. The boundary layer thicknesses used are

$$\delta^* = \int_0^\delta \left[1 - \frac{\rho u}{\rho_e u_e} \right] dy = displacement\ thickness \tag{15-2}$$

$$\theta = \int_0^\delta \frac{\rho u}{\rho_e u_e} \left[1 - \frac{u}{u_e} \right] dy = momentum\ thickness \tag{15-3}$$

Gruschwitz (1950) employs a general boundary layer velocity profile of the form

$$\frac{u}{u_e} = C_1 \eta + C_2 \eta^2 + C_3 \eta^3 + C_4 \eta^4 \tag{15-4}$$

where u = boundary layer velocity. The parameter η is defined by

$$\eta = \frac{1}{\delta'} \int_0^y \frac{\rho}{\rho_e} dy \tag{15-5}$$

$$\delta' = \int_0^\delta \frac{\rho}{\rho_e} dy \tag{15-6}$$

where δ = boundary layer thickness. By requiring the boundary layer profile to match the boundary layer edge velocity distribution, the coefficients in Eq. (15-2) are found to be

$$C_1 = 2 + \Lambda/6; C_2 = -\Lambda/2$$
$$C_3 = \Lambda/2 - 2; C_4 = 1 - \Lambda/6 \tag{15-7}$$

where Λ = a boundary layer shape factor defined by

$$\Lambda = \frac{\rho_e^2}{\rho_w} \frac{(\delta')^2}{\mu} \frac{du_e}{dx} \tag{15-8}$$

Gruschwitz defines two additional parameters convenient for the analysis as

$$b_0 = \frac{T_w}{T_e} = \frac{\rho_e}{\rho_w} \tag{15-9}$$

$$K = \Lambda \left(\frac{\theta}{\delta'} \right)^2 = b_0 \frac{\rho_e \theta^2}{\mu} \frac{du_e}{dx} \tag{15-10}$$

where μ = fluid viscosity; and the following boundary layer thicknesses are used

$$\delta_E = \int_0^\delta \frac{\rho u}{\rho_e u_e} \left[1 - \frac{u^2}{u_e^2} \right] dy = energy\ thickness \tag{15-11}$$

$$\delta_h = \int_0^\delta \frac{\rho u}{\rho_e u_e} \left[\frac{h}{h_e} - 1 \right] dy = enthalpy\ thickness \tag{15-12}$$

$$\delta_u = \int_0^\delta \left[1 - \frac{u}{u_e} \right] dy = velocity\ thickness \tag{15-13}$$

With these definitions, the basic boundary layer thicknesses are directly related to Λ by

$$\frac{\theta}{\delta'} = \frac{37}{315} - \frac{\Lambda}{945} - \frac{\Lambda^2}{9072} \tag{15-14}$$

$$\frac{\delta_E}{\delta'} = \frac{798048 - 4656\Lambda - 758\Lambda^2 - 7\Lambda^3}{4324320} \tag{15-15}$$

$$\frac{\delta_u}{\delta'} = \frac{3}{10} - \frac{\Lambda}{120} + \frac{Fu_e^2}{2c_p T_e} \tag{15-16}$$

$$F = 0.232912 - 0.831483 \left(\frac{\Lambda}{100} \right)$$

$$+ 0.650584 \left(\frac{\Lambda}{100} \right)^2 + 17.8063 \left(\frac{\Lambda}{100} \right)^3 \tag{15-17}$$

Gruschwitz's (1950) analysis treats adiabatic walls and any value of the gas Prandtl number. There is no real justification for the added complexity in the present application, since the analysis is only an approximation of the real flow. Hence, it will be assumed that the Prandtl number is unity (i.e., $T_w = T_t'$, the total temperature relative to the blade row frame of reference). Then

$$\delta_h = \frac{u_e^2}{2c_p T_e} \delta_E \tag{15-18}$$

where c_p = specific heat at constant pressure. From the above, Gruschwitz developed the universal functions

$$\delta^* = \delta_h + \delta_u \tag{15-19}$$

$$\tfrac{1}{2}c_f = \frac{\tau_w}{\rho_e u_e^2} = \frac{\mu}{\rho_e u_e \delta'} \left(2 + \frac{\Lambda}{6} \right) \tag{15-20}$$

$$b_0 = \frac{T'_t}{T_e} \tag{15-21}$$

The above equations are sufficient to solve the momentum integral equation. It can be seen that Λ assumes the role of a primary boundary layer profile shape factor in this analysis. In the process of integrating Eq. (15-1) along the blade surface, it is required that $-12 \le \Lambda \le 12$, with $\Lambda = -12$ corresponding to boundary layer separation. It can be shown that K and Λ are directly related by

$$K = \left(\frac{37}{315} - \frac{\Lambda}{945} - \frac{\Lambda^2}{9072} \right)^2 \Lambda \tag{15-22}$$

The solution consists of integrating Eq. (15-1) along the wall, starting with $\theta = 0$ at the first station (blade leading edge). At each subsequent station, the solution is iterative in nature, proceeding as follows:

1. Compute b_0 from Eq. (15-21) and estimate θ.
2. Compute K from Eq. (15-10).
3. Compute Λ from Eq. (15-22).
4. Compute δ' from Eq. (15-14).
5. Compute δ^* from Eq. (15-15) through (15-19).
6. Compute τ_w from Eq. (15-20).
7. Recompute θ from Eq. (15-1).
8. Repeat steps 2 through 7 until convergence on θ is acheived

Eventually, the boundary layer will undergo transition to turbulent flow. Many different criteria for transition have been suggested. The author prefers to use the following momentum thickness Reynolds number relation as the criterion for boundary layers to be turbulent:

$$Re_\theta = \frac{\rho_e u_e \theta}{\mu} > 250 \tag{15-23}$$

15.2 Two-Dimensional Turbulent Boundary Layer Analysis

The turbulent boundary layer is conveniently analyzed using the Entrainment Method of Head (Head, 1958; 1968; Cumpsty and Head, 1967; Green, 1968; Rotta, 1966; Summer and Shanebrook, 1971). This involves solving the equations of mass and momentum conservation in integral form, i.e., Eq. (15-1) and

$$\frac{d}{dx} [b\rho_e u_e(\delta - \delta^*)] = b\rho_e u_e E \tag{15-24}$$

where E is an empirical entrainment function governing the rate at which fluid is entrained into the boundary layer at the boundary layer edge. Equation (15-24) was derived in Chapter 3 as Eq. (3-40). The method reviewed here is based on the work of Head (1968) with Green's (1968) modifications for compressible flow. The kinematic shape factor, H_k, and a shape factor, S, are defined as

$$H_k = \frac{1}{\theta} \int_0^\delta \frac{\rho}{\rho_e} \left[1 - \frac{u}{u_e} \right] dy \qquad (15\text{-}25)$$

$$S = (\delta - \delta^*)/\theta \qquad (15\text{-}26)$$

H_k is used as the primary boundary layer shape factor. Empirical relations must be used to relate E and S to H_k. The relations used are

$$E = 0.025(H_k - 1) \qquad (15\text{-}27)$$

$$H_k = 1 + [0.9/(S - 3.3)]^{0.75} \qquad (15\text{-}28)$$

and are illustrated in Fig. 15-1. For incompressible flow, the well-known Ludwieg-Tillmann skin friction coefficient relation Ludwieg and Tillmann (1950) is

$$c_{fi} = 0.246 e^{-1.561 H_k} Re_\theta^{-0.268} \qquad (15\text{-}29)$$

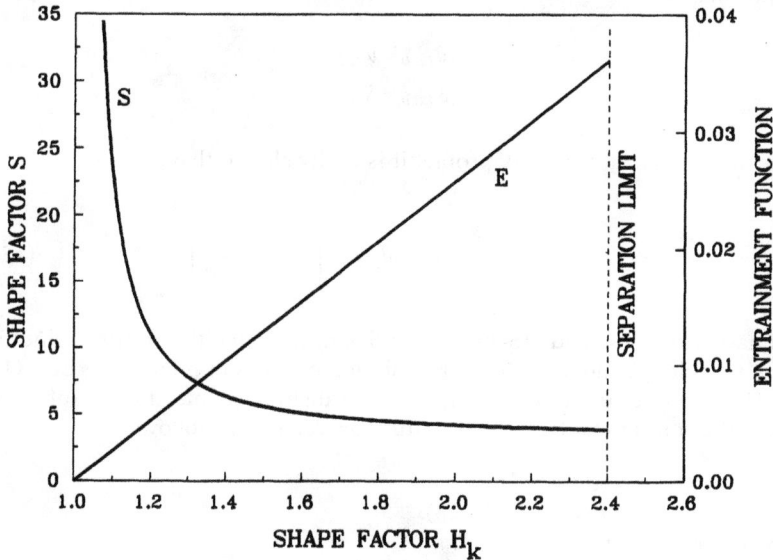

FIGURE 15-1. Boundary Layer Shape Factors

Green also considers the case of adiabatic walls with any value of the Prandtl number. As with the laminar case, the approximate nature of the boundary layer analysis for the present application does not warrant the added complexity. Hence, it will be assumed that the Prandtl number is unity, to obtain a simpler relation between H and H_k, i.e.

$$H = \delta^*/\theta = (H_k + 1)T'_t/T_e - 1 \tag{15-30}$$

Green corrects the skin friction coefficient for compressibility effects by

$$c_f = c_{fi}H_k(H_k + 1)/[2H_k + H(H_k - 1)] \tag{15-31}$$

Equation (15-1) and Eqs. (15-24) through (15-31) are sufficient to solve the two-dimensional turbulent boundary layer problem. To start the turbulent boundary layer analysis, conventional practice is to use the momentum thickness from the laminar solution at transition along with some assumption on how the shape factor, H, changes at transition. By using the Gruschwitz laminar solution procedure and the entrainment method for the turbulent case, a more fundamental approach can be used. Clearly, it is necessary that mass and momentum be conserved at the transition point. This requires that

$$(\delta - \delta^*)_{turb} = (\delta - \delta^*)_{lam}$$
$$(\delta - \delta^* - \theta)_{turb} = (\delta - \delta^* - \theta)_{lam} \tag{15-32}$$

or equivalently

$$\theta_{turb} = \theta_{lam}$$
$$S_{turb} = S_{lam} \tag{15-33}$$

Using the Gruschwitz velocity profile, it is easily shown that

$$(\delta - \delta^*) = \int_0^\delta \frac{\rho u}{\rho_e u_e}\, dy = \delta'\left[\frac{7}{10} + \frac{\Lambda}{120}\right] \tag{15-34}$$

Equations (15-14) and (15-34) are sufficient to define the starting conditions for the turbulent boundary layer. The solution consists of integrating Eqs. (15-1) and (15-24) along the wall, starting at the transition point. At each subsequent station, the solution is iterative in nature, proceeding as follows:

1. Estimate θ and $(\delta - \delta^*)$
2. Compute H_k from Eq. (15-28).
3. Compute E from Eq. (15-27).
4. Compute H and δ^* from Eq. (15-30).
5. Compute c_f from Eqs. (15-29) and (15-31).

6. Solve Eqs. (15-1) and (15-24) for θ and $(\delta - \delta^*)$.
7. Repeat steps 2 through 6 until convergence on θ and $(\delta - \delta^*)$ is achieved

Throughout the solution process, this author uses $H_k > 2.4$ as a boundary layer separation criterion. The analysis limits H_k to a maximum of 2.4 to avoid solution divergence when this situation is encountered. In this way, the analysis can continue through a separation zone. This is necessary for the centrifugal compressor application since a separated boundary layer may well reattach further downstream.

15.3 Blade Passage Profile Losses

Once the two-dimensional boundary layer analysis is accomplished along both blade surfaces, it is useful to convert the trailing edge boundary layer data into an equivalent total pressure loss coefficient for purposes of evaluating the design. Designers must recognize the limitations involved and use the results in a qualitative rather than quantitative sense. The boundary layer analysis conducted has ignored the important three-dimensional effects. In the real flow case, there is usually substantial secondary flow, i.e., boundary layer fluid migrates across the inviscid flow stream surfaces, usually causing significantly higher losses than indicated by the two-dimensional boundary layer analysis. Also, boundary layer separation is often encountered, which the basic boundary layer analysis cannot handle correctly. In the presence of separation, many of the fundamental assumptions of boundary layer theory cease to be valid. Nevertheless, the calculated loss coefficient is useful as a means of evaluating design alternatives with respect to each other.

Lieblein and Roudebush (1956) develops the following equation for the total pressure loss coefficient based on the trailing edge velocity pressure:

$$\overline{\omega} = \frac{\Delta p_t}{p_{t,id} - p} = \sum \frac{2\theta}{S \sin \beta} \left[1 + \frac{\theta}{2S} \frac{H^2}{\sin \beta} \right] \tag{15-35}$$

where now S = the spacing between the blades; $p_{t,id}$ = trailing edge ideal (no loss) total pressure; p = trailing edge static pressure; and the summation is for the boundary layers on both blade surfaces. If preferred, this is easily converted to a loss coefficient based on the leading edge velocity pressure simply by multiplying it with the ratio of the trailing-edge-to-leading-edge velocity pressure. In the case of rotating blades, this can also be used to compute an efficiency if the user prefers.

15.4 End-Wall Turbulent Boundary Layer Analysis

The analysis of boundary layers along the hub-and-shroud end walls of a centrifugal compressor component is a more complicated problem than the blade surface case. Almost all end-wall boundary layers are highly three-dimensional in nature. End walls can be rotating or stationary. In vaned components, blade forces act on the boundary layer fluid and must be considered in the analysis. With these complications, there is little merit in conducting two-dimensional boundary layer analyses in the end-wall region.

The axisymmetric, three-dimensional boundary layer model is about the only approximation that can yield meaningful results, yet require practical computer running times for supporting the iterative component design process. Indeed, for vaneless components this model yields excellent results, as shown in Aungier (1988b), Senoo et al. (1977) and Davis (1976). In vaned components, it is clear that the boundary layers are not axisymmetric. Yet this approximation makes some sense in terms of modeling an "average" boundary layer behavior, as discussed in Horlock (1970). Consequently, this is the model used by the author for this purpose. The fundamental integral equations governing the axisymmetric, three-dimensional turbulent boundary layer problem in a rotating coordinate system are developed in Chapter 3, Section 3.4. It is convenient to express these equations in a coordinate system fixed to the wall rather than to the blade. To maintain a nomenclature consistent with previous chapters, the velocity relative to the wall will be denoted as V, while absolute velocities continue to be denoted by C, and velocities relative to blade rows by W. Hence, if ω is the blade row angular velocity and ω_w is the wall angular velocity

$$V_m = W_m = C_m \tag{15-36}$$

$$V_U + r\omega_w = W_U + r\omega = C_U \tag{15-37}$$

Equations (3-44) through (3-53) are valid for any rotating coordinate system. So here they can be used directly by substituting V for W and ω_w for ω, i.e.

$$\frac{\partial}{\partial m}\left[r\rho_e V_{me}(\delta - \delta^*)\right] = r\rho_e V_e E \tag{15-38}$$

$$\frac{\partial}{\partial m}\left[r\rho_e V_{me}^2 \theta_{11}\right] + \delta_1^* r\rho_e V_{me}\frac{\partial V_{me}}{\partial m}$$

$$- \sin\alpha_C \rho_e V_{Ue}[V_{Ue}(\delta_2^* + \theta_{22}) + 2\omega_w r\delta_2^*]$$

$$= r\tau_{mw} + r\delta(f_{me} - f_m) \tag{15-39}$$

$$\frac{\partial}{\partial m}\left[r^2\rho_e V_{me}V_{Ue}\theta_{12}\right] + r\delta_1^*\rho_e V_{me}\left[r\frac{\partial V_{Ue}}{\partial m} + \sin\alpha_C(V_{Ue} + 2\omega_w r)\right]$$

$$= r^2\tau_{Uw} + r^2\delta(f_{Ue} - f_U) \tag{15-40}$$

where the various mass and momentum defect thicknesses are defined by

$$\rho_e V_{me} \delta_1^* = \int_0^\delta (\rho_e V_{me} - \rho V_m) dy \qquad (15\text{-}41)$$

$$\rho_e V_{me}^2 \theta_{11} = \int_0^\delta \rho V_m (V_{me} - V_m) dy \qquad (15\text{-}42)$$

$$\rho_e V_{me} V_{Ue} \theta_{12} = \int_0^\delta \rho V_m (V_{Ue} - V_U) dy \qquad (15\text{-}43)$$

$$\rho_e V_{Ue} \delta_2^* = \int_0^\delta (\rho_e V_{Ue} - \rho V_U) dy \qquad (15\text{-}44)$$

$$\rho_e V_{Ue}^2 \theta_{22} = \int_0^\delta \rho V_U (V_{Ue} - V_U) dy \qquad (15\text{-}45)$$

The last term on the right-hand side of Eqs. (15-39) and (15-40) are blade force defect terms. They account for variations in the blade force components through the boundary layer. The blade force components at the boundary layer edge can be computed from the boundary layer edge conditions using conservation of momentum, as outlined in Chapter 3, i.e.

$$f_{me} = \rho_e V_{me} \frac{\partial V_{me}}{\partial m} + \frac{\partial p_e}{\partial m} - \rho_e (V_{Ue} + r\omega_w)^2 \frac{\sin \alpha_C}{r} \qquad (15\text{-}46)$$

$$f_{Ue} = \rho_e V_{me} \left[\frac{\partial V_{Ue}}{\partial m} + \frac{\sin \alpha_C}{r} (V_{Ue} + 2r\omega_w) \right] = \frac{\rho_e V_{me}}{r} \frac{\partial r C_{Ue}}{\partial m} \qquad (15\text{-}47)$$

If the blade force components are constant through the boundary layer, the blade force defect terms in the momentum integral equations will be zero. In the case of vaneless components, normally all blade force terms will be zero, although Aungier (1988b) makes use of these force defect terms to account for the special cases of merged or separated boundary layers in a coupled viscous-inviscid flow analysis.

As with any turbulent boundary layer analysis, the solution requires additional empirical models. For this purpose, Aungier (1988b) developed empirical models that were shown to yield results in good agreement with experimental data for vaneless components. That method employs power-law velocity profiles as the basis for solution.

$$V_m = V_{me} \left(\frac{y}{\delta} \right)^n \qquad (15\text{-}48)$$

$$V_U = V_{Ue} \left(\frac{y}{\delta} \right)^m \qquad (15\text{-}49)$$

Substitution of the power-law profiles into the definitions of the defect thicknesses while assuming density is essentially constant, and yields

$$n = \theta_{11}/(\delta - \delta_1^* - 2\theta_{11}) \tag{15-50}$$

$$m = \theta_{12}(n + 1)^2/[\delta - \theta_{12}(n + 1)] \tag{15-51}$$

$$H_1 = \delta_1^*/\theta_{11} = 2n + 1 \tag{15-52}$$

$$\delta - \delta_1^* = 2H_1\theta_{11}/(H_1 - 1) \tag{15-53}$$

$$H_2 = \delta_2^*/\theta_{22} = 2m + 1 \tag{15-54}$$

$$\delta_2^*/\delta = m/(m + 1) \tag{15-55}$$

For the vaneless passage (Aungier, 1988b), these profiles and associated empirical equations are sufficient for most applications. But, analysis within blade rows in centrifugal compressors, with the presence of blade force effects, requires additional considerations. It is quite common for the tangential defect thicknesses to become negative. This happens when viscous effects reduce the angular momentum flux [the second term in Eq. (15-40)] such that the blade force is not balanced. As will be seen later in this chapter, it also occurs when the boundary layer moves from a rotating wall to a stationary one, or vice versa. This situation simply means that the boundary layer integrated tangential mass or momentum exceeds that of the free stream fluid. The governing equations can handle this situation, but the power-law profile models cannot. Note from Eq. (15-51) that a negative tangential momentum defect requirs $m < 0$, which invalidates the relations given in Eqs. (15-51), (15-54) and (15-55), since the velocity at the wall is infinite rather than zero when $m < 0$. Indeed, the power-law profile assumption is of questionable validity for profile power-law exponents less than about 0.05 ($H_2 < 1.1$).

For boundary layer analysis in axial-flow turbomachinery, this is a somewhat academic problem. But, in the centrifugal compressor application, the tangential velocity profile shape strongly influences the meridional momentum conservation, as seen from the third term in Eq. (15-39). To handle these situations, it is necessary to extend the tangential velocity profiles to account for the excess tangential momentum in the boundary layer. Figure 15-2 illustrates the extended profiles used by this author, along with the conventional power-law profiles for normal situations. The approach used is to employ an alternate profile whenever $m < 0.05$ is encountered. This extended profile form is

$$\frac{V}{V_e} = \left(\frac{y}{\delta}\right)^{0.05} + 0.1705(1 - 20m)\left(1 - \frac{y}{\delta}\right)^2\left(\frac{y}{\delta}\right)^{0.1} \tag{15-56}$$

which is chosen to match the power-law profile at $m = 0.05$, yield mass and momentum defect thicknesses of zero when $m = 0$, and extend the profiles in a plausible fashion when $m < 0$. Substitution of this profile assumption into the definitions for the defect thicknesses yields

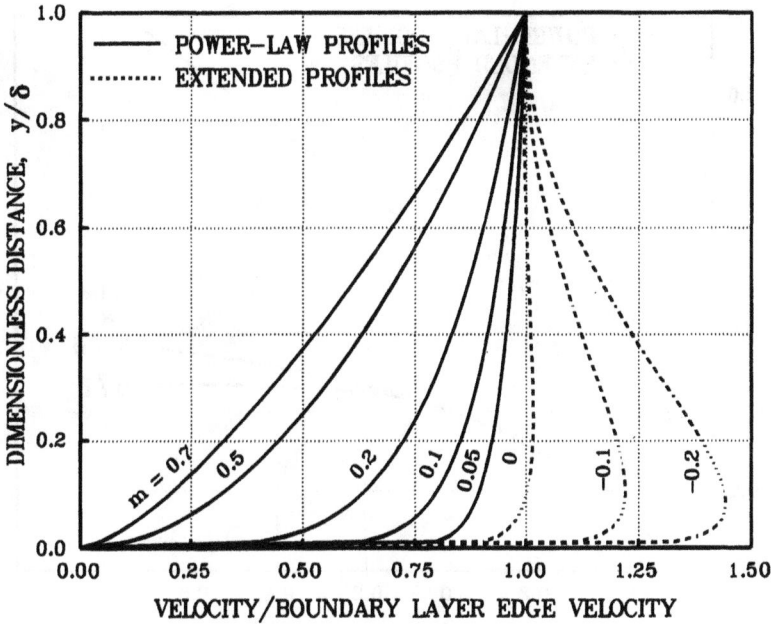

FIGURE 15-2. Boundary Layer Velocity Profiles

$$\delta_2^*/\delta = 20m/21 \tag{15-57}$$

$$\theta_{22}/\delta = 0.95m - 1.684m^2 \tag{15-58}$$

$$m = 0.05 + \frac{1}{6.82}\left[\frac{\theta_{12}}{\delta} - \frac{0.05}{(n+1)(n+1.05)}\right](n+1.1)(n+2.1)(n+3.1) \tag{15-59}$$

Figure 15-3 illustrates the functional form of the key boundary layer profile parameters as a function of m for the power-law profile and this extended profile, where $m = 0.05$ is the transition point between the two profiles.

The wall shear stress is assumed to be directed along the boundary layer edge streamline. Then, the wall shear stress components are given by

$$\tau_m = \tfrac{1}{2}c_f\rho_e V_e V_{me} \tag{15-60}$$

$$\tau_U = \tfrac{1}{2}c_f\rho_e V_e V_{Ue} \tag{15-61}$$

where c_f = a generalized Ludwieg-Tillmann skin friction coefficient based on the boundary layer profiles along the boundary layer edge streamline direction

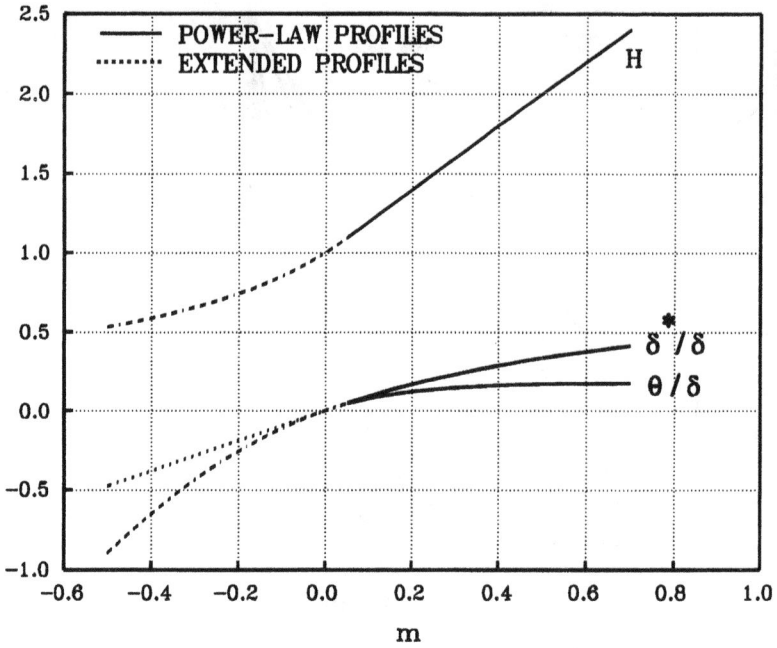

FIGURE 15-3. Boundary Layer Parameters

$$c_f = 0.246 \exp(-1.561 H_s)(\rho_e V_e \theta_s / \mu)^{-0.268} \qquad (15\text{-}62)$$

and δ_s^* and θ_s = displacement and momentum thicknesses, respectively, in the free stream direction; and $H_s = \delta_s^*/\theta_2$. Denoting the velocity component in the free stream direction as V_s,

$$V_s = V_m \sin \alpha_e + V_U \cos \alpha_e \qquad (15\text{-}63)$$

where α_e = boundary layer edge flow angle with respect to the tangential direction, the streamwise defect thicknesses are defined as

$$\rho_s V_{se} \delta_s^* = \int_0^\delta (\rho_e V_{se} - \rho V_s) dy \qquad (15\text{-}64)$$

$$\rho_e V_{se}^2 \theta_s = \int_0^\delta \rho V_s (V_{se} - V_s) dy \qquad (15\text{-}65)$$

By substituting Eq. (15-63) into Eqs. (15-64) and (15-65) and expanding, these terms can be expressed in terms of the defect thicknesses in Eqs. (15-41) and (15-45).

$$\delta_s^* = \sin^2 \alpha_e \delta_1^* + \cos^2 \alpha_e \delta_2^* \tag{15-66}$$

$$\theta_s = \sin^4 \alpha_e (\delta_1^* + \theta_{11}) + \cos^4 \alpha_e (\delta_2^* + \theta_{22})$$
$$+ 2 \sin^2 \alpha_e \cos^2 \alpha_e (\delta_1^* + \theta_{12}) - \delta_s^* \tag{15-67}$$

In evaluating Eqs. (15-67) and (15-66), normal practice is to limit δ_2^*, θ_{22} and θ_{12} to values no less than zero. Since α_e is normally quite small, use of negative values of these parameters can quickly produce values of δ_s^* and θ_s that invalidate Eq. (15-62). As further precaution, limits are imposed on the solution of Eq. (15-62). First, the momentum thickness Reynolds number can be no less than the transition value, i.e., $\rho_e V_e \theta_s / \mu \geq 250$. Also, the streamwise shape factor is limited to the separation value, i.e., $H_s \leq 2.4$.

The entrainment function is computed from

$$E = 0.025(H - 1) = 0.05n;$$
$$E \geq 0.025(H_s - 1) \tag{15-68}$$

Aungier (1988b) used only the first of the above relations, because of its vaneless component application, since the meridional profile shape factor is always dominant for that case. When the analysis is applied within blade passages, the streamwise profile shape may also govern the entrainment rate. Hence, the larger of the two estimates is used to cover both cases.

To apply this analysis within blade rows, it is necessary to consider the behavior of the blade force terms in the boundary layer. Early attempts to model end-wall boundary layers had limited success, largely due to the assumption that the blade force remains constant across the boundary layer. Since blade forces arise from blade surface pressure differences, that assumption is consistent with the usual boundary layer assumption that pressure is constant across the boundary layer. Mellor and Wood (1971) presented compelling arguments as to the necessity of blade force defects in the end-wall boundary layer. Their exceptional insight had a profound influence on the end-wall boundary layer analysis problem.

Smith (1970) provided experimental confirmation of blade force defects from an evaluation of tangential blade forces in a multistage compressor, considering hub-and-shroud boundary layers for both rotors and stators. He introduced the tangential defect thickness defined as

$$v_2 = \int_0^\delta [1 - fU/f U_e] dy \tag{15-69}$$

Smith's tangential force defect data indicates that the blade force defect is significant and always positive, but the data shows a great deal of scatter. Smith tentatively selected $v_2 = 0.65\delta_1^*$ as an estimate. To reduce the data scatter, Smith subsequently reworked his data (Koch and Smith, 1976) to develop average values of v_2 and δ_1^* for rotors and stators for both hub-and-shroud boundary layers. Figure 15-4 shows those results.

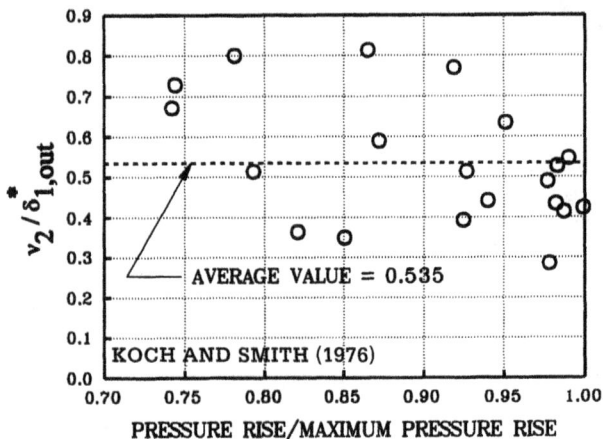

FIGURE 15-4. Expeirmental Force Defect Data

Common practice in axial-flow compressor end-wall boundary layer analysis is to predict boundary layer parameters between blade rows, without regard to the detailed development within the blade passages. In that context, relating the blade force defects to blade row discharge boundary layer parameters is sufficient. But it is of limited value when seeking to integrate the boundary layer equations through the blade row. In that case, it is necessary to relate the local blade force defect to local boundary layer parameters. The centrifugal compressor application introduces additional uncertainties. With centrifugal effects present, blade force terms assume greater significance. Also, the profile shape factor, H_1, shows substantially more variation than encountered in axial flow compressors. Strong coupling between the blade force terms and H_1 makes boundary layer displacement thickness a poor choice for a correlating parameter unless necessitated by the physics of the problem.

Hunter and Compsty (1982) report some very detailed experiments on an axial flow compressor rotor, concentrating on the shroud boundary layer with tip clearance effects. While their results are more limited in scope than Smith's data, they included detailed measurements for all boundary layer defect thicknesses. Figure 15-5 shows results from Hunter and Cumpsty (1982) normalized in the same manner as that used by Smith. Figure 15-6 indicates a different presentation of the same results. Here, the tangential defect thickness is normalized by the average of the blade row inlet and discharge momentum thicknesses, θ_{11}. Normalization by an average parameter will provide a correlation more suitable for local use within the blade passage. It was expected that boundary layer momentum thickness would be a better choice as a correlating parameter than displacement thickness. It can be seen from Figs. 15-5 and 15-6 that the momentum thickness shows similar significance. On average, $\nu_2/\theta_{11,ave}$ is about 60% larger than ν_2/δ_1^*. Using this relationship to adjust Smith's more comprehensive data set to this form of correlation leads to the following correlation for the local tangential blade force defect thickness:

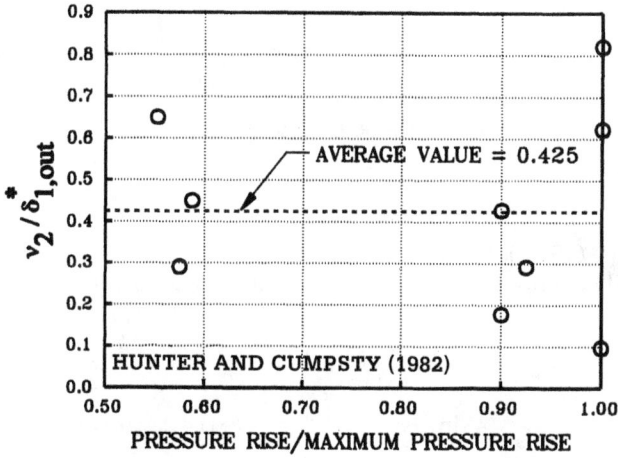

FIGURE 15-5. Experimental Force Defect Data

$$v_2/\theta_{11} \approx 1 \qquad (15\text{-}70)$$

Given the tangential force defect, the meridional force defect can be estimated. The blade force must be approximately normal to the mean streamline. It is reasonable to expect that the force defect in both directions is similar, i.e., we are dealing with specific components of the overall blade force defect. Following Smith, define

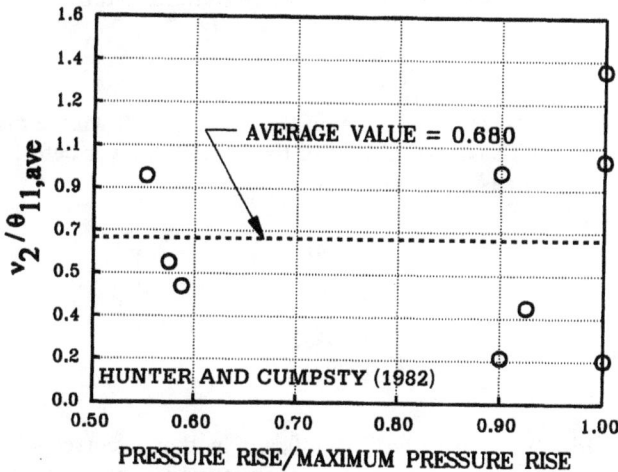

FIGURE 15-6. Experimental Force Defect Data

$$v_1 = \int_0^\delta [1 - f_m/f_{me}]dy \qquad (15\text{-}71)$$

the requirement that blade force be normal to the streamline yields

$$v_1 f_{me} = -v_2 f_{Ue} W_U/W_m \qquad (15\text{-}72)$$

Using Eqs. (15-46), (15-47), (15-69) and (15-71), the boundary layer momentum integral equations can be written in a more convenient form for solution

$$\frac{\partial}{\partial m} [r\rho_e V_{me}^2 \theta_{11}] + \delta_1^* r\rho_e V_{me} \frac{\partial V_{me}}{\partial m}$$

$$- \sin \alpha c \rho_e V_{Ue}[V_{Ue}(\delta_2^* + \theta_{22}) + 2\omega_w r \delta_2^*]$$

$$= r\tau_{mw} + r v_1 f_{me} \qquad (15\text{-}73)$$

$$\frac{\partial}{\partial m} [r^2 \rho_e V_{me} V_{Ue} \delta_{12}] + r^2 \delta_1^* f_{Ue} = r^2 \tau_{Uw} + r^2 v_2 f_{Ue} \qquad (15\text{-}74)$$

To complete the development of the analysis, consideration must be given to boundary layer development along surfaces where the wall is both rotating and stationary. This is not uncommon for centrifugal compressors, e.g., when analyzing the hub wall boundary layer through a rotating impeller and into the stationary vaneless diffuser which follows it. To develop the "jump" conditions at the abrupt change from a rotating wall to a stationary wall, consider this impeller-diffuser hub boundary layer problem. For clarity, different nomenclature will be used for the rotating and stationary frames of reference. In the impeller, the velocity relative to the wall is designated by W, while C is used for the absolute velocity appropriate to the diffuser wall. These velocities are related by

$$W_U = C_U - \omega r \qquad (15\text{-}75)$$

Similarly, boundary layer parameters on the rotating wall will be distinguished by a prime. From the basic definitions of the boundary layer defect thicknesses and Eq. (15-75), it is easily shown that

$$C_{Ue}\theta_{12} = W_{Ue}\theta_{12}' \qquad (15\text{-}76)$$

$$C_{Ue}\delta_2^* = W_{Ue}\delta_2^{*'} \qquad (15\text{-}77)$$

$$C_{Ue}^2\theta_{22} = W_{Ue}^2\theta_{22}' + \omega r W_{Ue}\delta_2^{*'} \qquad (15\text{-}78)$$

$$C_{Ue}^2\theta_{22}(1 + H_2) = W_{Ue}^2\theta_{22}'(1 + H_2') + 2\omega r W_{Ue}\delta_2^{*'} \qquad (15\text{-}79)$$

Equation (15-79) is particularly significant in that it corresponds to a key term in the meridional momentum integral equation and shows that this term is invariant between a rotating and stationary frame of reference. Note that most typical centrifugal compressor analysis problems have $W_U < 0$ and $C_U > 0$. Hence, the boundary layer defect thicknesses in Eqs. (15-76) and (15-77) will normally have opposite signs, i.e., a deficit in one frame of reference is an excess in the other.

This completes the theory and supporting empirical models required to solve the axisymmetric, three-dimensional turbulent boundary layer problem. Since only turbulent boundary layers are considered, initial upstream boundary layer data must be supplied to start the analysis. Convenient parameters for this purpose are H_1, H_2 and θ_{11} (or the momentum thickness Reynolds number based on θ_{11}). This author often starts the analysis with $H_1 = 1.4$. The classical 1/7th power-law profile is used for the tangential velocity profile, i.e., $m = 1/7$ or $H_2 = 1.286$. Typically, the momentum thickness Reynolds number based on θ_{11} is set to 250, a typical value for transition from laminar to turbulent flow. For centrifugal compressor components, the boundary layer analysis is rather insensitive to these starting conditions, except very close to the inlet station. The analysis procedure is basically identical to that used for the two-dimensional turbulent boundary layer analysis, except that two momentum integral equations must be integrated and blade force defects must be evaluated when the boundary layer is inside a blade passage.

To illustrate this analysis, Figs. 15-7 and 15-8 show results obtained by applying it to the same boundary layer edge flow data and contour geometry for the flow along an impeller shroud contour, extended into the entrance region of the diffuser. In Fig. 15-7, the impeller is assumed to be open (stationary wall) while for Fig. 15-8, a covered impeller (rotating wall) is assumed. The two boundary layer analyses yield quite different results, particularly near the impeller tip.

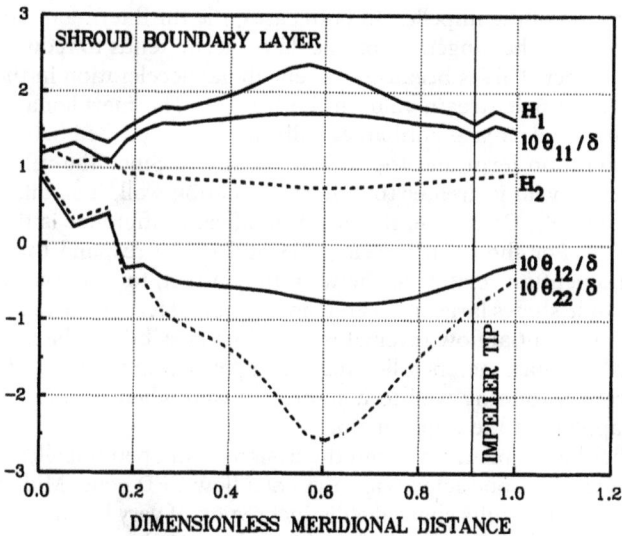

FIGURE 15-7. Open Impeller Boundary Layer

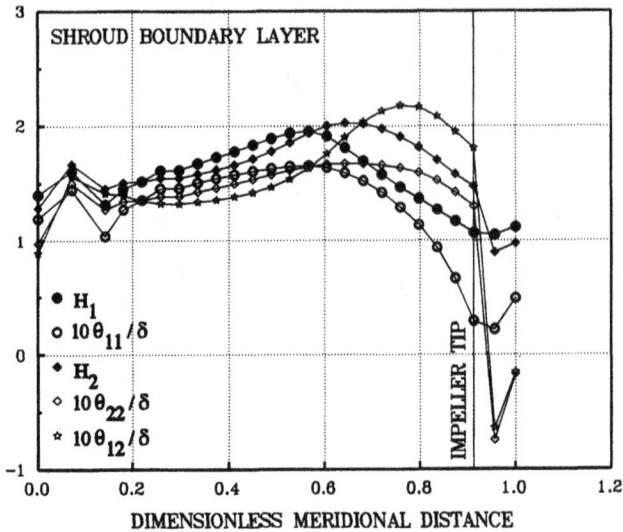

FIGURE 15-8. Covered Impeller Boundary Layer

From Eq. (15-74) it is expected that the tangential momentum integral equation should yield similar results for both cases. This follows from the fact that the blade force, f_{U_e}, is the same in both cases. If Eq. (15-76) is used to relate results from the two analyses to the same frame of reference, this is in fact the case. The meridional boundary layer parameters also show similar behavior in the region where the contour angle with the axial direction is small, but are quite different in the radial portion of the impeller. In both cases, the third term on the right-hand side of Eq. (15-73), the tangential profile defect term, tends to reduce the meridional profile defect. This is because the centrifugal acceleration in the boundary layer exceeds the free stream value in both cases. The meridional momentum thickness tends to decrease until an equilibrium is established between the centrifugal acceleration term and the wall shear stress. Since the velocity relative to the stationary wall is greater than for the rotating wall, it stabilizes at larger value of θ_{11} and H_1. Of course, the other significant difference is the tangential velocity profile. For the open impeller, θ_{22} and H_2 correspond to an excess in tangential momentum relative to the wall. In contrast, for the rotating wall, the tangential profile shows large mass and momentum defects relative to the wall, to a degree that tangential flow reversal is a distinct possibility. The abrupt change in boundary layer parameters following the impeller tip in Fig. 15-8 is due to the transition from a rotating wall to a stationary wall when Eqs. (15-76) through (15-78) are applied at the impeller tip.

This particular problem was from the design of an open impeller, for which a final evaluation was conducted with a viscous flow CFD code. Midway between the blades, the CFD results showed a distinct region of very low meridional velocity corresponding to the region in Fig. 15-7 where H_1 becomes large, but there is no sign of flow separation. Hence, the boundary layer analysis provided good design guidance in this case, such that no unexpected behavior was indicated by the final CFD evaluation. Although it cannot present a detailed evaluation of viscous effects, it does provide useful guidance in the design of end-wall contour and blade geometry to minimize the redesign activity needed to achieve satisfactory results in a final viscous flow CFD evaluation.

ANSWERS TO THE EXERCISES

1.1 The covered impeller will have a very wide tolerance on axial clearance, which is the most difficult problem in multi-stage machines. Also, in the event of a seal rub with the cover, damage is likely to be limited to the seal fins, which are relatively cheap to repair.

1.2 For similarity, curves of H versus. Q_0 at constant M_U are needed.

$M_U = \pi N r_2/(30 a_{0t})$ so N/a_{0t} is the desired speed parameter.
$\mu = H/U_2^2 = H/(M_U a_{0t})^2$ so $H/(a_{0t})^2$ is the desired head parameter
$\phi = Q_0/(\pi r^2 M_U a_{0t})$ so Q_0/a_{0t} is the desired flow parameter

1.3 $U_2 = N \pi r_2/30 = 9,000(0.25)\pi/30 = 235.62$ m/sec.
Equation (1-1): $\phi = 0.925/[\pi(.25)^2(235.62)] = 0.020$.
From Fig. 1-9; $\eta = 0.745$, $\mu = 0.515$.

1.4 The compressor stage of exercise 1.3 has $\phi = 0.02$, $\mu = 0.515$. From Fig. 1-9, $\phi = 0.032$ will give the 5% efficiency gain with $\mu = 0.525$.
Equation (1-2) (original): $H = 0.515(235.62)^2 = 28,590$ m^2/sec^2.
Equation (1-2) (new): $U_2 = [(28,590/0.525)]^{0.5} = 233.36$ m/sec.
Equation (1-1): $d_2 = 2\{0.925/[(0.032)(233.36)\pi]\}^{0.5} = 39.7$ cm.

1.5 Equation (1-2): $H = \mu U_2^2 = \mu(N \pi d_2/120)^2$, so reduce the impeller diameter to a factor of $(0.95)^{0.5}$ of the original to reduce the head by 5%.

1.6 From Fig. 1-9, a modest drop in efficiency can be expected, so the required head will not likely be achieved. Similitude should be useable, some reasonable corrections, since the modest change in the flow passage widths should not greatly change the character of the design. The impeller diameter can be increased to compensate for the reduced efficiency. Or the rotation speed can be increased for this purpose. But that will also increase the flow, per Eq. (1-1), so a passage width reduction greater than 10% would be needed if speed increase is used.

2.1 Equation (2-45):

$P_R = 1.5 : \ln(T_R) = 0.4^* \ln(1.5)/[1.4\eta_p];$ $\quad T_R = 1.1558.$
$P_R = 6.0 : \ln(T_R) = 0.4^* \ln(6)/[1.4\eta_p];$ $\quad T_R = 1.8963.$

Equation (2-44):

$$P_R = 1.5 : \eta_{ad} = (1.5^{(2/7)} - 1)/0.1558 = 0.7883.$$
$$P_R = 6.0 : \eta_{ad} = (6^{(2/7)} - 1)/0.8963 - 0.7458.$$

2.2 Equation (2-49): $p_4 = 170 + 0.65(200 - 170) = 189.5$ kPa.
Equation (2-40): $T_3 = 300[170/200]^{(2/7)} = 286.4°$K.
Equation (2-53): $p_{t4} = 200 - 0.1(200 - 170) = 197$ kPa.
Equation (2-40): $T_{4ad} = 300[189.5/200]^{(2/7)} = 295.4°$K.
Equation (2-40): $T_4 = 300[189.5/197]^{(2/7)} = 296.7°$K.
Equation (2-18): $\Delta h = c_p(296.7 - 286.4) = 10.3c_p$.
Equation (2-18): $\Delta h_{ad} = c_p(295.4 - 286.4) = 9c_p$.
Equation (2-48): $\eta_{diff} = 9/10.3 = 0.874$.

2.3 Equation (2-54): $\ln(100/4065) = 7(1 + 0.3254)(1 - 374.3/T)/3$
$T = 246.2°$K.

2.4 Definition of R_U: $R = R_U/M = 8314/28 = 296.93$ m^2/(sec^2 - °K).
Equation (2-17) and (2-22): $c_p = 7(296.93)/2 = 1039.15$ m^2/(sec^2 - °K).
Equation (2-40): $T_i = 300(85/100)^{(2/7)} = 286.4°$K.
$T_{ad} = 300(70/100)^{(2/7)} = 270.9°$K.
Equation (2-41): $C_i = [2(1039.15)(300 - 286.4)]^{0.5} = 168.1$ m/sec.
Equation (2-41): $C_{ad} = [2(1039.15)(300 - 270.9)]^{0.5} = 245.9$ m/sec
Equation (2-50): $C_d^2 = (168.1)^2 + 0.95[(245.9)^2 - (168.1)^2]$.
$C_d = 242.6$ m/sec.
Equation (2-41): $T_d = 300 - (242.6)^2/[2(1039.15)] = 271.7°$K.
Equation (2-40): $P_d = 70(300/271.7)^{(7/2)} = 99.0$ kPa.
Equation (2-53): loss coefficient = $(100 - 99)/(100 - 85) = 0.0667$.

3.1 All derivatives with respect to t and θ are zero; $\omega = 0$. Results are:

$$\frac{\partial r b \rho C_m}{\partial m} = 0$$

$$C_m \frac{\partial C_m}{\partial m} - \frac{\sin \alpha_C}{r} C_U^2 = -\frac{1}{\rho} \frac{\partial p}{\partial m}$$

$$\frac{\partial C_U}{\partial m} + \frac{\sin \alpha_C}{r} C_U = 0$$

$$\frac{\partial h_t}{\partial m} = 0$$

3.2 Same approach as exercise 3.1. Result is:

$$\frac{\partial r C_U}{\partial m} = 0$$

which is easily shown to be identical to the result in exercise 3.1 using

$$\sin \alpha_C = \frac{\partial r}{\partial m}$$

3.3 From Eqs. (3-32) and (3-34) it is easily shown that

$$\int_0^\delta \rho u \, dy = \rho_e u_e (\delta - \delta^*)$$

$$\int_0^\delta \rho u^2 \, dy = \rho_e u_e^2 (\delta - \delta^* - \theta)$$

3.4 Denote mixed-out flow with subscript m. Conservation of mass requires
Conservation of momentum requires

$$\rho u_e (b - 2\delta^*) = b\rho u_m$$

$$u_m = u_e (1 - 2\delta^*/b)$$

$$\rho u_m^2 = \rho u_e^2 (1 - 2\delta^*/b)^2$$

$$b p_e + \rho u_e^2 (b - 2\delta^* - 2\theta) = b p_m + b\rho u_m^2$$

Combining results and using the definition of p_t from the problem statement yields

$$P_{te} - P_{tm} = \rho u_e^2 [2\theta/b + 2(\delta^*/b)^2]$$

3.5 Use Eq. (3-22). Per assumptions given, all gradients along n are zero, and n is the radial coordinate. Then, W_m can be expressed

$$\frac{\partial W_m}{\partial r} = -k_{m1} W_m$$

$$W_m = W_{m1} \exp[k_{m1}(r_1 - r)]$$

Equation (3-1) gives W_U, and with rC_U specified constant,

$$\tan \alpha = W_m/W_U = W_{m1} \exp[k_{m1}(r_1 - r)]/[r_1 C_{U1}/r - \omega r]$$

4.1 Only C_t and C_c are functions of N. From Figs. (4-8) and (4-9),
Current seal: $C_t \approx 0.42$; $C_c \approx 1.8$.
Modified seal: $C_t \approx 0.295$; $C_c \approx 2.4$.
From Eq. (4-33) the ratio of leakage for the modified to the current seal is $(0.295)(2.4)/[(0.42)(1.8)] = 0.937$. With only about a 6% reduction in leakage, doubling the manufacturing cost would be very questionable.

4.2 From Eqs. (4-2), (4-5) and (3-1) and the definition of ϕ_2, the velocity components are:

$C_{U2}/U_2 = 0.88$, $W_{U2}/U_2 = -0.12$, $C_{m2}/U_2 = 0.3$, so the velocity diagram is easily constructed.

4.3 Using Eqs. (4-22) to (4-25), the four torque coefficients are:

$C_{M1} = 0.00157$, $C_{M2} = 0.00559$, $C_{M3} = 0.00726$, $C_{M4} = 0.00600$, so C_{M3} is the relevant value and the flow regime is turbulent, merged boundary layers.

4.4 From Eq. (4-27), the Reynolds number where roughness first has significance is:

$$Re_s = 1,100(0.001)^{-0.4}/(0.00726)^{0.5} \approx 204,600.$$

So the disk is already hydraulically smooth and there is no purpose in additional surface polishing.

4.5 From Eqs. (1-6) and (2-10): $I = (0.52)/(0.85) = 0.6118$.
From Eq. (4-1): $I_B = I - 0.02 = 0.5918$.
From Eq. (4-5): $\cot \beta_2 = [1 - 0.5918)/0.88]/[1.1(0.3)]$ or $\beta_2 = 45.2°$.

4.6 The table below gives data for the sketch in this problem and the next one, computed from Eq. (4-5), the specified head coefficient drop and the definition of hydraulic efficiency.

	– – – –90 Degree – – – –			– – – –50 Degrees – – – –		
ϕ_2	I_B	μ	η	I_B	μ	η
0.2000	0.8800	0.6550	0.7443	0.7176	0.4926	0.6864
0.2100	0.8800	0.6640	0.7545	0.7094	0.4934	0.6955
0.2200	0.8800	0.6710	0.7625	0.7013	0.4923	0.7020
0.2300	0.8800	0.6760	0.7682	0.6932	0.4892	0.7057
0.2400	0.8800	0.6790	0.7716	0.6851	0.4841	0.7066
0.2500	0.8800	0.6800	0.7727	0.6769	0.4769	0.7046
0.2600	0.8800	0.6790	0.7716	0.6688	0.4678	0.6995
0.2700	0.8800	0.6760	0.7682	0.6607	0.4567	0.6912
0.2800	0.8800	0.6710	0.7625	0.6526	0.4436	0.6797
0.2900	0.8800	0.6640	0.7545	0.6444	0.4284	0.6648
0.3000	0.8800	0.6550	0.7443	0.6363	0.4113	0.6464

For a radially bladed impeller, peak hydraulic efficiency always occurs at peak head and so is usually at the flow where surge can be expected. Normally, it will not be an "operating" point.

4.7 See the table above for the 50° blade angle impeller sketch. Now, peak efficiency is well away from the peak head flow and could be used as a regular "operating" point. The coworker does not realize that the backward leaning impeller design would normally be run at a higher tip speed to get the required (absolute) work input. If absolute loss levels were similar for the two impellers, the "head coefficient loss" would be lower for the higher speed impeller. So, we can't use the assumed loss coefficient form for the two impeller types for any meaningful comparison beyond observation of the relative location of peak head and peak efficiency for both.

5.1 A performance analysis can give the fluid dynamic data to solve Eq. (5-4) for surface finish when roughness effects start to become significant. Then, for any component, you can determine what surface finish is needed to avoid roughness effects on friction losses.

5.2 Should be obvious.

5.3 Friction losses will be large when the path length divided by the passage width is large, which is characteristic of low flow coefficient stages (e.g., see Fig. 1-8). Diffusion losses become higher when the parameter, D, in Eq. (5-50) increases, which is more significant for wide passages, i.e., higher flow coefficient stages.

5.4 The inlet swirl velocity for a vaneless diffuser is set by the impeller. Hence, designing with lower flow angles will produce lower meridional velocities, requiring wider passages to conserve mass. So, designers should favor low flow angles for low flow coefficient stages and higher flow angles for higher flow coefficient stages. This conclusion will be reasonably valid for other component types since the passage width influences friction losses in all components in a similar way.

5.5 Under the conditions stated in the problem, the only parameter affecting stall is h_{th} in Eq. (5-64). It will have to become smaller as the number of vanes is increased to maintain a constant overall throat area, which will directly increase the flow angle at the vaned diffuser stall, moving stall to higher flow coefficients. So, for a wide stable operating range (low stall flow angle), fewer vanes should always be favored.

6.1 Revise the specifications for efficiency, parasitic work and impeller distortion factor based on the performance analysis, and repeat the preliminary design. Repeat process again if necessary until agreement is satisfactory.

6.2 The various stage components may not be adequately matched to one another if these design specifications are not realistic.

6.3 Increase the vaned diffuser design incidence angle.

6.4 Change the design impeller blade loading parameter in Eq. (6-22). Also could increase the impeller axial length or change impeller style to produce longer blades, per Eq. (4-42).

6.5 (a) Increase impeller inlet relative flow angle to increase inlet meridional velocity.
(b) Reduce the impeller tip flow angle (or tip flow coefficient).
(c) Increase impeller axial length or reduce passage widths by using higher impeller flow angles.
(d) Increase impeller design incidence or use splitter blades to reduce throat metal blockage.

(e) Reduce impeller tip and diffuser exit flow angles to obtain wider passages.

7.1 $X_1 = 0$, $X_2 = 1$, $Y_1 = 0.1$, $Y_2 = 0.2$, $\alpha_2 = 0$, $L_1 = 0$, $L_2 = 0.5$, and α_1 is available to change the shape of the curve.

7.2 Hub: $X_1 = 0$, $X_2 = 0$, $Y_1 = 15$, $\alpha_1 = 90°$, $\alpha_2 = 90°$, $L_1 = 0$, $L_2 = 0$.
Shroud: $X_1 = -1$, $X_2 = -0.5$, $Y_1 = 15$, $\alpha_1 = 80°$, $\alpha_2 = 90°$, $L_1 = 0$, $L_2 = 0$.
And Y_2 is undefined. This allows you to adjust that parameter to obtain an acceptably smooth curve.

7.3 Fit these curves with a Bezier polynomial curve.

7.4 The capability to invert the geometry when exporting it to the quasi-three-dimensional flow analysis is the only change needed (the inlet in the gas path design system becomes the discharge in the quasi-three-dimensional flow analysis and vice versa).

8.1 Designing to locate the minimum loss at the design flow does not necessarily result in the lowest achievable loss at this flow. This practice usually fails in its objective for very low flow coefficient stages, where locating the minimum loss at a flow greater than the design flow via wider passage widths will usually reduce the loss at the design flow. The major benefit of this practice is to increase the flow range between the design flow and the diffuser stall flow.

8.2 Diffuser #1 achieves the highest pressure recovery, which is about the only good thing that can be said about it. Diffuser #4 achieves lower loss and higher discharge flow angles for flow less than design flow, so it would be a good choice if a wide stable operating flow range were desired. Diffuser #4 would not be a good choice for a volute stage since its high discharge velocity (low pressure recovery) would result in much higher volute losses.

8.3 Matching the volute will almost certainly yield a better efficiency than optimizing the diffuser loss. Reducing volute losses is really the only reason that a diffuser is included. But, the designer will have to be careful to achieve an adequate stable operating range from design flow to stall flow.

9.1 The match achieved for a vaned diffuser with its downstream component will be approximately the same for all flow rates. Hence, it will directly affect performance for the entire operating range of the compressor.

9.2 (a) It can achieve greater flow diffusion or pressure recovery than a vaneless diffuser. This is most important in the high flow coefficient range where diffusion losses are the largest contributor to the overall diffuser loss.
(b) It can allow operation at lower flow angles (or with wider passages) without encountering diffuser stall. This is most important in the low flow coefficient range where friction losses are the largest contributor to the overall diffuser loss.

9.3 (a) How high is the impeller tip flow angle? Vaned diffusers are not very effective at high flow angles; thus it is difficult to use a vaned diffuser inlet angle less than the impeller tip flow angle.

(b) What are the requirements for matching the downstream component? Increased diffusion won't do much good if that causes increased loss in the downstream component due to severe mismatching. (c) What is the stage flow coefficient? The probability of a performance improvement with a vaned diffuser is greater at lower flow coefficients where lower diffuser friction losses can produce substantial efficiency gains.

9.4 (a) Can the diffuser be matched to the impeller at a fairly low flow angle? Essentially a "*must*" for a wide stable operating range. (b) What are the restrictions (i.e., mechanical) on the number of vanes? A low number of vanes is also essential for wide stable operating range. (c) How high is the impeller tip Mach number? Choking in the vaned diffuser may severely reduce the maximum flow achievable.

10.1 Direct substitutions stated yields equations. Blade loading terms are:

$$K = \frac{Y - 1}{\cot \beta_y - \cot \beta_6} \frac{\partial \cot \beta}{\partial y}$$

$$C_s - C_p = \frac{2\pi \sin \beta}{z} \frac{\partial (rC_r \cot \beta)}{\partial y}$$

10.2 For the same basic bend curvatures, a shorter axial-length return system is achieved; therefore, the stage axial length is reduced.

10.3 For low flow coefficient stages, low flow angles will be preferred to permit use of a wider flow passage to reduce the dominant friction losses. For high flow coefficient stages, the reverse is true since narrower passages reduce the more dominant curvature and diffusion losses.

10.4 (a) Specifying b_6 is best for high flow coefficient stages since it directly controls the overall flow diffusion (area ratio) across the bend. Specifying the crossover exit angle helps most for low flow coefficient stages where using the lowest practical flow angle (or widest passage) is the most important consideration.

(b) The parameter F_A is most appropriate for higher flow coefficient stages where control of diffusion rate within the bend is important. Specifying r_5 is most appropriate for low flow coefficient stages since it directly sets the diffuser exit radius, and diffusion losses have only a modest impact on performance.

(c) Specifying b/R is most appropriate for high flow stages, and is an important consideration for the wider passages involved. Specifying R_C is most appropriate for low flow coefficient stages since it gives more control over the bend axial length, and the b/R parameter has little impact on performance for the narrower passages involved.

11.1 $R^2 [0.75\pi + 1] = [R + r_5]\theta b_5 \tan \alpha_5$.

11.2 $R^2 [0.75\pi + 1] = r_5\theta b_5 \tan \alpha_5 + R(b_5 + t).$

11.3 $H[W - \frac{1}{2}\theta b_5 \tan \alpha_5] = \theta r_5 b_5 \tan \alpha_5.$

11.4 $H[W - \frac{1}{2}(b_5 + t)] = \theta r_5 b_5 \tan \alpha_5.$

11.5 Under ideal flow conditions, the semi-external will achieve no static pressure recovery between Stations 5 and 6. So, external volute achieves the higher "internal" static pressure recovery. This has no real significance since the two volutes should achieve the same full-collection plane velocity (kinetic energy), thus imposing the same flow diffusion requirements on the exit cone.

11.6 Under ideal flow conditions, the semi-external should achieve the lower overall loss. The additional diffusion achieved in the extended diffuser passage (larger r_5) will reduce the meridional velocity head dump loss and produce as much diffusion of the tangential velocity head before the flow enters the volute as would be achieved within the external volute. Thus, the semi-external volute should achieve the higher static pressure at the full-collection plane even though it offers no internal static pressure recovery. This is due to the higher static pressure achieved by the extended diffuser passage.

REFERENCES

Aungier, R. H. (1968). "A Time-Dependent Numerical Method for Calculating the Flow About Blunt Bodies." *Technical Report AFWL-TR-68-52*, Air Force Weapons Laboratory, Kirtland AFB, NM.

Aungier, R. H. (1970). "A Computational Method for Exact, Direct and Unified Solutions for Axisymmetric Flow Over Blunt Bodies of Arbitrary Shape (Program BLUNT)." *Technical Report AFWL-TR-70-16*, Air Force Weapons Laboratory, Kirtland AFB, NM.

Aungier, R. H. (1971a). "A Computational Method for Unified Solutions to the Inviscid Flow Field About Blunt Bodies." *The Entry Plasma Sheath and Its Effects on Space Vehicle Electromagnetic Systems (Proceedings, Fourth Plasma Sheath Symposium), NASA SP-252*, 241–260.

Aungier, R. H. (1971b). "A Computational Method for Two-Dimensional, Axisymmetric and Three-Dimensional Blunt Body Flows (Program ATTACK)." *Technical Report AFWL-TR-70-124*, Air Force Weapons Laboratory, Kirtland AFB, NM.

Aungier, R. H. (1988a). "A Systematic Procedure for the Aerodynamic Design of Vaned Diffusers." *Flows in Non-Rotating Turbomachinery Components*, ASME, Vol. 69, 27–34.

Aungier, R. H. (1988b). "A Performance Analysis For the Vaneless Components Of Centrifugal Compressors." *Flows in Non-Rotating Turbomachinery Components*, ASME, Vol. 69, 35–43.

Aungier, R. H. (1990). "Aerodynamic Performance Analysis of Vaned Diffusers." *Fluid Machinery Components*, ASME, Vol. 101, 27–44.

Aungier, R. H. (1993a). "Aerodynamic Design and Analysis of Vaneless Diffusers and Return Channels." *Paper No. 93-GT-101*, ASME, New York.

Aungier, R. H. (1993b). "Mean Streamline Aerodynamic Performance Analysis of Centrifugal Compressors," *Proceedings Rotating Machinery Conference and Exposition (ROCON'93), Vol. I*. (Also in *Transactions ASME, Journal of Turbomachinery*, July 1995, 360–366.)

Aungier, R. H. (1994). "A Fast, Accurate Real Gas Equation of State for Fluid Dynamic Analysis Applications." *Contributed Papers In Fluids Engineering 1994*, ASME, Vol. 182, 1–6. (Also in *Transactions ASME, Journal of Fluids Engineering*, June 1995, 277–281).

Aungier, R. H. (1995). "Centrifugal Compressor Preliminary Aerodynamic Design and Component Sizing." *ASME Paper No. 95-GT-78*, ASME, New York.

Aungier, R. H. (1998). "Thermodynamic State Relations." *The Handbook of Fluid Dynamics*, R. W. Johnson, ed. CRC Press LLC, Boca Raton, Florida, 4-29–4-34.

Balje, O. E. (1981). *Turbomachines—A Guide to Design, Selection and Theory*, Wiley & Sons, New York.

Barnes, F. J. (1973) PhD thesis, Department of Chemical Engineering, University of California, Berkeley.

Benedict, R. P., Carlucci, N. A., and Swetz, S. D. (1966). "Flow Losses in Abrupt Enlargements and Contractions." *Transactions ASME, Journal of Engineering for Power*, Jan., 73–81.

Benedict, M., Webb, G. B., and Rubin, L. C. (1951). "An Empirical Equation for Thermodynamic Properties of Light Hydrocarbons and Their Mixtures." *Chemical Engineering Progress*, Vol. 47, 419–422.

Bishnoi, P. R., Miranda, R. D., and Robinson, D. B. (1974). "BWR Applied to NG/SNG Needs." *Hydrocarbon Processing*, 197–201.

Brown, W. B. and Bradshaw, G. R. (1949). "Design and Performance of a Family of Diffusing Scrolls with Mixed Flow Impellers and Vaneless Diffusers." *NACA Report 936*.

Busemann, A. (1928). "The Head Ratio of Centrifugal Pumps with Logarithmic Spiral Blades." *ZAMM*, Vol. 8, Oct., 372–384 (in German).

Casey, M. V. (1983). "A Computational Geometry for the Blades and Internal Flow Channels of Centrifugal Compressors." *ASME Journal of Engineering for Power*, Vol. 105, 288–295.

Casey, M. V. (1985). "The Effects of Reynolds Number on the Efficiency of Centrifugal Compressor Stages." *Transactions ASME, Journal of Engineering for Gas Turbines and Power*, Vol. 107, Apr., 541–548.

Chen, N. H. (1965). *Journal of Chemical Engineering Data*, Vol. 10, 207.

Colwill, W. H. (1980). "Impeller Performance Prediction Using Three-Dimensional Flow Analysis." *Performance Prediction of Centrifugal Pumps And Compressors*, ASME, New York, 125.

Conrad, O., Raif, K., and Wessels, M. (1979). "The Calculation of Performance Maps for Centrifugal Compressors with Vane-Island Diffusers." *Performance Prediction of Centrifugal Pumps and Compressors*, ASME, New York, 135–147.

Cooper, H. W. and Goldfrank, J. C. (1967). "B-W-R Constants and New Correlations." *Hydrocarbon Processing*, Vol. 46, 140–146.

Courant, R., Friedricks, K. O., and Lewy, H. (1928). "Uber die Partiellen Differenzengleichungen der Mathematischen Physik." *Math. Ann.*, 100, 32.

Cumpsty, N. A. (1989). *Compressor Aerodynamics*, Longman Scientific and Technical, Essex, United Kingdom.

Cumpsty, N. A., and Head, M. R. (1967). "The Calculation of Three-Dimensional Turbulent Boundary Layers, Parts I and II." *Aero Quarterly*, Vol. 18, 55–84 and 150–164.

Daily, J. W. and Nece, R. E. (1960a). "Chamber Dimension Effects on Induced

Flow and Frictional Resistance of Enclosed Rotating Disks." *Transactions ASME, Journal of Basic Engineering*, Mar., 217–232.

Daily, J. W. and Nece, R. E. (1960b). "Roughness Effects on Frictional Resistance of Enclosed Rotating Disks." *Transactions ASME, Journal of Basic Engineering*, Sept., 553–562.

Davis, W. R. (1976). "Three-Dimensional Boundary-Layer Computation on the Stationary End-Walls of Centrifugal Turbomachinery." *Transactions ASME, Journal of Fluids Engineering*, Sept., 431–442.

Denton, J. D. (1982). "An Improved Time-Marching Method for Turbomachinery Flow Calculation." *Paper No. 82-GT-239*, ASME, New York.

Dunevant, J. C., Emery, J. C., Walsh, H. C., and Westphal, W. R. (1955). "High-Speed Cascade Tests of The NACA 65-($12A_{10}$)-10 and NACA 65-($12A_2I_{8b}$)-10 Compressor Blade Sections," *NACA Research Memorandum RM-L55I08*.

Egli, A. (1935). "The Leakage of Steam Through Labyrinth Glands." *Transactions ASME*, Vol. 57, 115–122.

Ferguson, T. B. (1963). *The Centrifugal Compressor Stage*, Butterworth & Co. Ltd., London.

Fister, W., Zahn, G., and Tasche, (1982). "Theoretical and Experimental Investigations about Vaneless Return Channels of Multi-Stage Radial Flow Turbomachines." *Paper No. 82-GT-209*, ASME, New York.

Forrest, A. R. (1972). "Interactive Interpolation and Approximation by Bezier Polynomials." *Computer Journal*, Vol. 15, 71–79.

Gopalakrishnan, S. and Bozzola, R. (1973). "Numerical Representation of Inlet and Exit Boundary Conditions in Transient Cascade Flow." *Paper No. 73-GT-55*, ASME, New York.

Green, J. E. (1968). "The Prediction of Turbulent Boundary Layer Development in Compressible Flow." *Journal of Fluid Mechanics*, Vol. 31, 753.

Gruschwitz, E. (1950). "Calcul Approche de la Couche Limite Laminaire en Ecoulement compressible Sur Une Paroi Non-conductrice de la Chaleur." *Publication No. 47*, Office National d'Etudes et de Recherche Aeronautiques (ONERA), Paris, France.

Gunn, R. D. and Yamada. T. (1971). *AICHE Journal*, Vol. 17, 1341.

Head, M. R. (1958). "Entrainment in the Turbulent Boundary Layer." *Aeronautical Research Council, R&M 3152*.

Head, M. R. (1968). "Cambridge Work on Entrainment." *Proceedings of Computation of Turbulent Boundary Layers*, Thermosciences Division, Stanford University, California, 188–194.

Herrig, L. J., Emery, J. C., and Erwin, J. R. (1957). "Systematic Two-Dimensional Cascade Tests of NACA 65-Series Compressor Blades at Low Speeds." *NACA TN 3916*.

Hohlweg, W. C. (1987). "Correlation and Application of Centrifugal Compressor Return System Losses." *Fluid Machinery for the Petrochemical and Related Industries, Proceedings of the IMechE*, 97–103.

Hohlweg, W. C., Direnzi, G. L., and Aungier, R. H. (1993). "Comparison of Conventional and Low Solidity Vaned Diffusers." *Paper No. 93-GT-98*, ASME, New York.

Horlock, J. H. (1970). "Boundary Layer Problems in Axial Turbomachines." *Flow Research on Blading*, Dring, L.S., ed., Elsevier Publishing, Amsterdam, 322–371.

Howell, A. R. (1947). "Development of the British Gas Turbine Unit." Lecture: Fluid Dynamics of Axial Compressors, ASME Reprint, New York.

Hunter, I. H. and Cumpsty, N. A. (1982). "Casing Wall Boundary-Layer Development Through an Isolated Compressor Rotor." *Transactions ASME, Journal of Engineering for Power*, Oct., 805–818.

Huntington, R. A. (1985). "Evaluation of Polytropic Calculation Methods for Turbomachinery Performance." *Paper No. 85-GT-13*, ASME, New York.

Jansen, W. (1964). "Steady Fluid Flow in a Radial Vaneless Diffuser." *Transactions ASME, Journal of Basic Engineering*, Sept., 607–619.

Japikse, D. (1982). "Advanced Diffusion Levels in Turbocharger Compressors and Component Matching." *Proceedings Conference on Turbocharging and Turbochargers, Paper C45/82*, Institute of Mechanical Engineering.

Japikse, D. and Osborne, C. (1982). *Vaneless Diffuser, Return Bend and Return Channel Investigation*, Creare Inc., TN 346, (Proprietary).

Johnsen, I. A. and Bullock, R. O., eds. (1965). "Aerodynamic Design of Axial Flow Compressors." *NASA SP-36*, Fig. 177.

Johnston, J. P. and Dean, R. C. (1966). "Losses in Vaneless Diffusers of Centrifugal Compressors and Pumps." *Transactions ASME, Journal of Engineering for Power*, Vol. 88, 49–62.

Katsanis, T. (1968). "Computer Program for Calculating Velocities and Streamlines on a Blade-To-Blade Stream Surface of a Turbomachine." *NASA TN D-4525*.

Katsanis, T. (1969). "Fortran Program for Calculating Transonic Velocities on a Blade-To-Blade Stream Surface of a Turbomachine." *NASA TN D-5427*.

Kenny, D. (1979). "A Novel Correlation of Centrifugal Compressor Performance for Off-Design Prediction." *Proceedings AIAA/ASME/SAE 15th Joint Propulsion Conference, Paper No. 79-1159*.

King, C. J. (1980). *Separation Processes*, 2nd Ed., McGraw-Hill, New York, 64–80.

Koch, C. C. and Smith, L. H. Jr. (1976). "Loss Sources and Magnitudes in Axial-Flow Compressors." *Transactions ASME, Journal of Engineering for Power*, Vol. 98, 411–424.

Kosuge, H., Ito, T., and Nakanishi, K. (1982). "A Consideration Concerning Stall and Surge Limitations Within Centrifugal Compressors." *Transactions ASME, Journal of Engineering for Power*, Vol. 104, Oct. 782–787.

Lax, P. D. (1954). "Weak Solutions of Nonlinear Hyperbolic Equations and their Numerical Computation." *Commun. Pure and Appl. Math.*, Vol. 7, 159–193.

Lax, P. D. and Wendroff, B. (1964). "Differencing Schemes for Hyperbolic Equations with High Order of Accuracy." *Commun. Pure and Appl. Math.*, Vol. 17, 381–398.

Lieblein, S. (1959). "Loss and Stall Analysis of Compressor Cascades." *ASME Transactions, Journal of Basic Engineering* Sept., 387–400.

Lieblein, S. and Roudebush, W. H. (1956). "Theoretical Loss Relations for Low-Speed Two-Dimensional-Cascade Flow." *NACA TN 3662.*

Ludwieg, H. and Tillmann, W. (1950). "Investigations of the Wall-Shearing Stress in Turbulent Boundary Layers." *NACA TM 1285.*

Mallen, M. and Saville, G. (1977). "Polytropic Processes in the Performance Prediction of Centrifugal Compressors." *Paper No. C183/77*, Institute of Mechanical Engineers, 89–96.

Mellor, G. L. and Wood, G. M. (1971). "An Axial Compressor End-Wall Boundary Layer Theory." *Transactions AMSE, Journal of Basic Engineering*, June, 300–316.

Mischina, H. and Gyobu, I. (1978). "Performance Investigations of Large Capacity Centrifugal Compressors." *Paper No. 78-GT-3*, ASME, New York.

Moretti, G. and Abbett, M. (1966). "A Time-Dependent Computational Method for Blunt Body Flows." *AIAA Journal*, Vol. 4, 2136–2141.

Moussa, Z. M. (1978). "Impeller Casing Clearance Gap Flow Prediction Program." *Internal Technical Report 9-1030-05, No. 9 (Proprietary)*, Carrier Corporation. Syracuse, NY.

Nelson, L. C. and Obert, E. F. (1954). "Generalized PVT Properties of Gases." *Transactions ASME*, Vol. 76, 1057–1066.

Nikuradse, J. (1930). "Laws of Resistance and Velocity Distribution for Turbulent Flow of Water in Smooth and Rough Pipes." *Proceedings 3rd International Congress for Applied Mechanics*, 239–248.

Northern Research and Engineering Corporation. (1981). "An Interactive Graphics System for the Design of Radial Turbomachinery (COMIG)," Woburn, Mass.

Novak, R. A. (1967). "Streamline Curvature Computing Procedures for Fluid-Flow Problems." *Transactions ASME, Journal of Engineering for Power*, Oct., 478–490.

Novak, R. A. and Hearsey, R. M. (1973). "Axisymmetric Computing System for Axial Flow Turbomachinery." *ASME Fluid Dynamics of Turbomachinery*, Lecture Notes, ASME short course, Iowa State University.

Novak, R. A. and Hearsey, R. M. (1976). "A Nearly Three-Dimensional Intrablade Computing System for Turbomachinery—Part I: General Description." *Paper No. 76-FE-19*, ASME, New York.

Nykorowytsch, P., ed. (1983). *Return Passages of Multi-Stage Turbomachinery*, ASME, Vol. 3.

Osborne, C. and Sorokes, J. (1988). "The Application of Low Solidity Diffusers in Centrifugal Compressors." *Flows in Non-Rotating Turbomachinery Components*, ASME, Vol. 69, 89–101.

Pampreen, R. C. (1972). "The Use of Cascade Technology in Centrifugal Compressor Vaned Diffuser Design." *Transactions ASME, Journal of Engineering for Power*, July, 187–192.

Pitzer, K. S., Lippmann, D. Z., Curl, R. F., Huggins, C. M., and Peterson, D. E.

(1955). "The Volumetric and Thermodynamic Properties of Fluids. II Compressibility Factor, Vapor Pressure and Entropy of Vaporization." *American Chemical Society*, Vol. 77, 3427–3440.

Pohlhausen, K. (1921). "Zur Naherungsweisen Integration der Differential-Gleichung der Laminare Reibungsschicht, *ZAMM* Vol. 1, 235.

Redlich, O. and Kwong, J. (1949). *Chemical Review* 233.

Reneau, L., Johnston, J., and Kline, S. (1967). "Performance and Design of Straight Two-Dimensional Diffusers." *Transactions ASME, Journal of Basic Engineering*, Mar., 141–150.

Ried, R. C. and Sherwood, T. K. (1966). *The Properties of Gases and Liquids*, McGraw-Hill, New York.

Ried, R. C., Prausnitz, J. M., and Sherwood, T. K. (1977). *The Properties of Gases and Liquids*, McGraw-Hill, New York.

Ried, R. C., Prausnitz, J. M. and Poling, B. E. (1987), *The Properties of Gases and Liquids*, Fourth edition, McGraw-Hill, New York.

Rogers, C. (1982). "The Performance of Centrifugal Compressor Channel Diffusers." *Paper No. 82-GT-10*, ASME, New York.

Rotta, J. C. (1966). "Recent Developments in Calculation Methods for Turbulent Boundary Layers with Pressure Gradients and Heat Transfer." *Transactions ASME, Journal of Applied Mechanics*, Vol. 88, 429.

Runstadler, P. and Dean, R. (1969). "Straight Channel Diffuser Performance at High Inlet Mach Numbers." *Transactions ASME, Journal of Basic Engineering*, Vol. 91, Sept., pp 397–422.

Runstadler, P., Dolan, F., and Dean, R. (1975). "Diffuser Data Book." Creare R&D Inc., TN 186, Hanover, NH.

Schlichting, H. (1968). *Boundary-Layer Theory*, 6th Ed., McGraw-Hill, New York.

Schlichting, H. (1979). *Boundary-Layer Theory*, 7th Ed., McGraw-Hill, New York.

Schultz, J. M. (1962). "The Polytropic Analysis of Centrifugal Compressors." *Transactions ASME, Journal of Engineering for Power*, Jan., 69–82.

Schumann, L. F. (1985). "A Three-Dimensional Axisymmetric Calculation Procedure for Turbulent Flows in a Radial Vaneless Diffuser." *Paper No. 85-GT-133*, ASME, New York.

Selby, S. M., ed. (1965). *Standard Mathematical Tables*, 14th Ed., The Chemical Rubber Company, Cleveland.

Senoo, Y. (1981). "Low Solidity Circular Cascade for Wide Flow Range Blower." *Proceedings, Advanced Concepts in Turbomachinery*, Fluid Dynamics Institute, Hanover, NH.

Senoo, Y. and Kinoshita, Y. (1978). "Limits of Rotating Stall and Stall in Vaneless Diffusers of Centrifugal Compressors." *Paper No. 78-GT-19*, ASME, New York.

Senoo, Y. and Nakase, Y. (1972). "An Analysis of Flow Through a Mixed Flow Impeller." *Transactions ASME, Journal of Engineering for Power*, 43–50.

Senoo, Y., Kinoshita, Y., and Ishida, M. (1977). "Axisymmetric Flow in Vaneless Diffusers of Centrifugal Blowers." *Transactions ASME, Journal of Fluids Engineering*, Mar., 104–114.

Senoo, Y., Hayami, H., and Uski, H. (1983). "Low-Solidity Tandem Cascade Diffusers for Wide-Flow-Range Centrifugal Blowers." *Paper No. 83-GT-3*, ASME, New York.

Sheets, H. E. (1950). "The Flow Through Centrifugal Compressors and Pumps." *Transactions ASME*, Vol. 72, 1009–1015.

Sheppard, D. G. (1956). *Principles of Turbomachinery*, Macmillan, New York.

Smith, L. H. Jr. (1970). "Casing Boundary Layers in Multi-Stage Axial-Flow Compressors." *Flow Research on Blading*, Dring, L. S., ed., Elsevier Publishing, Amsterdam, 275–304.

Smith, D. J. L. and Frost, D. H. (1969). "Calculation of the Flow Past Turbomachine Blades." *Proceedings Institution of Mechanical Engineering*, Vol. 184, Paper 27.

Soave, G. (1972). "Equilibrium Constants from a Modified Redlich-Kwong Equation of State." *Chemical Engineering Science*, Vol. 27, 1197–1203.

Sokolnikoff, I. S. and Redheffer, R. M. (1958). *Mathematics of Physics and Modern Engineering*, McGraw-Hill, New York.

Stanitz, J. D. (1952). "One-Dimensional Compressible Flow in Vaneless Diffusers of Radial and Mixed-Flow Centrifugal Compressors, Including Effects of Friction, Heat Transfer and Area Change." *NACA-TN-2610*.

Stodola, A. (1927). *Steam and Gas Turbines*, McGraw-Hill, New York.

Summer, W. J. and Shanebrook, J. R. (1971). "Entrainment Theory for Compressible Turbulent Boundary Layers on Adiabatic Walls." *AIAA Journal*, Vol. 9, 330–332.

Van den Braembussche, R. (1987). "Rotating Stall in Centrifugal Compressors." *Preprint 1987-16*, von Karman Institute for Fluid Dynamics.

Vavra, M. H. (1960). *Aero-Thermodynamics and Flow in Turbomachines*, Wiley, New York.

Von Neumann, J. and Richtmyer, R. D. (1950). "A Method for the Numerical Calculation of Hydrodynamic Shocks." *Journal of Applied Physics*, Vol. 21, 232–237.

Walsh, J. L., Ahlberg, J. H., and Nilson, E. N. (1962). "Best Approximation Properties of the Spline Fit." *Journal of Mathematics and Mechanics*, Vol. 11, 225–234.

Weber, C. R. and Koronowski M. E. (1986). "Meanline Performance Prediction of Volutes in Centrifugal Compressors." *Paper No. 86-GT-216*, ASME, New York.

Whitfield, A. and Baines, N. C. (1990). *Design of Radial Turbomachines*, Longman Scientific and Technical, Essex, United Kingdom.

Whitfield, A. and Roberts, D. V. (1983). "Alternative Vaneless Diffusers and Collecting Volutes for Turbocharger Compressors." *Paper No. 83-GT-32*, ASME, New York.

Whitney, W. J., Szanca, E. M., Moffitt, T. P., and Monroe, D. E. (1967). "Cold-Air Investigation of a Turbine for High-Temperature-Engine Application." *Technical Note TN D-3751*, NASA.

Wiesner, F. J. (1967). "A Review of Slip Factors for Centrifugal Impellers." *Transactions ASME, Journal of Engineering for Power*, Oct., 558–572.

Wiesner, F. J. (1979). "A New Appraisal of Reynolds Number Effects on Centrifugal Compressor Performance." *Transactions ASME, Journal of Engineering for Power*, Vol. 101, July, 384–396.

Wilson, G. M. (1966). "Calculation of Enthalpy Data from a Modified Redlich-Kwong Equation of State." *Advances in Cryogenic Engineering*, Vol. 11, 392.

Wislicenus, G. F. (1947). *Fluid Dynamics of Turbomachinery*, McGraw-Hill, New York.

Wu, C. H. (1952). "A General Theory of Three-Dimensional Flow in Subsonic and Supersonic Turbomachines of Axial-, Radial- and Mixed-Flow Types." *NACA TN 2604*.

Yoshinaga, Y., Gyobu, I., Mischina, H., Koseki, F., and Nishida, H. (1980). "Aerodynamic Performance of a Centrifugal Compressor with Vaned Diffusers." *Transactions ASME, Journal of Fluids Engineering*, Dec., 486–493.

Zemansky, M. W. (1957). *Heat and Thermodynamics*, McGraw-Hill, New York.

ABOUT THE AUTHOR

Mr. Aungier is the Manager of Product Development for the Elliott Turbomachinery Co. Inc. in Jeannette, Pennsylvania. He has been active in fluid mechanics research and development for more than 33 years, including over 29 years in turbomachinery aerodynamics, specializing in centrifugal compressors. He has numerous publications in this field, primarily through the American Society of Mechanical Engineers. He is a graduate of Cornell University, where he received a masters degree in Aerospace Engineering and a bachelors degree in Engineering Physics.

Mr. Aungier started his career in 1966 as an officer in the U.S. Air Force, conducting research in hypersonic re-entry vehicle aerodynamics at the Air Force Weapons Laboratory in Albuquerque, New Mexico. He is the author of numerous Air Force and NASA publications, some of which are the basis for one of the analysis techniques described in this book. In 1970, Mr. Aungier joined the Research Division of Carrier Corporation in Syracuse, New York, where he spent 11 years managing and conducting applied research on the fluid dynamics of turbomachinery and air handling equipment. Most of his individual research was in support of The Elliott Company (then a division of Carrier). He developed performance analysis techniques for axial-flow compressors, centrifugal compressors and radial turbines, which formed the basis for methods still used at Elliott and Carrier. In 1981, Mr. Aungier transferred to Elliott Company as the Manager of Compressor Development, where he started focusing on the development of systematic and efficient techniques for the aerodynamic design of centrifugal compressors. His responsibilities were expanded to include turbine aerodynamic development in 1983 and mechanical design and analysis in 1987. He continues to be an active contributor to turbomachinery aerodynamic technology, including the aerodynamic design of new centrifugal compressor stages.

INDEX

www.ingramcontent.com/pod-product-compliance
Lightning Source LLC
Chambersburg PA
CBHW050454190326
41458CB00005B/1279